New Series of Latin School Books.

2. Latin Grammar. A Grammar of the Latin Language, for the use of Schools and Colleges. By Professors E. A. ANDREWS and S. STODDARD. This work, which for many years has been the text-book in the department of Latin Grammar in a large portion of American schools and colleges, and which claims the merit of having first introduced into the schools of this country the subject of grammatical analysis, which now occupies a conspicuous place in so many grammars of the English language, has been recently revised and carefully corrected in every part.

3. Questions on the Grammar. This little volume is intended to aid the student in preparing his lessons, and the teacher in conducting his recitations.

4. A Synopsis of Latin Grammar, comprising the Latin Paradigms, and the Principal Rules of Latin Etymology and Syntax. The few pages composing this work contain those portions of the Grammar to which the student has occasion to refer most frequently in the preparation of his daily lessons.

5. Latin Reader. The Reader, by means of two separate and distinct sets of notes, is equally adapted for use in connection either with the First Latin Book or the Latin Grammar.

6. Viri Romæ. This volume, like the Reader, is furnished with notes and references, both to the First Latin Book and to the Latin Grammar. The principal difference in the two sets of notes found in each of these volumes consists in the somewhat greater fulness of those which belong to the smaller series.

7. Latin Exercises. This work contains exercises in every department of the Latin Grammar, and is so arranged that it may be studied in connection with the Grammar through every stage of the preparatory course. It is designed to prepare the way for original composition in the Latin language, both in prose and verse.

8. A Key to Latin Exercises. This Key, in which all the exercises in the preceding volume are fully corrected, is intended for the use of teachers only.

9. Cæsar's Commentaries on the Gallic War, with a Dictionary and Notes. The text of this edition of Cæsar has been formed by reference to the best German editions. The Notes are principally grammatical. The Dictionary, which, like all the others in the series, was prepared with great labor, contains the usual significations of the words, together with an explanation of all such phrases as might otherwise perplex the student.

10. Sallust. Sallust's Jugurthine War and Conspiracy of Catiline, with a Dictionary and Notes. The text of this work, which was based upon that of Cortius, has been modified by reference to the best modern editions, especially by those of Kritz and Geriach; and its orthography is, in general, conformed to that of Pottier and Planche. The Dictionaries of Cæsar and Sallust connected with this series are original works, and, in connection with the Notes in each volume, furnish a very complete and satisfactory apparatus for the study of these two authors.

11. Ovid. Selections from the Metamorphoses and Heroides of Ovid, with Notes, Grammatical References, and Exercises in Scanning These selections from Ovid are designed as an introduction to Latin poetry. They are accompanied with numerous brief notes explanatory of difficult phrases, of obscure historical or mythological allusions, and especially of grammatical difficulties. To these are added such Exercises in Scanning as serve fully to introduce the student to a knowledge of Latin prosody, and especially of the structure and laws of hexameter and pentameter verse.

New Series of Latin School Books.

ANDREWS AND STODDARD'S LATIN GRAMMAR has long since been introduced into the LATIN SCHOOL OF THE CITY OF BOSTON, and into most of the other principal Classical Schools in this country. It is adopted by all the Colleges in New England, viz., HARVARD, YALE, DARTMOUTH, AMHERST, WILLIAMS, BOWDOIN, WATERVILLE, MIDDLEBURY, BURLINGTON, BROWN UNIVERSITY at Providence, WESLEYAN UNIVERSITY at Middletown, and WASHINGTON COLLEGE at Hartford; also at HAMILTON COLLEGE, New York, NEW YORK UNIVERSITY, city of New York, CINCINNATI COLLEGE and MARIETTA COLLEGE, Ohio, RANDOLPH MACON COLLEGE, Virginia, MOUNT HOPE COLLEGE, near Baltimore, MARYLAND INSTITUTE OF INSTRUCTION and ST. MARY'S COLLEGE, Baltimore, and the UNIVERSITIES OF MICHIGAN and ALABAMA; and has been highly recommended by Professors Kingsley, Woolsey, Olmstead, and Gibbs, of Yale College; Professor Beck, of Harvard College; President Penney and Professor North, of Hamilton College; Professor Packard, of Bowdoin College; Professor Holland, of Washington College; Professor Fisk, of Amherst College, and by Professor Hackett, of Brown University;—also by Messrs. Dillaway and Gardner, of the Boston Latin School; Rev. Lyman Colman, of the English High School, Andover; Hon. John Hall, Principal of the Ellington School, Conn.; Mr. Shaler, Principal of the Connecticut Literary Institution, at Suffield; Simeon Hart, Esq., Farmington, Conn.; Professor Cogswell, of Round Hill School, Northampton; President Shannon, of Louisiana College, and by various periodicals.

As a specimen of the communications received from the above sources, the following extracts are given:—

It gives me great pleasure to bear my testimony to the superior merits of the Latin Grammar lately edited by Professor Andrews and Mr. Stoddard. I express most cheerfully, unhesitatingly, and decidedly, my preference of this Grammar to that of Adam, which has, for so long a time, kept almost undisputed sway in our schools.—*Dr. C. Beck, Professor of Latin in Harvard University.*

I know of no grammar published in this country, which promises to answer so well the purposes of elementary classical instruction, and shall be glad to see it introduced into our best schools.—*Mr. Charles K. Dillaway, Master of the Public Latin School, Boston.*

Your new Latin Grammar appears to me much better suited to the use of students than any other grammar I am acquainted with.—*Professor William M. Holland, Washington College, Hartford, Conn.*

I can with much pleasure say that your Grammar seems to me much better adapted to the present condition and wants of our schools than any one with which I am acquainted, and to supply that which has long been wanted—a good Latin grammar for common use.—*Mr. F. Gardner, one of the Masters Boston Lat. Sch.*

The Latin Grammar of Andrews and Stoddard is deserving, in my opinion, of the approbation which so many of our ablest teachers have bestowed upon it. It is believed that, of all the grammars at present before the public, this has greatly the advantage, in regard both to the excellence of its arrangement, and the accuracy and copiousness of its information; and it is earnestly hoped that its merits will procure for it that general favor and use to which it is entitled.—*H. B. Hackett, Professor of Biblical Literature in Newton Theol. Sem.*

The universal favor with which this Grammar is received was not unexpected. It will bear a thorough and discriminating examination. In the use of well-defined and expressive terms, especially in the syntax, we know of no Latin or Greek grammar which is to be compared to this.—*Amer. Quarterly Register.*

The Latin Grammar of Andrews and Stoddard I consider a work of great merit. I have found in it several principles of the Latin language correctly explained which I had myself learned from a twenty years' study of that language, but had never seen illustrated in any grammar. Andrews's First Lessons I con sider a valuable work for beginners, and in the sphere which it is designed to occupy, I know not that I have met its equal. — *Rev. James Shannon, President of College of Louisiana.*

These works will furnish a series of elementary publications for the study of Latin altogether in advance of any thing which has hitherto appeared, either in this country or in England. — ***American Biblical Repository.***

We have made Andrews and Stoddard's Latin Grammar the subject both of reference and recitation daily for several months, and I cheerfully and decidedly bear testimony to its superior excellence to any manual of the kind with which I am acquainted. Every part bears the impress of a careful compiler. The *principles* of syntax are happily developed in the rules, whilst those relating to the moods and tenses supply an important deficiency in our former grammars. The rules of prosody are also clearly and fully exhibited. — *Rev. Lyman Coleman, Principal of Burr Seminary, Manchester, Vt.*

I have examined Andrews and Stoddard's Latin Grammar, and regard it as superior to any thing of the kind now in use. It is what has long been needed, and will undoubtedly be welcomed by every one interested in the philology of the Latin language. We shall hereafter use it as a text-book in this institution — *Mr. Wm. H. Shaler, Principal of the Connecticut Lit. Institution at Suffield.*

This work bears evident marks of great care and skill, and ripe and accurate scholarship in the authors. It excels most grammars in this particular, that, while by its plainness it is suited to the necessities of most beginners, by its fulness and detail it will satisfy the inquiries of the advanced scholar, and will be a suitable companion at all stages of his progress. We cordially commend it to the student and teacher. — *Biblical Repository.*

Your Grammar is what I expected it would be — an excellent book, and just the thing which was needed. We cannot hesitate a moment in laying aside the books now in use, and introducing this. — *Rev. J. Penney, D. D., President of Hamilton College, New York.*

Your Grammar bears throughout evidence of original and thorough investigation and sound criticism. I hope, and doubt not, it will be adopted in our schools and colleges, it being, in my apprehension, so far as simplicity is concerned, on the one hand, and philosophical views and sound scholarship on the other, far preferable to other grammars; a work at the same time highly creditable to yourselves and to our country. — *Professor A. Packard, Bowdoin College, Maine.*

This Grammar appears to me to be accommodated alike to the wants of the new beginner and the experienced scholar, and, as such, well fitted to supply what has long been felt to be a great desideratum in the department of classical learning. — *Professor S. North, Hamilton College, New York.*

From such an examination of this Grammar as I have been able to give it, I do not hesitate to pronounce it superior to any other with which I am acquainted. I have never seen, any where, a greater amount of valuable matter compressed within limits equally narrow. — *Hon. John Hall, Prin. of Ellington School, Conn.*

We have no hesitation in pronouncing this Grammar decidedly superior to any now in use. — *Boston Recorder.*

I am ready to express my great satisfaction with your Grammar, and do not hesitate to say, that I am better pleased with such portions of the syntax as I have perused, than with the corresponding portions in any other grammar with which I am acquainted. — *Professor N. W. Fiske, Amherst College, Mass.*

I know of no grammar in the Latin language so well adapted to answer the purpose for which it was designed as this. The book of Questions is a valuable attendant of the Grammar. — *Simeon Hart, Esq., Farmington, Conn.*

This Grammar has received the labor of years, and is the result of much reflection and experience, and mature scholarship. As such, it claims the attention of all who are interested in the promotion of sound learning. — *N. Y. Obs.*

This Grammar is an original work. Its arrangement is philosophical, and its rules clear and precise, beyond those of any other grammar we have seen — *Portland Christian Mirror.*

PALMER'S ARITHMETIC.

ARITHMETIC,

ORAL AND WRITTEN,

PRACTICALLY APPLIED

BY MEANS OF

SUGGESTIVE QUESTIONS.

BY

THOMAS H. PALMER,

AUTHOR OF THE PRIZE ESSAY ON EDUCATION, ENTITLED THE "TEACHER'S MANUAL," "THE MORAL INSTRUCTOR," ETC.

Genuine education exercises deeper powers than the memory. That mind only is truly educated, which can hang up, as it were, a subject before it, view it distinctly in all its bearings, compare them carefully, weigh them justly, and then form a *correct decision for itself.* PROF. SHEDD.

BOSTON:
PUBLISHED BY CROCKER & BREWSTER,
NO. 47 WASHINGTON STREET.
1854.

STEREOTYPED BY
HOBART & ROBBINS,
NEW ENGLAND TYPE AND STEREOTYPE FOUNDERY,
BOSTON.

PREFACE.

I HAD three main objects in the preparation of this Treatise, namely, 1. So to simplify the arrangement of the subject as to give a clear general view, to enable the student to grasp the science as a whole; 2. To save his time by the introduction of shorter and more rapid processes; and, 3. To develop his reasoning powers by constant practice.

I. *Simplicity of arrangement.* The most inattentive observer can hardly have failed to notice the present confused classification of Arithmetic. Who among the brightest of our scholars can tell us the principles on which it is arranged? can, at one view, take a clear and definite survey of the whole ground? In the present work, the science has been analyzed into a few simple principles, traced back, indeed, to one grand element, into which all others can be resolved; being, in fact, nothing more than mere abbreviations, by the omission of steps become superfluous by practice.

II. *Rapid computation.* — The present method of calculating in our schools has been aptly characterized as an awkward mode of *spelling* figures by taking them singly, in place of *reading* them in groups. The pupil is so shackled with the multitude of words SUPPOSED *to be indispensable*, that his progress is necessarily slow and limping. If this work is used agreeably to the directions in p. viii. and elsewhere, these incumbrances will be almost entirely avoided *from the outset*, and a degree of rapidity quickly attained, which, at first sight, would hardly appear credible. This rapidity of operation, too, is much increased by a very great *diminution in the number of figures* employed, amounting, in many cases, to more than a half.

III. *Intellectual culture.* — One of the principal objects in the study of the mathematics is the mental culture it affords; and undoubtedly arithmetic cannot be acquired at all without *some* improvement of the mind. The difference in this respect between the present work and all others is simply this: In the latter, the pupil reads or commits to memory the reasonings of *another* mind; in the former, his rules are the result of *his own* mental processes. He is led forward by appropriate questions; but he cannot take a single step without active intellectual employment, thus continually eliciting energy of thought, clearness of expression, and fruitfulness of invention.

In the *Oral Arithmetic*, the lessons should be short, and the questions read slowly, till the pupils become prompt with their answers. Every section should be repeated till it is thoroughly mastered. Answers are occasionally given in the book; but this is merely to indicate the *form* of expression. None of the exercises will be found too difficult when taken in regular order. Obstacles can only arise from omissions.

In the *Written Arithmetic*, one question is generally answered by the adjoining one. At other times the answer is given, unless rendered unnecessary by the mode of proof being pointed out. The exemplifications *may* be studied on the slate, when the teacher is unable to find time to use the black-board; but such a course is by no means to be recommended. A few of the illustrations may appear obscure when read unconnected with the computations. But all such obscurities will vanish when they are read in their proper connection.

The paragraphs within brackets, [ ], are intended chiefly for the teacher.

In p. 199, l. 16 from the bottom, a question will be found leading to the formation of a formal rule by the pupil. This, or a similar question, can be repeated wherever the teacher may consider such formulas of any importance. But, where the pupil has acquired clear ideas of the *principles* of Arithmetic, which he cannot fail to do if he studies this book properly, formal rules will rarely, if ever, be found necessary. See p. 203 and 308—14, for another method of forming rules.

When the pupil is at a loss at a computation, the teacher should neither work it out for him, nor directly instruct him how to proceed. He should merely ask one or more leading questions, or refer him to the proper passage in the book, in order to elicit thought, and lead the pupil to rely on his own resources.

See p. viii. for particular instructions as to the method of using the book.

CONTENTS.

CHAPTER III.

FRACTIONS OF UNITY.

CHAPTER IV.

DETERMINATE FRACTIONS, OR FRACTIONS EXPRESSED IN CONCRETE WORDS, OR BY SIGNS.

CHAPTER V.

PART II. — WRITTEN ARITHMETIC.

CHAPTER I.

CHAPTER II.

THE SHORTENED PROCESSES OF INCREASE AND DECREASE OF INTEGERS AND DECIMAL FRACTIONS.

CHAPTER III.

THE SHORTENED PROCESSES OF INCREASE AND DECREASE APPLIED TO COMMON FRACTIONS AND DETERMINATE FRACTIONS.

CHAPTER IV.

CHAPTER V.

PRACTICAL APPLICATION OF ARITHMETIC BY MEANS OF RATIO AND PROPORTION.

SUPPLEMENT.

CONTRACTED MULTIPLICATION AND DIVISION OF DECIMAL FRACTIONS.

METHOD OF USING THIS BOOK.

Beginners should commence by daily practice on the *Improved Numeral Frame*, described in p. 11, followed by the exercises in *Oral Arithmetic*, Chap. I., pp. 15—66. When the learner, or class, has become familiar with a few sections in this chapter, he may commence the study of the first chapter in *Written Arithmetic*, pp. 108—124. The oral and written Exercises should now proceed simultaneously; the former clearing the way and facilitating the operations in the latter. Chaps. II. and III. Oral Arithmetic, should be fully mastered before the pupil commences Chap. III. Written Arithmetic.

Those who have already made some progress in Arithmetic should pursue pretty much the same course, omitting, or passing rapidly over such parts as may seem familiar to them, if any such there be. It is believed, however, that this will seldom be found to be the case.

See the Preface for the general Object of the work.

The following paragraphs on "Exchange" have been omitted by mistake in their proper place at p. 248.

Sterling may be changed to Canada money by adding $\frac{1}{9}$ to its amount; Why? [See Table of Provincial Currencies, p. 227.] It may be changed to New England by adding $\frac{1}{3}$; Why? To New York by adding $\frac{7}{9}$; Why? To Pennsylvania by adding $\frac{2}{3}$; Why? To South Carolina by adding $\frac{1}{27}$; Why?

Canada may be changed to New England money by adding $\frac{1}{5}$; to New York by adding $\frac{3}{5}$; to Pennsylvania by adding $\frac{1}{2}$.

New England may be changed to New York by adding $\frac{1}{3}$; to Pennsylvania, $\frac{1}{4}$.

Pennsylvania may be changed to New York by adding $\frac{1}{15}$.

South Carolina may be changed to Canada by adding $\frac{1}{14}$; to New England by adding $\frac{2}{7}$; to New York by adding $\frac{5}{7}$; to Pennsylvania by adding $\frac{17}{28}$.

Every one of these operations may be reversed by *subtracting* instead of *adding* the proportionate part, after changing the respective fraction by adding the numerator to the denominator for a new denominator, and allowing the numerator to remain as before. Thus in changing Canada to Sterling money, the fraction $\frac{1}{9}$ must be changed to $\frac{1}{10}$; while New York to Sterling requires $\frac{7}{9}$ to be changed to $\frac{7}{16}$. Why is this so? The reason will be readily discovered by attentively operating with a fraction that has 1 for numerator, and then enlarging it to 2, 3, &c.

The same principles are applicable to *Foreign* Exchange.

ARITHMETIC.

ADDRESSED TO TEACHERS.

Arithmetic is the science of numbers, or that branch of mathematics which teaches us to *combine and separate numbers* with ease and rapidity. The propriety of this definition appears evident, when it is considered that there are only two operations in arithmetic, *increase* and *decrease*. A number may be increased by one or more additions. It may be diminished by one or more subtractions. Such is the whole sum and substance of arithmetic. Now both these operations may be performed by Numeration, that is, by the successive addition or subtraction of a unit at a time. If, for instance, a child wishes to know the amount of 4 and 3, he enumerates *five, six, seven;* if he wants to know how many of 7 apples will remain after parting with three, he also enumerates, *six, five, four*.

But practice makes perfect. These slow though sure movements seem every day more tedious. The child's numeration, therefore, is transformed into Addition and Subtraction, by the *omission of superfluous steps*. Four and 3 at *one* step make 7; 3 from 7, in the same manner, make 4.

One more progressive movement completes the system. When several *equal* numbers are to be added or subtracted, the same desire for economy in time and space prompts to a similar *omission of superfluous steps*. In place of *three* steps to reckon 3, and 3, and 3, and 3, *one* is made to suffice, by saying 4 *times* 3; and the operation is reversed, by finding *at once* how many *times* 3 is contained in 12, in place of subtracting at *three* different times. Thus another important improvement is introduced, by substituting, when the numbers are identical, Multiplication for addition, Division for subtraction; and the whole secret lies, as before, in the *omission of superfluous steps*.

Involution and Evolution are merely slight modifications of

multiplication and division. The former differs in no other respect from multiplication, save that of the factors being identical. In the latter, the divisor and quotient being also equal factors, both are required to be found by *analysis of the dividend.*

Such, in all its beautiful simplicity, is our system of Decimal Arithmetic. The whole science may be considered a mere enumeration of numbers, with wider and more rapid steps, as the subject becomes more and more familiar. We shall endeavor to follow up this *natural mode* of improving the science, by suggesting to the pupil from time to time very many other cases in which both time and labor may be saved by the *omission of superfluous steps.*

For the sake of convenience, arithmetic is commonly arranged under two heads, viz., Oral Arithmetic and Written Arithmetic.

PART I.

ORAL ARITHMETIC.

INTRODUCTION,

ADDRESSED TO TEACHERS.

Although oral and written arithmetic occupy separate places in this book, it is intended that they should be taught simultaneously. By this means, while the pupil is acquiring a knowledge of Notation and Numeration in written arithmetic, he is mentally practising the elementary operations, preparatory to the larger processes on the slate; and thus, if care be taken to proceed with equal steps in both parts of the science, the oral operations precede, and smooth, and facilitate the progress of the pupil in *written* arithmetic, and enable him wholly to dispense with *artificial rules*, or, should these be thought necessary, to form them for himself.

Arithmetic is suited to the capacity of the youngest child that attends school. At a very early age children understand,

and delight to operate with, abstract as well as concrete numbers. Indeed, within certain limits, the effort is less in young than in older children. Experience shows that, of two boys of equal capacity, the one of seven, the other of nine or ten years of age, when commencing arithmetic, their progress is more nearly in an inverse than in a direct ratio to their respective ages. Nor is this all. The faculty of *attention*, one of the most important powers of the mind, is thus surely, yet easily and gradually, developed by oral arithmetic when properly taught. Indeed, children in the habitual practice of such operations, acquire the art of *reading* and *orthography* much more rapidly than those who delay the study of arithmetic to a later period.

This work commences with the most simple elements. The very first lessons may not, therefore, be necessary to all. The teacher should use a proper discretion in this respect. His judgment, indeed, must be relied on throughout the oral division of the subject, as well in the supply of additional questions for a dull class, as in the omission of such as may be superfluous for brighter scholars. No treatise can be formed exactly to suit every degree of mental capacity. It is believed, however, that it will be beneficial for all to proceed regularly through the book, and more especially through the fundamental exercises on the frame given below. Probably no other means would convey to the mind of childhood such clear and exact conceptions of the nature of our decimal arithmetic.

Description of the Improved Numeral Frame.

A numeral frame is an indispensable tool for a good teacher of arithmetic. No contrivance could be better adapted to convey clear ideas of the first principles of that science. But, as commonly arranged, with twelve wires, and twelve beads on each wire, its worth, if it has any, must be exceedingly limited. To fit it for our *decimal* system, it should be pierced for eleven wires only, ten of which should be at equal distances, the eleventh farther apart. In one already formed, the eleventh wire may be taken out, which will leave the others properly arranged. The beads, as in the old-fashioned frame, should be of two dissimilar colors, such as black and white, or blue and yellow. There should be *ten* beads on each wire, arranged as follows: three dark, two light, and two dark, three light. By this disposition of the beads, any number, not exceeding eleven hundred, can be *read* from the frame *at a glance.* Of the upper 100 beads, each stands for a single unit, while each of the 10 beads on the lower wire represents 100. By this simple arrangement into *units, tens,* and *hundreds,* our decimal system of arithmetic is distinctly presented to the *eye,* and is readily comprehended by the youngest pupil, and read off as easily and rapidly as by the use of figures.

The first lesson for beginners should be as follows: Let the teacher hold the frame so that all the beads fall to one side, and, passing one of those on the upper wire across, say, "There is one bead. Repeat after me, *one bead* [passing another across], *two beads,*" &c., till the first ten beads are all passed across and named.

The chief object of the second lesson is to qualify the pupil to read off any number, from one to ten, upon the frame *at a glance,* for which the arrangement of the beads upon the wires (3 and 2, and 2 and 3) renders it eminently qualified. The teacher should commence by a repetition of the first lesson, and, when that is perfectly known, proceed further, thus: Pass three beads across separately, and name as before, adding, "Now try to recollect three." Then pass those across at once on a different wire, and ask the number. If the child does not know, let this part of the lesson be repeated till the number three on the frame becomes familiar to the eye. Use four, five, six, and seven beads, till these numbers also can be readily named on the frame, *without counting them.* For eight, nine, and ten, direct attention to the opposite end of the wire; eight being known at first by two beads opposite, nine by one, ten by none.

When the pupil or class has become familiar with the first ten numbers, and able to name them on the frame at a glance, the difficulty of the nomenclature is nearly surmounted; the names of the others being chiefly a repetition of the first ten. It will be found most convenient and useful to give these larger numbers their *original* appellation before introducing their *common* or *contracted* names, as the former explains and simplifies the whole *system* of arithmetic. Let the class, therefore, be informed that *ten* has three different names, viz.

I. Ten, standing by itself, is called . . . . ten.
II. Ten, added to another number, ... teen.
III. In the plural, i. e., more than one ten, . . ty.

Applying this to the frame, when the beads on the first wire are passed across, we may say, "There is one ten." Then, passing one across on the second wire, "There is *oneteen;* another will make *twoteen;* three, *threeteen;* four, *fourteen;*" &c., to nineteen; and passing the last one across, "Now we have *twoty.*" Again, by passing the beads of the third wire singly across we shall have twoty-one, twoty-two, &c., to *threety;* and, continuing the operation, *fourty, fivety, sixty,* &c., to *ninety-nine,* the last bead giving *tenty,* or a hundred.

When the class has thus become familiar with the nomenclature, the abbreviations, which are few and simple, should be explained. The most difficult are the first two, *oneteen, twoteen,* which somehow have been changed to *eleven, twelve.* The rest are mere abbreviations. Thus, *three*teen and *three*ty have been shortened to *thir*teen and *thir*ty; *five*teen and *five*ty to *fif*teen and *fif*ty; lastly, *two*ty (originally *twain*ty) is now *twen*ty, and *four*ty is *for*ty.

This exercise is chiefly intended for those who have no knowledge of arithmetic. But it would be well for the whole school to go over the frame once or twice, as few children have such a clear and abiding idea of the meaning of *teen* or *ty* as may be thence derived.

When a child is at a loss for a reply to a question, it is not advisable to give him direct assistance. Such a course is apt to foster a habit of leaning on the teacher instead of his own resources. It is far better to aid him by a few suggestive questions. Suppose, for instance, he is at a loss for the amount of 52 and 27, it would be appropriate to ask, "How many are 50 and 20?" "Well, then, how many are 52 and 27?" Again, if he cannot answer, "Twelve are four times what number?" he might be asked, "How many fours in twelve?" "Well, then, twelve are four times what number?" Such instances, however, will rarely occur where the attention of a class is kept fully engaged during these exercises.

Every class in oral arithmetic should clearly understand that, during recitations, no question is to be repeated; a rule, indeed, that should never be broken. At the commencement of the study it might be proper for one of the class by turns to repeat each question after the teacher. But even this practice should not be continued too long. It is more important that children should be *good listeners* than *apt arithmeticians.* Let them learn, then, to listen with sufficient attention to hold on to what they hear, and they will acquire a habit that will exert a beneficial influence on the whole of after life.*

The questions should not be addressed to an individual, but to a class, so that every mind may be steadily engaged during the whole recitation. Children should never be fatigued with *long* lessons, but, while they *are* engaged, they should be *thoroughly* engaged. Let no habits of dreamy wandering of mind be *acquired* in school. If already acquired, they cannot too soon be broken up. Every week's delay will add to the difficulty. Let each question, then, be solved mentally by every individual in the class, and let the teacher wait till all have given the signal, before any one is called on to state the result, and this call should be made *promiscuously*, not in regular order, the more certainly to ensure the attention of all.

When an answer is given, the class should be asked if all agree; if not, every one who differs should give the result of his solution. Let the teacher then ask the first who spoke, "How do you know it to be so and so?" requiring an explanation of the *manner* in which it has been solved; those who differ being called on for the same purpose. The children should be required to give these explanations in a distinct and lucid style. Suppose, for instance, the question to be this: John has 6 nuts, Mary has 2, and James 4; how many nuts have the three children in all? The answer, "12 nuts," being given, and a child called on to state why it is so, may say, "If John has 6 nuts, and Mary 2, John and Mary together have 8, because 6 and 2 make 8; and, if we add James's 4, there will be 12, because 8 and 4 make 12." Such minuteness of detail may to some seem tedious and unnecessary; but, if the inappreciable

* At a convention of teachers, where the subject of oral arithmetic in schools was in discussion, a clergyman rose to give his experience in the matter. All are aware, said he, of the difficulty of maintaining the undivided attention of an auditory, even on subjects on which they feel a deep interest. The opening of a door, the restlessness of a child, the fall of a book: either of these is sufficient to distract the thoughts of a whole congregation. Now a course of oral arithmetic, properly conducted, presents a sure remedy for this serious evil. To classes thus trained, he continued, he could talk or read for half an hour at a time, and receive unwavering attention, although persons were going out and in, backwards and forwards, nearly the whole time.

Surely, when we consider how prone are mankind to wandering thoughts, how common it is for all "to hear a little, and guess the rest," we can scarcely overvalue so simple a cure for this vicious habit, which obstructs education alike in the primary school and the college, and exhibits its baneful effects in every grade of social life.

value of habits of close attention and of clearness of diction be taken into view, the labor will not be considered in vain. Let this system, then, of questioning by the teacher, and demonstration by the class, be steadily persevered in wherever practicable.

Finally, let this plan be thoroughly executed, without hurry, and success may be relied on. What man *has* done, man *can* do. A child taught thus will possess an immense, an obvious superiority over his fellows. In such practice, the mind is continually on the alert. That sluggish mental inactivity, which is the bane of our schools, and which nullifies most of their good effects, is completely banished. The powers of thought are kept constantly awake, steadily in employment. The pupil is not merely made an expert mathematician, but, what is of infinitely more importance, he is preparing the way for *self-education*, — he is acquiring the *control over his own mind*, one of the rarest and most valuable of all acquisitions.

N. B. None of the questions in this part of the book are to be worked out on the slate. It is to be, strictly speaking, *oral* arithmetic, with occasional illustrations by the teacher on the Frame and Black-board.

CHAPTER I.

INCREASE AND DECREASE OF INTEGERS, OR WHOLE NUMBERS.

EXERCISES ON THE NUMERAL FRAME.

SECTION I. *Increase and Decrease by Unity.*

1. ONE bead and one bead are how many beads? Two beads and one bead? Three beads and one bead? &c. [Continue to nine beads and one bead.]

2. If one bead be taken from ten beads, how many are left? One bead from nine beads, how many? One bead from eight beads, how many? &c. [Continue to one bead from one bead.]

3. One bead and one bead, how many ? One bead and two beads ? One bead and three beads ? &c. [Continue to one bead and nine beads.]

[Repeat the above till it becomes sufficiently familiar without the frame. Then repeat it again in *abstract* numbers, by omitting the word *bead*, till equally familiar. To a dull class, the word bead may be occasionally used in the following lessons.]

4. One and one are how many ? Take one from two, and how many remain ? Two from two ? How many are two times one ? [Here separate the two beads a little.] One time two ? Are two times one and one time two the same number, then ? How many ones in two ? Twos in two ?

5. Two and one are how many ? Take one from three, and how many remain ? Three from three ? How many are three times one ? One time three ? Are three times one and one time three the same, then ? How many ones in three ? Twos in three ? [One two and one over.] Threes in three ?

6. Three and one, how many ? Take one from four, how many remain ? Two from four ? How many are two and two, then ? Four from four ? How many are four times one ? One time four ? Are four times one and one time four the same, then ? How many ones in four ? Twos in four ? Threes in four ? [Separate the beads, or rather let one of the class separate them, at all such questions as the last two, till the class can separate them by the eye.] Fours in four ?

7. Four and one, how many ? Take one from five, how many ? Two from five ? Three from five ? How many are two and three, then ? Three and two ? Four from five ? Five from five ? How many are five times one ? One time five ? Are five times one and one time five the same, then ? How many ones in five ? Twos ? Threes ? Fours ? Fives ?

8. Five and one, how many ? One from six, how many ? Two from six ? How many are two and four, then ? How many are four and two ? Are two and four, then, the same as four and two ? Take three from six, how many ? Three and three, how many, then ? How many threes in six, then ? Take four from six ? Five from six ? Six from six ? How many are six times one ? One time six ? Are six times one and one time six the same, then ? How many ones in six ? Twos ? Threes ? Fours ? Fives ? Sixes ?

9. Six and one, how many ? One from seven, how many ? Two from seven ? How many are two and five, then ? Five

and two? Are two and five the same as five and two, then? Three from seven, how many? How many are three and four, then? Four and three? Are three and four, then, the same as four and three? Take four from seven? Five from seven? Six from seven? Seven from seven? How many ones in seven? Twos? Threes? Fours? Fives? Sixes? Sevens?

10. Seven and one, how many? Take one from eight, how many? Two from eight? How many are six and two, then? Two and six? Take three from eight? How many are five and three, then? Three and five? Take four from eight? How many are four and four, then? Two fours? Take five from eight? Six from eight? Seven from eight? Eight from eight? How many ones in eight? Twos? Threes? Fours? Fives? Sixes? Sevens? Eights?

11. Eight and one, how many? Take one from nine, how many? Two from nine? How many are seven and two, then? Two and seven? Take three from nine? How many are three and six, then? Six and three? Take four from nine? How many are four and five, then? Five and four? Take six from nine? Seven? Eight? Nine? How many ones in nine? Twos? Threes? Fours? Fives? Sixes? Sevens? Eights? Nines?

12. Nine and one, how many? Take one from ten? Two from ten? How many are eight and two, then? Two and eight? Take three from ten? How many are three and seven, then? Seven and three? Take four from ten? How many are six and four, then? Four and six? Take five from ten? How many fives in ten, then? Six from ten? Seven from ten? Eight? Nine? Ten? How many ones in ten? Twos? How many are five twos, then? Two fives? How many threes in ten? Fours? Fives? Sixes? Sevens? Eights? Nines? Tens?

13. Ten and one, how many? One from eleven, how many? Two from eleven? How many are nine and two, then? Two and nine? Three from eleven? How many are eight and three, then? Three and eight? Four from eleven? How many are seven and four? Four and seven? Five from eleven? How many are six and five, then? Five and six? Six from eleven? Seven from eleven? Eight? Nine? Ten? Eleven? How many ones in eleven? Twos? Threes? Fours? Fives? Sixes? Sevens, &c., to Elevens?

14. Eleven and one, how many? One from twelve? Two

from twelve? Ten and two? Three from twelve? Nine and three? Three and nine? Four from twelve? Eight and four? Four and eight? Five from twelve? Seven and five? Five and seven? Six from twelve? Six and six? How many sixes in twelve? How many twos? Seven from twelve? Eight from twelve? Nine? Ten? Eleven? Twelve? How many ones in twelve? Threes? Fours? Fives? Sevens? Eights? Nines? Tens? Elevens? Twelves?

15. Twelve and one, how many? Take one from thirteen? Take two? Eleven and two, then? Two and eleven? Take three from thirteen? Ten and three, then? Four from thirteen? Nine and four, then? Four and nine? Five from thirteen? Eight and five? Five and eight, then? Nine from thirteen? Nine and four, then? Four and nine? Ten from thirteen? Three and ten, then? Ten and three? Eleven from thirteen? Two and eleven, then? Twelve from thirteen? Thirteen from thirteen? How many ones in thirteen? Twos? Threes? Fours? Fives? Sixes? Sevens, &c., to Thirteens?

16. Thirteen and one, how many? One from fourteen? Two from fourteen? Twelve and two, then? Two and twelve? Three from fourteen? Eleven and three, then? Three and eleven? Four from fourteen? Ten and four? Four and ten? Five from fourteen? Nine and five? Five and nine? Six from fourteen? Eight and six? Six and eight? Seven from fourteen? Seven and seven, then? How many sevens in fourteen? Eight from fourteen? Nine? Ten? Eleven? Twelve? Thirteen? Fourteen? How many ones in fourteen? Twos? Sevens, then? Threes? Fours? Fives? Sixes? Eights? Nines? Tens? Elevens? Twelves? Thirteens? Fourteens?

17. How many are fourteen and one? Two from fifteen? Thirteen and two, then? Two and thirteen? Three from fifteen? Twelve and three, then? Three and twelve? Four from fifteen? Eleven and four, then? Four and eleven? Five from fifteen? Ten and five then? Five and ten? Six from fifteen? Nine and six, then? Six and nine? Seven from fifteen? Eight and seven? Seven and eight? Eight from fifteen? Nine from fifteen? Ten? Eleven? Twelve? Thirteen? Fourteen? Fifteen? How many ones in fifteen? Fifteen times one, then? How many twos in fifteen? Threes? Five times three, then? Three times five? Is three times five the same as five times three? Sixes in fifteen? Sevens, &c., to fifteens in fifteen?

18. Fifteen and one, how many? Take one from sixteen? Two? Fourteen and two, then? Two and fourteen? Three from sixteen? Thirteen and three, then? Three and thirteen? Four from sixteen? Twelve and four, then? Four and twelve? Five from sixteen? Eleven and five, then? Five and eleven? Six from sixteen? Ten and six, then? Six and ten? Seven from sixteen? Nine and seven, then? Seven and nine? Eight from sixteen? Eight and eight, then? How many eights in sixteen? Twos in sixteen, then? Nine from sixteen? Ten, &c., to sixteen from sixteen? How many ones in sixteen? Twos? Threes? Fours? How many are four times four, then? Fives, &c., to sixteens in sixteen?

19. Sixteen and one, how many? One from seventeen? Two? How many are two and fifteen, then? Fifteen and two? Three from seventeen? Fourteen and three, then? Three and fourteen? Four from seventeen? Thirteen and four, then? Four and thirteen? Five from seventeen? Twelve and five, then? Five and twelve? Six from seventeen? Eleven and six, then? Six and eleven? Seven from seventeen? Ten and seven, then? Seven and ten? Eight from seventeen? Nine and eight, then? Eight and nine? Ten from seventeen? Eleven? Twelve, &c., to seventeen from seventeen? How many ones in seventeen? Twos? Threes? Fours, &c., to seventeens?

20. Seventeen and one, how many? One from eighteen? Two? Sixteen and two, then? Two and sixteen? Three from eighteen? Fifteen and three, then? Three and fifteen? Four from eighteen? Fourteen and four, then? Four and fourteen? Five from eighteen? Thirteen and five, then? Five and thirteen? Six from eighteen? Twelve and six, then? Six and twelve? Seven from eighteen? Eleven and seven, then? Seven and eleven? Eight from eighteen? Ten and eight, then? Eight and ten? Nine from eighteen? Nine and nine, then? How many nines in eighteen? How many are two times nine, then? Ten from eighteen? Eleven? Twelve, &c., to eighteen from eighteen? How many ones in eighteen? Twos, &c., to eighteens in eighteen?

[The last seventeen lessons should now be repeated, over and over, without the frame, till perfectly familiar. The instrument, however, should always be at hand for illustration. For, let it be steadily borne in mind, that all the answers are to be results of the workings of the child's own mind, though, occasionally, the teacher may put them into a *better form of words* for him, some instances of which appear in this book. The frame will still be useful in many of the following lessons, especially when they are first recited. It should not, however, be used more frequently than is absolutely necessary. Perhaps it would be well that the members of the class should manage the frame by turns during the recitations.]

SECTION II. *Explanatory.*

1. WHAT is the meaning of *teen* in the words fourteen, sixteen, &c.* What does fourteen signify, then? *Ans.* Four *and* ten. Sixteen? Eighteen? What is the contracted. or common name for oneteen? For twoteen? Threeteén? Fiveteen? What is the meaning of *ty*, in the words sixty, seventy,* &c.? What is the contracted name for twoty, or twainty? For threety? For fivety? How many tens in fifty, then? In twenty? Forty? Ninety? Thirty? [Repeat the above at the commencement of each lesson, till perfectly familiar.]

2. How many are twenty and ten? [Show 20 on the upper two wires, and 10 on the fourth wire of the frame.] Twenty and twenty? [Show 20 on first and second, and 20 on ninth and tenth wires.] Twenty and thirty? Twenty and fifty? Thirty and ten? Thirty and fifty? Thirty and twenty? Thirty and sixty? Forty and twenty? Forty and ten? Forty and thirty? Fifty and thirty? Fifty and ten? Sixty and twenty? Ten and fifty? Thirty and fifty? Ten from twenty, how many? Twenty from forty? Thirty from seventy? Fifty from sixty? One hundred and two hundred? Two hundred and two hundred? Two hundred and five hundred? Four hundred and two hundred? Three hundred and four hundred? One hundred and eight hundred? Two hundred from eight hundred? Three hundred from seven hundred? Two hundred from one thousand? Two hundred from seven hundred?

3. Show ten on the frame. [Let the class show this by turns.] Twenty. Forty. Fifty. Seventy. Ninety. Sixteen. Eleven. Thirteen. Fifteen. Fourteen. Twelve. A hundred and sixteen. A hundred and twenty-three. Two hundred and

* See Introduction, page 12, near the bottom.

thirty-four. Three hundred and twenty-six. Six hundred and forty-five. [Continue and vary this exercise as far as necessary.]

SECTION III. *Increase and Decrease by a Small Number, without causing a change in the* TY *or tens.*

1. SIX and one, how many? Sixteen and one? Twenty-six and one? Sixty-six and one? Thirty-six and one? A hundred and twenty-six and one? Take one from seven; from seventeen; from thirty-seven; twenty-seven; fifty-seven; a hundred and thirty-seven.

2. Five and three, how many? Twenty-five and three? Fifteen and three? Fifty-five and three? A hundred and twenty-five and three? Two hundred and thirty-five and three? Take three from eight. Three from eighteen. Three from thirty-eight. Three from fifty-eight. Three from a hundred and twenty-eight. Three from a hundred and eight. Three from five hundred and sixty-eight. Three from two hundred and eighteen.

3. Four and two, how many? Fourteen and two? Fifty-four and two? Ninety-four and two? A hundred and four and two? Two hundred and fourteen and two? A thousand and four and two? Two from six? Two from sixteen? From thirty-six? From sixty-six? From a hundred and sixteen?

4. Two and two, how many? Twelve and two? Seventy-two and two? Thirty-two and two? Fifty-two and two? Twenty-two and two? A hundred and forty-two and two? Three hundred and forty-two and two? A thousand and forty-two and two? Two *from* four? From fourteen? From seventy-four? From ninety-four? From twenty-four? From a hundred and forty-four?

5. Two and three, how many? Twelve and three? Sixty-two and three? Forty-two and three? Eighty-two and three? A hundred and fifty-two and three? Three hundred and twenty-two and three? Two *from* five? From fifteen? From sixty-five? From forty-five? From a hundred and fifty-five? From three hundred and twenty-five?

6. Three and three, how many? Thirteen and three? Fifty-three and three? Ninety-three and three? Twenty-three

and three? Three hundred and sixty-three and three? Take three *from* six? From sixteen? From thirty-six? From seventy-six? From a hundred and six? From a hundred and sixteen? From a thousand and sixteen?

7. One and four, how many? Eleven and four? Forty-one and four? Twenty-one and four? Fifty-one and four? Ninety-one and four? A hundred and one and four? Six hundred and eleven and four? Take four *from* five? From fifteen? From forty-five? From two hundred and fifteen? From two hundred and sixty-five?

8. Six and two, how many? Sixteen and two? Eighty-six and two? Twenty-six and two? Five hundred and six and two? Take two *from* eight? From eighteen? From forty-eight? From three hundred and eighteen?

9. Two and seven, how many? Twelve and seven? Thirty-two and seven? Seventy-two and seven? Four hundred and two and seven? Seven hundred and twelve and seven? Five hundred and thirty-two and seven? Take seven *from* nine? From nineteen? From thirty-nine? From a hundred and nine? From five hundred and nineteen?

10. Two and five, how many? Twelve and five? Twenty-two and five? Eighty-two and five? Sixty-two and five? A hundred and two and five? Two hundred and twelve and five? Take five *from* seven? From seventeen? From twenty-seven? From fifty-seven? From a hundred and seventeen? From five hundred and seventeen?

11. One and seven, how many? Eleven and seven? Thirty-one and seven? Sixty-one and seven? Two hundred and twenty-one and seven? Five hundred and eleven and seven? Take seven *from* eight? From eighteen? From seventy-eight? From three hundred and eighteen?

12. Five and four, how many? Fifteen and four? Fifty-five and four? Two hundred and fifteen and four? Take four *from* nine? From nineteen? From forty-nine? From two hundred and nineteen?

13. One and eight, how many? Eleven and eight? Forty-one and eight? Seventy-one and eight? Ninety-one and eight? Three hundred and one and eight? Four hundred and eleven and eight? Take eight *from* nine? From nineteen? From ninety-nine? From two hundred and nine? From five hundred and nineteen?

SECTION IV. *Increase and Decrease by a Small Number, causing a change in the ty or tens.*

[LET the Teacher show here, by means of the frame, that in this, as in other successive lessons, the sum of the units *increases* the number of tens *by one;* and that, in the subtraction, the number of tens is, for a similar reason, *decreased* by one.]

1. Four and six are how many? Fourteen and six? Twenty-four and six? Forty-four and six? Sixty-four and six? Two hundred and thirty-four and six? Take four from ten? From twenty? From fifty? From a hundred? From two hundred and fifty? From two hundred and ten? From a thousand and ten?

2. Four and seven, how many? Fourteen and seven? Twenty-four and seven? A hundred and fifty-four and seven? Five hundred and thirty-four and seven? Nine hundred and sixty-four and seven? Take four from eleven? Four from twenty-one? Thirty-one? Fifty-one? Eighty-one? Sixty-one? A hundred and eleven? Two hundred and twenty-one? Three hundred and forty-one? Five hundred and fifty-one? One thousand and twenty-one?

3. Nine and five, how many? Nineteen and five? Twenty-nine and five? Thirty-nine and five? Fifty-nine and five? Seventy-nine and five? A hundred and nine and five? Three hundred and nineteen and five? Take five from fourteen? From twenty-four? From forty-four? Sixty-four? A hundred and four? Two hundred and thirty-four? Three hundred and fourteen? A thousand and fourteen?

4. Eight and three, how many? Eighteen and three? Twenty-eight and three? Fifty-eight and three? Eighty-eight and three? Two hundred and seventy-eight and three? Nine hundred and forty-eight and three? One thousand and eighteen and three? Take three from eleven? Three from twenty-one? From thirty-one? Sixty-one? Eighty-one? A hundred and twenty-one? A thousand and twenty-one?

5. Four and nine, how many? Fourteen and nine? Twenty-four and nine? Forty-four and nine? Eighty-four and nine? Three hundred and twenty-four and nine? Nine hundred and eighty-four and nine? Take four from thirteen? Four from twenty-three? Thirty-three? Sixty-three? A hundred and thirteen? A thousand and twenty-three?

6. Six and eight, how many? Sixteen and eight? Twenty-

six and eight? Fifty-six and eight? Eighty-six and eight? Sixty-six and eight? Nine hundred and seventy-six and eight? Take six from fourteen? Six from twenty-four? Thirty-four? Eighty-four? Fifty-four? A hundred and fourteen? Seven hundred and twenty-four? Eight hundred and sixty-four? Nine hundred and twenty-four? A thousand and fourteen?

SECTION V. *Increase and Decrease by a Small Number, causing a change in the Tens and the Hundreds.*

1. FIVE and seven are how many? Twenty-five and seven? Fifteen and seven? Forty-five and seven? Eighty-five and seven? Fifty-five and seven? Thirty-five and seven? Ninety-five and seven? Two hundred and ninety-five and seven? Seven hundred and ninety-five and seven? Nine hundred and ninety-five and seven? Take seven from twelve? Seven from twenty-two? Seven from fifty-two? Seven from ninety-two? Seven from a hundred and two? From four hundred and two? A thousand and two? Eight hundred and two?

2. Two and nine, how many? Thirty-two and nine? Twelve and nine? Fifty-two and nine? Eighty-two and nine? Ninety-two and nine? A hundred and forty-two and nine? Five hundred and ninety-two and nine? Nine hundred and ninety-two and nine? A thousand and twelve and nine? Take nine from eleven? Nine from a hundred and one? Nine from twenty-one? From fifty-one? From two hundred and one? From a thousand and one? From five hundred and twenty-one?

3. Six and four, how many? Thirty-six and four? Sixteen and four? Ninety-six and four? Nine hundred and ninety-six and four? Eighty-six and four? A thousand and six and four? Two hundred and sixteen and four? Take four from ten? Four from twenty? From a hundred? From three hundred? From two hundred and sixty? From a thousand? From five hundred and ten? From nine hundred and eighty?

4. Seven and five, how many? Ninety-seven and five? Two hundred and seventeen and five? Three hundred and ninety-seven and five? Five hundred and seventeen and five? Eight hundred and seven and five? Seven hundred and seven and five? Take five from twelve? From twenty-two? From

a hundred and two? From a thousand and two? From two hundred and twenty-two? From eight hundred and ninety-two?

5. Eight and seven, how many? Twenty-eight and seven? Eighteen and seven? Forty-eight and seven? Ninety-eight and seven? A hundred and eight and seven? Eight hundred and eight and seven? Nine hundred and eighteen and seven? Nine hundred and ninety-eight and seven? A thousand and eight and seven? Take seven from fifteen? From twenty-five? From ninety-five? From a hundred and five? From three hundred and fifteen? From six hundred and five? From four hundred and twenty-five? From a thousand and five? From a thousand and ninety-five?

6. Eight and nine, how many? Forty-eight and nine? Eighteen and nine? Ninety-eight and nine? Sixty-eight and nine? Eighty-eight and nine? A hundred and eight and nine? Two hundred and twenty-eight and nine? Six hundred and fifty-eight and nine? A thousand and eight and nine? Nine hundred and ninety-eight and nine? Take nine from seventeen? From eighty-seven? From twenty-seven? From a hundred and seven? From two hundred and twenty-seven? From three hundred and seven? From eight hundred and eighty-seven? From a thousand and seven? From a thousand and twenty-seven?

7. Seven and four, how many? Forty-seven and four? Seventeen and four? Ninety-seven and four? A hundred and seven and four? Eighty-seven and four? Three hundred and ninety-seven and four? Nine hundred and ninety-seven and four? A thousand and seventeen and four? Take four from eleven? From twenty-one? From sixty-one? From a hundred and one? From a thousand and one? From six hundred and eleven? From nine hundred and twenty-one? From three hundred and one?

SECTION VI. *Increase and Decrease by Larger Numbers, the Units causing no change in the Tens or Hundreds.*

1. Two and two, how many? Twelve and twelve? [Let the units occupy a wire in the frame between the two series of wires representing the tens.] How many are twelve and

twenty-two? Twenty-two and twenty-two? Thirty-two and twenty-two? Thirty-two and fifty-two? A hundred and twenty-two and forty-two? Two hundred and thirty-two and two hundred and forty-two? Take two from four? Twelve from twenty-four? Twelve from thirty-four? Twenty-two from thirty-four? Twenty-two from forty-four? Twenty-two from fifty-four? Thirty-two from fifty-four? Thirty-two from eighty-four? Fifty-two from eighty-four? A hundred and twenty-two from a hundred and sixty-four? Forty-two from a hundred and sixty-four? Two hundred and thirty-two from four hundred and seventy-four? Two hundred and forty-two from four hundred and seventy-four?

2. One and two, how many? Eleven and two? Eleven and twelve? Eleven and fifty-two? Eleven and twenty-two? Eleven and forty-two? Eleven and sixty-two? Eleven and eighty-two? Eleven and a hundred and twelve? Eleven and a thousand and two? Eleven from thirteen? Two from thirteen? Eleven from twenty-three? Twelve from twenty-three? Eleven from sixty-three? Fifty-two from sixty-three? Eleven from thirty-three? Twenty-two from thirty-three? Eleven from fifty-three? Forty-two from fifty-three? Eleven from seventy-three? Sixty-two from seventy-three? Eleven from a hundred and twenty-three? A hundred and twelve from a hundred and twenty-three? Eleven from a thousand and thirteen? A thousand and two from a thousand and thirteen?

3. Four and three, how many? Fourteen and thirteen? Thirty-four and thirteen? Thirty-four and fifty-three? A hundred and twenty-four and a hundred and fifty-three? Fourteen from twenty-seven? Thirteen from twenty-seven? Thirteen from forty-seven? Thirty-four from forty-seven? Thirty-four from eighty-seven? Fifty-three from eighty-seven? A hundred and twenty-four from two hundred and seventy-seven? A hundred and fifty-three from two hundred and seventy-seven?

4. Two and six, how many? Twelve and sixteen? Twelve and thirty-six? Twenty-two and fifty-six? Forty-two and fifty-six? Thirty-two and forty-six? A hundred and twenty-two and two hundred and forty-six? Six from eight? Two from eight? Twelve from twenty-eight? Sixteen from twenty-eight? Twelve from forty-eight? Thirty-six from forty-eight? Twenty-two from seventy-eight? Fifty-six from seventy-eight? Forty-two from ninety-eight? Fifty-six from ninety-eight?

Thirty-two from seventy-eight? Forty-six from seventy-eight? A hundred and twenty-two from three hundred and sixty-eight? Two hundred and forty-six from three hundred and sixty-eight?

5. Three and six, how many? Thirteen and sixteen? Twenty-three and thirty-six? Twenty-six and fifty-three? Forty-three and forty-six? A hundred and thirty-three and forty-six? Two hundred and twenty-six and three hundred and fifty-three? Three from nine? Six from nine? Thirteen from twenty-nine? Sixteen from twenty-nine? Twenty-three from fifty-nine? Thirty-six from fifty-nine? Twenty-six from seventy-nine? Fifty-three from seventy-nine? Forty-three from eighty-nine? Forty-six from eighty-nine? Forty-six from a hundred and seventy-nine? A hundred and thirty-three from a hundred and seventy-nine? Two hundred and twenty-six from five hundred and seventy-nine? Three hundred and fifty-three from five hundred and seventy-nine?

6. Two and three, how many? Twelve and thirteen? Twelve and twenty-three? Thirty-two and forty-three? Twenty-two and fifty-three? A hundred and thirty-two and forty-three? Two hundred and seventy-three and twelve? Two from five? Three from five? Twelve from twenty-five? Thirteen from twenty-five? Twelve from thirty-five? Twenty-three from thirty-five? Thirty-two from seventy-five? Forty-three from seventy-five? Twenty-two from seventy-five? Fifty-three from seventy-five? A hundred and thirty-two from a hundred and seventy-five? Forty-three from a hundred and seventy-five? Two hundred and seventy-three from two hundred and eighty-five? Twelve from two hundred and eighty-five?

7. Eight and one, how many? Eighteen and eleven? Twenty-eight and thirty-one? Seventy-eight and eleven? Two hundred and eleven and three hundred and eight? Two hundred and twenty-one and eighteen? Eight from nine? One from nine? Eighteen from twenty-nine? Eleven from twenty-nine? Twenty-eight from fifty-nine? Thirty-one from fifty-nine? Seventy-eight from eighty-nine? Eleven from eighty-nine? Two hundred and eleven from five hundred and nineteen? Three hundred and eight from five hundred and nineteen? Two hundred and twenty-one from two hundred and thirty-nine? Eighteen from two hundred and thirty-nine?

8. Four and two, how many? Fourteen and twelve? Twenty-four and twelve? Twenty-two and seventy-four? A hun-

dred and twelve and two hundred and fourteen? Two hundred and two and eight hundred and forty-four? Two from six? Four from six? Fourteen from twenty-six? Twelve from twenty-six? Twenty-four from thirty-six? Twelve from thirty-six? Twenty-two from ninety-six? Seventy-four from ninety-six? A hundred and twelve from three hundred and twenty-six? Two hundred and fourteen from three hundred and twenty-six? Two hundred and twenty-two from a thousand and sixty-six? Eight hundred and forty-four from a thousand and sixty-six?

9. Five and four, how many? Fifteen and fourteen? Thirty-four and fifteen? Twenty-four and thirty-five? Two hundred and twenty-five and three hundred and fourteen? Three hundred and seventy-five and seven hundred and fourteen? Five from nine? Four from nine? Fifteen from twenty-nine? Fourteen from twenty-nine? Thirty-four from forty-nine? Fifteen from forty-nine? Twenty-four from fifty-nine? Thirty-five from fifty-nine? Two hundred and twenty-five from five hundred and thirty-nine? Three hundred and fourteen from five hundred and thirty-nine? Three hundred and seventy-five from a thousand and eighty-nine? Seven hundred and fourteen from a thousand and eighty-nine?

10. Two and five, how many? Twelve and fifteen? Twenty-two and thirty-five? Sixty-two and fifteen? Two hundred and thirty-two and fifteen? Four hundred and twelve and six hundred and fifteen? Two from seven? Five from seven? Twelve from twenty-seven? Fifteen from twenty-seven? Twenty-two from fifty-seven? Thirty-five from fifty-seven? Sixty-two from seventy-seven? Fifteen from seventy-seven? Two hundred and thirty-two from two hundred and forty-seven? Fifteen from two hundred and forty-seven? Four hundred and twelve from a thousand and twenty-seven? Six hundred and fifteen from a thousand and twenty-seven?

11. Five and three, how many? Fifteen and thirteen? Twenty-five and thirty-three? A hundred and fifteen and a hundred and fifty-three? Two hundred and forty-five and eight hundred and thirteen? Five from eight? Three from eight? Fifteen from twenty-eight? Thirteen from twenty-eight? Twenty-five from fifty-eight? Thirty-three from fifty-eight? A hundred and fifteen from two hundred and sixty-eight? A hundred and fifty-three from two hundred and sixty-eight? Two hundred and forty-five from a thousand and

sixty-eight? Eight hundred and thirteen from a thousand and sixty-eight?

Section VII. *Practical Questions.*

[This section is the first that requires very close attention on the part of the pupils. The class, therefore, should once more be warned that the questions are not to be repeated, one of the main objects of the course being to make good and correct listeners.]

1. John picked up an apple in the orchard. His father gave him another. How many apples had he then? *Ans.* John had then two apples. How do you know? Because he picked up one in the orchard, and his father gave him one, and one and one make two. [Let the Practical Questions be resolved in this manner throughout. Use the frame when the child is at a loss.]

2. William's father gave him two plums, and his mother gave him two. How many plums did they both give him? *Ans.* They both gave him plums. How do you know?

3. If an orange cost five cents, and an apple two cents, how many cents will both cost? How do you know? How many will the orange cost more than the apple? How do you know? *Ans.* Because the orange cost five cents, and the apple two, and the difference between five and two is three.

4. Robert had six nuts, and gave two of them to his sister. How many had he left? Why?

5. If you had seven nuts in the one hand, and four in the other, how many would you have in the one more than in the other? How many in both hands? [Repeat "Why?" after every question.]

6. A man had four apples, which he divided equally between his two boys. How many did he give them apiece?

7. A lady gave two apples to each of her three children. How many did she give to them all?

8. A man bought six peaches, and divided them equally among his three children. How many did they get apiece?

9. William gave two nuts to each of his three brothers. How many did he give to them all?

10. If you had six cents in one pocket, and five cents in another, how many would you have in both?

11. If you had eleven cents, and were to pay away three, how many would you have remaining?

12. There were four boys, each of whom had three cents. How many had they altogether?

13. A lady, who had five children, wished to divide fifteen apples equally among them. How many would they get apiece?

14. A man, who had twenty-eight dollars in his pocket, paid away five of them for a barrel of flour. How much had he left?

15. John had sixteen marbles in a bag, and four in one of his pockets. How many were in both? How many more in the bag than in the pocket?

16. James gathered nineteen apples, and put them in a basket to carry them home; but, when he got there, he found only fifteen in the basket. How many had he lost out?

17. A man owing thirty-seven dollars, paid all but seven dollars. How much did he pay?

18. A man bought three calves for six dollars each. What did they cost him? If he were to sell them for twenty dollars, how much would he gain?

19. A merchant bought a firkin of butter for twelve dollars; but, as it was found to be damaged, he had to sell it for eight dollars. How much did he lose?

20. Three girls one day counted their needles, and put them into one cushion: one had five, another four, and the third eight. How many needles had they amongst them?

21. When the girls had finished sewing, they found that six of the needles had been broken, and agreed to share the loss equally among them. How many did each of them get, and how many had they in the whole?

22. A boy had twenty apples, which he divided among his companions as follows: to one he gave three; to another two; to another four; and to another five. How many did he give away, and how many had he left?

23. A man went to a provision store, and bought three pounds of beef for eighteen cents, and four pounds of mutton for twenty cents. He gave the man fifty cents. How much change should he receive?

24. A man bought a cabbage for five cents, some turnips for eight cents, some carrots for six cents, and a head of celery for five cents, and gave the owner a twenty-five cent piece to pay for them. What was his change?

25. A man bought a sleigh for fifteen dollars, and gave nine dollars to have it repaired and painted. He hired it to one of his neighbors for a few days for a dollar, and to another for a month for four dollars. He then sold it for twenty dollars. Did he gain or lose by the bargain, and how much?

26. Dick had twenty-five plums? He gave seven of them to Harry, and half of the rest to John. How many had each of them then?

27. A boy had twenty-five cents. He bought two oranges at six cents each, four apples at one cent each, and a lemon for four cents. How much money had he left?

28. Dick had ten peaches, Harry twelve, and Charles thirteen: Dick gave three to Stephen, Harry gave him six, and Charles gave him five. How many had Stephen, and how many had each left?

29. A boy, having received fifty cents for his work, bought a slate for ten cents, two pencils for a cent, a book of arithmetic for fifteen cents, and a book of geography for twenty cents. How many cents had he left?

30. John, having received fifty cents from his father, bought one of the Rollo books for twenty cents. His mother then gave him twenty-five cents, after which he bought for his sister one of the Lucy books for twenty cents, and a ribbon for six cents. How many cents had he left, and how many more did he spend for his sister than for himself?

SECTION VII. — *Increase and Decrease by Large Numbers, the Units causing a change in the Tens.*

1. FOUR and six, how many? Fourteen and sixteen? [Show, on the frame, that, when the units amount to ten or more, the number of ty or tens is increased by one.] Thirty-four and sixteen? Twenty-four and twenty-six? Thirty-four and twenty-six? Fifty-four and twenty-six? Forty-four and thirty-six? Thirty-four and twenty-six? Eighty-six and fourteen? A hundred and twenty-four and sixteen? Five hundred and fourteen and a hundred and sixteen? Fourteen from thirty? [Here show, on the frame, that, when there are not a sufficiency of units for the subtraction, one of the tens must be broken. Repeat illustrations of this kind on the frame wher-

ever necessary.] Sixteen from thirty? Sixteen from fifty? Thirty-four from fifty? Twenty-six from fifty? Thirty-four from sixty? Twenty-six from eighty? Forty-four from eighty? Thirty-four from ninety? Eighty-six from a hundred? A hundred and twenty-four from a hundred and forty? A hundred and sixteen from six hundred and thirty?

2. Eight and five, how many? Eighteen and fifteen? Twenty-eight and thirty-five? Forty-eight and forty-five? Twenty-eight and sixty-five? Sixty-eight and fifteen? A hundred and eighteen and a hundred and fifteen? Five hundred and twenty-eight and a hundred and fifteen? Five from thirteen? Eight from thirteen? Eighteen from thirty-three? Twenty-eight from sixty-three? Forty-five from ninety-three? Twenty-eight from ninety-three? Fifteen from eighty-three? A hundred and eighteen from two hundred and thirty-three? A hundred and fifteen from six hundred and forty-three?

3. Seven and four, how many? Seventeen and fourteen? Twenty-seven and fifty-four? Thirty-seven and fifty-four? Thirty-seven and twenty-four? Sixty-seven and twenty-four? A hundred and seventeen and thirty-four? A hundred and fifty-seven and thirty-four? Two hundred and thirty-seven and three hundred and fourteen? Fourteen from thirty-one? Twenty-seven from eighty-one? Thirty-seven from sixty-one? Twenty-four from sixty-one? Twenty-four from ninety-one? A hundred and seventeen from a hundred and ninety-one? A hundred and seventeen from a hundred and fifty-one? Thirty-four from a hundred and ninety-one? Two hundred and thirty-seven from five hundred and fifty-one? Three hundred and fourteen from five hundred and fifty-one?

4. Six and six, how many? Sixteen and twenty-six? Sixteen and sixteen? Thirty-six and forty-six? Fifty-six and twenty-six? Sixteen and fifty-six? A hundred and sixteen and a hundred and twenty-six? Three hundred and sixteen and two hundred and fifty-six? Six from twelve? Sixteen from forty-two? Forty-six from eighty-two? Fifty-six from eighty-two? Sixteen from seventy-two? A hundred and twenty-six from two hundred and forty-two? Three hundred and sixteen from five hundred and seventy-two? Two hundred and fifty-six from five hundred and seventy-two?

5. Seven and eight, how many? Seventeen and eighteen? Seventeen and forty-eight? Twenty-seven and fifty-eight? A hundred and twenty-seven and eighteen? Two hundred and

seventeen and five hundred and thirty-eight? Five hundred and thirty-seven and three hundred and forty-eight? Seven from fifteen? Eighteen from thirty-five? Seventeen from sixty-five? Fifty-eight from eighty-five? Twenty-seven from eighty-five? Eighteen from a hundred and forty-five? Two hundred and seventeen from seven hundred and fifty-five? Five hundred and thirty-seven from eight hundred and eighty-five? Three hundred and forty-eight from eight hundred and eighty-five?

6. Nine and four, how many? Nineteen and fourteen? Nineteen and twenty-four? Thirty-nine and thirty-four? A hundred and twenty-nine and a hundred and fourteen? Two hundred and nineteen and three hundred and thirty-four? Nine from thirteen? Four from thirteen? Nineteen from thirty-three? Twenty-four from forty-three? Thirty-nine from seventy-three? Thirty-four from seventy-three? A hundred and fourteen from two hundred and forty-three? Two hundred and nineteen from five hundred and fifty-three? Three hundred and thirty-four from five hundred and fifty-three?

7. Six and eight, how many? Sixteen and eighteen? Sixteen and thirty-eight? Seventy-six and eighteen? A hundred and sixteen and nine hundred and eighteen? Two hundred and thirty-six and two hundred and thirty-eight? Six hundred and sixteen and a hundred and twenty-eight? Six from fourteen? Eight from fourteen? Sixteen from thirty-four? Sixteen from fifty-four? Eighteen from ninety-four? Seventy-six from ninety-four? A hundred and sixteen from a thousand and thirty-four? Two hundred and thirty-eight from four hundred and seventy-four? Six hundred and sixteen from seven hundred and forty-four? A hundred and twenty-eight from seven hundred and forty-four?

8. Nine and seven, how many? Nineteen and seventeen? Forty-nine and seventeen? Sixty-nine and twenty-seven? Thirty-nine and forty-seven? Two hundred and nineteen and a hundred and forty-seven? Four hundred and nineteen and five hundred and seventeen? Nine from sixteen? Seven from sixteen? Nineteen from thirty-six? Seventeen from thirty-six? Seventeen from sixty-six? Sixty-nine from ninety-six? Forty-seven from eighty-six? A hundred and forty-seven from three hundred and sixty-six? Four hundred and nineteen from nine hundred and thirty-six? Five hundred and seventeen from nine hundred and thirty-six?

9. Two and nine, how many? Twelve and nineteen? Thirty-two and nineteen? Sixty-two and twenty-nine? A hundred and twelve and nineteen? Two hundred and fifty-two and two hundred and thirty-nine? Three hundred and thirty-two and three hundred and thirty-nine? A hundred and twelve and nine hundred and nineteen? Two from eleven? Nine from eleven? Nineteen from thirty-one? Thirty-two from fifty-one? Sixty-two from ninety-one? A hundred and twelve from a hundred and thirty-one? Two hundred and fifty-two from four hundred and ninety-one? Three hundred and thirty-nine from six hundred and seventy-one? A hundred and twelve from a thousand and thirty-one? Nine hundred and nineteen from a thousand and thirty-one?

10. Five and nine, how many? Fifteen and nineteen? Twenty-five and thirty-nine? Sixty-five and twenty-nine? A hundred and five and three hundred and nineteen? Six hundred and fifteen and two hundred and fifty-nine? Five from fourteen? Nine from fourteen? Fifteen from thirty-four? Thirty-nine from sixty-four? Twenty-nine from ninety-four? A hundred and five from four hundred and twenty-four? Six hundred and fifteen from eight hundred and seventy-four? Two hundred and fifty-nine from eight hundred and seventy-four?

Section VIII. — *Addition Circles.*

[Explanation.— Addition circles are formed by adding a number continually to itself, dropping the hundreds, till we return to the number with which we commenced. Take, for instance, 9, 18, 27, 36, 45, 54, 63, 72, 81, 90, 99, 108, then omit the hundreds, 17, 26, &c. Each circle should be recited again and again, till it can be repeated as fast as the words can be spoken. The circles for 7 and 3 may be formed in the same manner as for 9. The even numbers and 5 must be managed somewhat differently, to insure sufficient variety. As the repetition of 5 will only give fives and ciphers in the unit's place, after repeating the circle sufficiently with that number, commence the circle again with the other digits. Thus, commencing with 1 gives 6, 11, 16, &c., with 2 gives 7, 12, 17, &c., and so with the other digits. In forming circles with the even numbers, in order that there may be sufficient variety, it will be necessary to commence at least twice, namely, once with an odd figure, and once with an even one. The exercises that follow increase in difficulty, nearly in the order in which they are arranged. They will be formed with more ease, if the pupil observes that the addition of 9, 19, 29, &c., to any number diminishes the number of units by 1; the addition of 8, 18, &c., by 2, and 7, &c., by 3. It may also be noticed that the addition of 1, 11, 21, &c., 2, 12, 22, &c., 3, 13, 23, &c., increase the units by 1, 2, and 3, respectively.]

1. Form an addition circle with 9; with 7; with 3; dropping the hundreds; and continuing the process till the pupil again arrives at the number 9, or 7, or 3, with which he commenced.

2. Form a circle with 5, commencing with that number. Form it again, commencing with 6; with 7; with 8; with 9.

3. Form addition circles with 2, 4, 6, 8; commencing, the first time, with these numbers severally; afterwards, with an additional 1, 3, 5, or 7.

4. Form addition circles with 11, 12, 13, 19, 18, 17, 16, 15, in a similar manner as with the single digits.

5. Form circles with the numbers 21 to 29, varying the commencement by the addition of other numbers as above, especially the even numbers and 25.

[These exercises ought to be repeated till the circles can be recited rapidly and without hesitation.]

SECTION IX. — *Increase and Decrease by Large Numbers, the Tens causing a change in the Hundreds.*

1. SEVENTY and seventy, how many? [A hundred and forty. Why? Because *seven*ty and *seven*ty make *fourteen*ty, and as ten ty make a hundred, fourteen ty are a hundred and forty. Repeat this idea in the following questions, till the principle is sufficiently familiar to the class, when the questions may be answered in the *usual terms.*] Ninety and forty? [*Ans.* A hundred and thirty. Why, &c.? Some pupils will add numbers like these more rapidly by *subtracting* a sufficient number of *ty* from the smaller number to increase the larger number to 100. Thus, 80+70=170—20=150; 90+40=140—10=

130; 84+42=146—20=126; 365+253=668—40=618. Observe, however, that *one* of the *two steps* in each of the above operations should be dropped as superfluous after a little practice.] Sixty and fifty? Eighty and ninety? Forty-two and eighty-four? Thirty-six and eighty-two? Fifty-four and sixty-two? Seventy-eight and eighty-one? A hundred and fifty and seventy? [Explain again, if any hesitation here.] Two hundred and eighty and a hundred and sixty? Three hundred and sixty-five and two hundred and fifty-three?

2. Seventy from a hundred and forty? [A hundred and forty being the same as fourteen *ty*, this question is, in fact, *seven*ty from *fourteen*ty, which is *seven*ty.] Ninety from a hundred and thirty? Forty-two from a hundred and twenty-six? Fifty-four from a hundred and sixteen? Seventy-eight from a hundred and fifty-nine? A hundred and sixty from four hundred and forty? Three hundred and sixty-four from six hundred and eighteen?

3. Sixty and eighty, how many? Seventy-three and forty-five? Seventy-two and seventy-six? Thirty-six and ninety-three? Five hundred and forty-six and two hundred and ninety-one? Five hundred and forty-six and two hundred and ninety-two? [Seven hundred and *thirteen*ty-eight, or, &c.] Three hundred and eighty-eight and two hundred and forty-one? [Five hundred and *twelve*ty-nine, or, &c.] Sixty from a hundred and forty? [or *fourteen*ty.] Forty-five from a hundred and eighteen? [or *eleven*ty-eight.] Seventy-two from a hundred and forty-eight? Thirty-six from a hundred and twenty-nine? [twelvety-nine.] Two hundred and ninety-two from eight hundred and thirty-eight? [seven hundred and thirteenty-eight.] Three hundred and eighty-eight from six hundred and twenty-nine? [five hundred and twelvety-nine.]

4. Ninety and eighty, how many? Seventy-three and sixty-two? Eighty-three and ninety-six? A hundred and fifty-four and two hundred and eighty-two? Three hundred and twenty-nine and two hundred and ninety? Four hundred and twenty-two and five hundred and thirty-three? Ninety from a hundred and seventy? [seventeenty.] Eighty from a hundred and seventy? Sixty-two from a hundred and thirty-five? [thirteenty-five.] Eighty-three from a hundred and seventy-nine? [seventeenty-nine.] Two hundred and eighty-two from four hundred and thirty-six? [three hundred and thirteenty-six.] Two hundred and ninety from six hundred and nine-

teen? [five hundred and eleventy-nine.] Four hundred and twenty-two from a thousand and fifty-five? [nine hundred and fifteenty-five.]

5. Seventy and sixty, how many? Sixty-two and eighty-two? Two hundred and thirty-six and a hundred and eighty-two? Two hundred and seventy-one and two hundred and eighty-four? Six hundred and ninety-five and three hundred and seventy-three? Seventy from a hundred and thirty? [thirteenty.] Sixty-two from a hundred and forty-four? Two hundred and eighty-two from four hundred and eighteen? [three hundred and eleventy-eight.] Two hundred and seventy-one from five hundred and fifty-five? Six hundred and ninety-five from a thousand and sixty-eight? [nine hundred and sixteenty-eight.]

6. Sixty-three and seventy-two, how many? Two hundred and fifty-four and three hundred and sixty-five? Four hundred and seventy-two and two hundred and fifty-three? Eight hundred and fifteen and a hundred and ninety-four? Sixty-three from a hundred and thirty-five? Two hundred and fifty-four from six hundred and nineteen? Four hundred and seventy-two from seven hundred and twenty-five? A hundred and ninety-five from a thousand and nine? Eight hundred and fifteen from a thousand and nine?

7. Forty-two and ninety-three, how many? Two hundred and sixty-seven and three hundred and sixty-one? Four hundred and forty-three and three hundred and ninety-four? Five hundred and seventy-eight and three hundred and sixty-one? Forty-two from a hundred and thirty-five? Two hundred and sixty-seven from six hundred and twenty-eight? Three hundred and ninety-four from eight hundred and thirty-seven? Five hundred and seventy-eight from nine hundred and thirty-nine?

8. Seventy-six and forty-three, how many? Two hundred and fifty-six and six hundred and eighty-three? Four hundred and ninety-six and two hundred and sixty-three? Eight hundred and seventy-six and ninety-three? Two hundred and forty-six and six hundred and eighty-three? Seventy-six from a hundred and nineteen? Forty-three from a hundred and nineteen? Six hundred and eighty-three from nine hundred and thirty-nine? Four hundred and ninety-six from seven hundred and fifty-nine? Ninety-three from nine hundred and sixty-nine? Six hundred and eighty-three from nine hun-

dred and thirty-nine? Two hundred and forty-six from nine hundred and thirty-nine?

Section X. — *Increase and Decrease by Large Numbers, the Units causing a change in the Tens, and the Tens in the Hundreds.*

1. Seventy-nine and sixty-five, how many? Two hundred and thirty-nine and three hundred and eighty-five? A hundred and fifty-nine and three hundred and eighty-five? Three hundred and seventy-nine and two hundred and eighty-five? Four hundred and fifty-nine and four hundred and fifty-five? Seventy-nine from a hundred and forty-four? Sixty-five from a hundred and forty-four? Three hundred and eighty-five from six hundred and twenty-four? Three hundred and eighty-five from five hundred and forty-four? Two hundred and eighty-five from six hundred and sixty-four? Four hundred and fifty-five from nine hundred and fourteen? Four hundred and fifty-nine from nine hundred and fourteen?

2. Sixty-three and eighty-seven, how many? Two hundred and seventy-three and ninety-seven? Five hundred and twenty-three and two hundred and eighty-seven? Seven hundred and fifty-three and two hundred and sixty-seven? Six hundred and forty-three and two hundred and eighty-seven? Seven hundred and fifty-three and two hundred and sixty-seven? Six hundred and forty-three and two hundred and eighty-seven? Five hundred and thirty-three and four hundred and sixty-seven? Sixty-three from a hundred and fifty? Eighty-seven from a hundred and fifty? Two hundred and seventy-three from three hundred and seventy? Two hundred and eighty-seven from eight hundred and ten? Two hundred and sixty-seven from a thousand and twenty? Six hundred and forty-three from nine hundred and thirty? Four hundred and sixty-seven from a thousand?

3. Fifty-six and sixty-seven, how many? Two hundred and sixty-six and a hundred and eighty-seven? Three hundred and thirty-six and two hundred and eighty-seven? Four hundred and forty-six and two hundred and ninety-seven? Eight hundred and eighty-six and fifty-seven? Four hundred and seventy-six and three hundred and eighty-seven? Fifty-six from a

hundred and twenty-three? Sixty-seven from a hundred and twenty-three? A hundred and eighty-seven from four hundred and fifty-three? Three hundred and thirty-six from six hundred and forty-three? Two hundred and ninety-seven from seven hundred and forty-three? Fifty-seven from nine hundred and forty-three? Three hundred and eighty-seven from eight hundred and sixty-three? Four hundred and seventy-six from eight hundred and sixty-three?

4. Eighty-nine and thirty-six, how many? Two hundred and fifty-nine and two hundred and fifty-six? Five hundred and nineteen and three hundred and eighty-six? Seven hundred and thirty-nine and two hundred and ninety-six? Six hundred and sixty-nine and two hundred and thirty-six? Five hundred and forty-nine and four hundred and eighty-six? Thirty-six from a hundred and twenty-five? Eighty-nine from a hundred and twenty-five? Two hundred and fifty-six from five hundred and fifteen? Three hundred and eighty-six from nine hundred and five? Two hundred and ninety-six from a thousand and thirty-five? Six hundred and sixty-nine from nine hundred and five? Four hundred and eighty-six from a thousand and thirty-five? Five hundred and forty-nine from a thousand and thirty-five?

5. Twenty-seven and ninety-four, how many? Three hundred and thirty-seven and three hundred and sixty-four? Five hundred and seventy-seven and three hundred and fifty-four? Two hundred and seven and two hundred and ninety-four? Three hundred and sixty-four and two hundred and forty-seven? Ninety-four from a hundred and twenty-one? Twenty-seven from a hundred and twenty-one? Three hundred and thirty-seven from seven hundred and one? Three hundred and fifty-four from nine hundred and thirty-one? Two hundred and ninety-four from five hundred and one? Two hundred and forty-seven from six hundred and eleven? Three hundred and sixty-four from six hundred and eleven?

6. Thirty-four and eighty-eight, how many? Two hundred and sixty-four and a hundred and fifty-eight? Three hundred and ninety-four and two hundred and sixty-eight? A hundred and seventy-four and two hundred and seventy-eight? A hundred and fourteen and six hundred and eighty-eight? Thirty-four from a hundred and twenty-two? Eighty-eight from a hundred and twenty-two? Two hundred and sixty-four from four hundred and twenty-two? Three hundred and ninety-

four from six hundred and sixty-two? Two hundred and seventy-eight from four hundred and fifty-two? A hundred and fourteen from eight hundred and two?

7. Thirty-four and nineteen, how many? Ninety-four and ninety-nine? Two hundred and sixty-four and two hundred and thirty-nine? Five hundred and twenty-four and three hundred and eighty-nine? Two hundred and seventy-four and three hundred and sixty-nine? Six hundred and eighty-four and two hundred and forty-nine? A hundred and four and three hundred and ninety-nine? Thirty-four from fifty-three? Nineteen from fifty-three? Ninety-four from a hundred and ninety-three? Two hundred and sixty-four from five hundred and three? Three hundred and sixty-nine from six hundred and forty-three? Two hundred and forty-nine from nine hundred and thirty-three? A hundred and four from five hundred and three? Three hundred and ninety-nine from five hundred and three?

8. Sixty-five and forty-seven, how many? Thirty-five and a hundred and eighty-seven? Two hundred and sixty-five and three hundred and seventy-seven? Four hundred and ninety-five and five hundred and forty-seven? A hundred and fifteen and two hundred and ninety-seven? Three hundred and forty-five and a hundred and sixty-seven? Two hundred and eighty-five and six hundred and forty-seven? Sixty-five from a hundred and twelve? Forty-seven from a hundred and twelve? Two hundred and sixty-five from six hundred and forty-two? Four hundred and ninety-five from a thousand and forty-two? Three hundred and forty-five from five hundred and twelve? Six hundred and forty-seven from nine hundred and thirty-two? Two hundred and eighty-five from nine hundred and thirty-two?

9. Thirty-six and eighty-eight, how many? Two hundred and fifty-six and seventy-eight? Three hundred and forty-six and a hundred and seventy-eight? A hundred and ninety-six and four hundred and eighteen? Six hundred and forty-six and three hundred and fifty-eight? Two hundred and sixteen and three hundred and eighty-eight? Five hundred and twenty-six and three hundred and ninety-eight? Thirty-six from a hundred and twenty-four? Eighty-eight from a hundred and twenty-four? Two hundred and fifty-six from three hundred and thirty-four? Three hundred and forty-six from five hundred and twenty-four? A hundred and ninety-six from six

hundred and fourteen? Three hundred and fifty-eight from a thousand and four? Three hundred and eighty-eight from six hundred and four? Three hundred and ninety-eight from nine hundred and twenty-four? Five hundred and twenty-six from nine hundred and twenty-four?

10. Forty-five and seventy-five, how many? A hundred and twenty-five and a hundred and ninety-five? Sixty-five and eight hundred and eighty-five? Two hundred and thirty-five and three hundred and seventy-five? Six hundred and fifty-five and three hundred and forty-five? Eight hundred and sixty-five and a hundred and seventy-five? Four hundred and seventy-five and two hundred and eighty-five? Forty-five from a hundred and twenty? Eighty-five from a hundred and twenty? A hundred and twenty-five from three hundred and twenty? Sixty-five from nine hundred and fifty? Two hundred and thirty-five from six hundred and ten? Six hundred and fifty-five from a thousand? Eight hundred and sixty-five from a thousand and forty? Four hundred and seventy-five from seven hundred and sixty? Two hundred and eighty-five from seven hundred and sixty?

SECTION XI.—*Miscellaneous.*

1. NINETY-SEVEN and six, how many? A hundred and forty-four and two hundred and forty-three? Two hundred and sixty-five and three hundred and seventy-two? Five hundred and ninety-seven and two hundred and forty-eight? Two hundred and sixty-five and six hundred and forty-seven? Two hundred and fifty-eight and four hundred and fifty-seven? Six from a hundred and three? Ninety-seven from a hundred and three? A hundred and forty-four from three hundred and eighty-seven? Three hundred and seventy-two from six hundred and thirty-seven? Two hundred and forty-eight from eight hundred and forty-five? Six hundred and forty-seven from nine hundred and twelve? Four hundred and fifty-seven from seven hundred and fifteen? Two hundred and fifty-eight from seven hundred and fifteen?

2. Seventy-five and eighty-seven, how many? Two hundred and five and three hundred and ninety-seven? Three hundred and fifty-four and five hundred and twenty-three? Two hun-

dred and sixty-eight and seven hundred and thirty-six? Three hundred and thirty-two and two hundred and fifty-four? Four hundred and twenty-nine and three hundred and eighty-eight? Seventy-five from a hundred and sixty-two? Eighty-seven from a hundred and sixty-two? Three hundred and ninety-seven from six hundred and two? Five hundred and twenty-three from eight hundred and seventy-seven? Seven hundred and thirty-six from a thousand and four? Two hundred and fifty-four from five hundred and eighty-six? Three hundred and eighty-eight from eight hundred and seventeen? Three hundred and twenty-nine from eight hundred and seventeen?

3. Thirty-six and ninety-eight, how many? Four hundred and eleven and two hundred and seventeen? Five hundred and thirty-six and two hundred and eighty-eight? Three hundred and five and six hundred and eighteen? Five hundred and fifty-two and four hundred and eighty-nine? Six hundred and thirty-seven and two hundred and eighty-eight? Thirty-six from a hundred and thirty-four? Ninety-eight from a hundred and thirty-four? Two hundred and seventeen from six hundred and twenty-eight? Two hundred and eighty-eight from eight hundred and twenty-four? Six hundred and eighteen from nine hundred and twenty-three? Four hundred and eighty-nine from a thousand and forty-one? Two hundred and eighty-eight from nine hundred and twenty-five? Six hundred and thirty-seven from nine hundred and twenty-five?

4. Ninety-eight and seventy-four, how many? Three hundred and fifty-seven and three hundred and sixty-two? Two hundred and forty-eight and three hundred and eighty-five? Five hundred and twenty-six and four hundred and twenty-one? Seven hundred and thirty-eight and two hundred and eighty-nine? Seventy-four from a hundred and seventy-two? Ninety-eight from a hundred and seventy-two? Three hundred and sixty-two from seven hundred and nineteen? Three hundred and eighty-five from six hundred and forty-three? Four hundred and twenty-one from nine hundred and forty-seven? Two hundred and eighty-nine from a thousand and twenty-seven? Seven hundred and thirty-eight from a thousand and twenty-seven?

5. Seventy-two and ninety-nine, how many? A hundred and fifty-six and a hundred and forty-two? A hundred and fifty-six and two hundred and eighty-seven? Two hundred and seventy-six and three hundred and forty-eight? Eight hundred

and seventeen and two hundred and fifteen? A hundred and forty-six and two hundred and twenty-two? Three hundred and fifty-seven and two hundred and forty-nine? Ninety-nine from a hundred and seventy-one? Seventy-two from a hundred and seventy-one? A hundred and forty-two from two hundred and ninety-eight? Two hundred and eighty-seven from four hundred and forty-three? Three hundred and forty-eight from six hundred and twenty-four? Two hundred and fifteen from a thousand and thirty-two? A hundred and forty-six from three hundred and sixty-eight? Three hundred and fifty-seven from six hundred and six?

6. Sixty-five and thirty-four, how many? A hundred and twenty-six and two hundred and thirty-seven? Two hundred and fifty-four and three hundred and sixty-seven? Four hundred and thirty-eight and two hundred and twenty-one? Six hundred and forty-five and three hundred and twenty-seven? Two hundred and fifty-four and six hundred and seventy-one? Sixty-five from ninety-nine? Thirty-four from ninety-nine? A hundred and twenty-six from three hundred and sixty-three? Two hundred and fifty-four from six hundred and twenty-one? Four hundred and thirty-eight from six hundred and fifty-nine? Six hundred and forty-five from nine hundred and seventy-two? Two hundred and fifty-four from nine hundred and twenty-five? Six hundred and seventy-one from nine hundred and twenty-five?

7. Twenty-seven and thirty-three, how many? A hundred and fifty-six and a hundred and fifty-nine? Two hundred and thirty-eight and three hundred and seventy-nine? Two hundred and sixty-five and a hundred and twenty-one? Four hundred and forty-six and three hundred and thirty-two? Five hundred and twenty-seven and two hundred and eighty-eight? Six hundred and ninety-two and a hundred and seventeen? Twenty-seven from sixty? Thirty-three from sixty? A hundred and fifty-six from three hundred and nine? Two hundred and thirty-eight from six hundred and seventeen? Two hundred and sixty-five from three hundred and eighty-six? Four hundred and forty-six from seven hundred and seventy-eight? Five hundred and twenty-seven from eight hundred and fifteen? Six hundred and ninety-two from eight hundred and nine?

8. Thirty-nine and fifty-eight? A hundred and sixty-five and two hundred and thirty-nine? Five hundred and twenty-

three and three hundred and sixty-six? Two hundred and forty-five and a hundred and eighty-nine? Six hundred and twenty and four hundred and twenty-six? Five hundred and eight and four hundred and ninety-nine? A hundred and sixty-two and three hundred and eighty-nine? Thirty-nine from ninety-seven? Fifty-eight from ninety-seven? A hundred and sixty-five from four hundred and four? Five hundred and twenty-three from eight hundred and eighty-nine? Two hundred and forty-five from four hundred and thirty-four? Six hundred and twenty from eight hundred and forty-six? Five hundred and eight from a thousand and seven? A hundred and sixty-two from five hundred and fifty-one? Three hundred and eighty-nine from five hundred and fifty-one?

SECTION XII.—*Increase and Decrease by more than one Number.*

[IN the questions requiring subtraction below, the teacher can either direct his pupils to take the smaller number from the sum of the larger, or to subtract them separately. The former is the easier method; the latter will insure more mental discipline.]

1. Two, and three, and five, how many? Five, and four, and eight? Six, and nine, and ten? Fourteen, and nine, and five? Sixteen, and fourteen, and eight, less six? Twenty, and eight, less five, and thirteen? From ten take five and three? From ten take two and three? From ten take two and five? From seventeen take eight and four? From seventeen take five and four? From seventeen take eight and five? From twenty-five take nine and ten? From twenty-five take six and nine? From twenty-five take six and ten? From twenty-eight take nine and five? From twenty-eight take fourteen and nine? From twenty-eight take fourteen and five? From thirty-two take fourteen and eight less six? From thirty-two take sixteen and fourteen? From thirty-two take sixteen and eight less six? From thirty-six take eight less five, and thirteen? From thirty-six take twenty, and eight less five? From thirty-six take twenty and thirteen?

2. Four, and five, and three, and eight, how many? Six, and seven, and four, and thirteen less nine? Seventeen less

three, and five, and eight less four? Nine, and eighteen less seven, and eight, and six? From twenty, take five, and three, and eight? From twenty, take four, and five, and three? From twenty, take four, and five, and eight? From twenty, take four, and three, and eight? From twenty-one, take seven, and four, and thirteen less nine? From twenty-one, take six, and seven, and four? From twenty-one, take six, and seven, and four? From twenty-one, take six, and seven, and thirteen less nine? From thirty-one, take five, and eight, and four? From thirty-one, take seventeen less three, and five, and eight? From thirty-one, take seventeen less three, and five, and four? From thirty-one, take seventeen less three, and eight, and four? From thirty-four, take eighteen less seven, and eight, and six? From thirty-four, take nine, and eighteen less seven, and eight? From thirty-four, take nine, and eighteen less seven, and six? From thirty-four, take nine, and eight, and six?

3. Seventeen, and eight, and twenty-four less nine, how many? Seventeen, and six, and thirty-five less seven? Twenty-three, and thirteen, and fifteen less seven? A hundred and five, and eight, and seventeen less three? From forty, take seventeen, and eight? From forty, take eight, and twenty-four less nine? From forty, take seventeen, and twenty-four less nine? From fifty-one, take seventeen, and six? From fifty-one, take six, and thirty-five less seven? From fifty-one, take seventeen, and thirty-five less seven? From forty-four, take twenty-three, and thirteen? From forty-four, take thirteen, and fifteen less seven? From forty-four, take twenty-three, and fifteen less seven? From a hundred and twenty-seven, take eight, and seventeen less three? From a hundred and twenty-seven, take a hundred and five, and eight? From a hundred and twenty-seven, take a hundred and five, and seventeen less three?

4. Fourteen, and eighteen, and fifteen less four, how many? Thirty-four, and twenty-six less five, and seventeen? Forty-seven, and fourteen less three, and thirty-two? Fifteen, and twenty-four, and six, and eight? From forty-three, take fourteen, and eighteen? From forty-three, take eighteen, and fifteen less four? From forty-three, take fourteen, and fifteen less four? From seventy-two, take thirty-four, and twenty-six less five? From seventy-two, take twenty-six less five, and seventeen? From seventy-two, take thirty-four and seventeen? From ninety, take forty-seven, and fourteen less three? From

ninety, take forty-seven and thirty-two? From ninety, take fourteen less three, and thirty-two? From fifty-three, take fifteen, and twenty-four, and six? From fifty-three, take twenty-four, and six, and eight? From fifty-three, take fifteen, and six, and eight? From fifty-three, take fifteen, and twenty-four, and eight?

5. Seventeen, and four, and sixteen, and five, how many? Twenty-three less eight, and twenty-nine less seven, and thirteen? Seventeen less four, and eighteen less two, and eight? Fourteen, and fifteen, and seventeen, and nine? From forty-two, take seventeen, and four, and sixteen? From forty-two, take four, and sixteen, and five? From forty-two, take sixteen, and five, and seventeen? From forty-two, take five, and seventeen, and four? From forty-two, take seventeen, and sixteen, and five? From forty-two, take seventeen, and four, and five? From fifty, take twenty-three less eight, and twenty-nine less seven? From fifty, take twenty-nine less seven, and thirteen? From fifty, take thirteen, and twenty-three less eight? From thirty-seven, take seventeen less four, and eight less two? From thirty-seven, take eighteen less two, and eight? From thirty-seven, take seventeen less four, and eight? From fifty-five, take fourteen, and fifteen, and seventeen? From fifty-five, take fifteen, and seventeen, and nine? From sixty-five, take seventeen, and nine, and fourteen?

6. Thirteen, and sixteen, and twenty-five less six, how many? Fourteen, and eighteen, and three, and nine? Twenty-seven, and thirty-two, and fifty-one, and four? Sixteen, and fourteen less five, and four, and seven? From forty-eight, take thirteen, and sixteen? From forty-eight, take sixteen, and twenty-five less six? From forty-eight, take thirteen, and twenty-five less six? From forty-four, take fourteen, and eighteen, and three? From forty-four, take eighteen, and three, and nine? From forty-four, take three, and nine, and fourteen? From a hundred and fourteen, take twenty-seven, and thirty-two, and fifty-one? From a hundred and fourteen, take thirty-two, and fifty-one, and four? From a hundred and fourteen, take fifty-one, and four, and twenty-seven? From a hundred and fourteen, take four, and twenty-seven, and thirty-two? From thirty-six, take sixteen, and fourteen less five, and four? From thirty-six, take fourteen less five, and four, and seven? From thirty-six, take four, and seven, and sixteen? From thirty-six, take seven, and sixteen, and fourteen less five?

SECTION XIII.—*Practical Questions.*

1. JOHN had four cents; his father gave him six more; each of his two brothers gave him three; and he then bought some apples for five cents. How many cents had he left? How do you know? [Let the pupil explain the process for each question.]

2. A man, who had twenty-five dollars, bought a barrel of flour for seven dollars, some sugar for three, and some coffee for two. How much had he left?

3. From Brandon to Pittsford is eight miles; from Pittsford to Rutland, eight miles; from Rutland to Clarendon, six miles; from Clarendon to Wallingford, three miles. How many miles from Brandon to Wallingford?

4. A man travelled from Washington to Rockville, fourteen miles; from Rockville to Seneca Creek, six miles; from Seneca Creek to Little Seneca, four miles; and thence back to Seneca Creek. How many miles did he travel, and how far was he then from Washington?

5. A boy bought a box for twenty cents; he paid six cents to have it varnished, and then sold it for twenty-nine cents. Did he gain or lose by his bargains, and how much?

6. William bought twenty peaches for twenty-five cents; he sold twelve of them for two cents apiece, and the rest for one cent apiece. Did he gain or lose by his bargains, and how much?

7. A boy bought a peck of apples, and found there were just twenty-five. He gave four to each of his two brothers, put three in his pocket, and divided the remainder between his two sisters. How many did each of the sisters get?

8. Three brothers went to an orchard for apples. John got three; William, five; and James, ten. Their mother told James to give a part of his apples to his brothers, so that each might have the same number. How many had they altogether? how many did each have after they were equally divided? and how many did James give to John, and how many to William?

9. A man went out to collect some debts, and to make some purchases. He got fifty dollars from Mr. A., and thirty-five dollars from Mr. B. He then bought a barrel of flour for seven dollars, fifty pounds of sugar for four dollars, and a quarter of beef for six dollars. How many dollars had he left?

10. John had twenty plums. He gave six to each of his two brothers, and six to his cousin Edward. He then went

into the house, and got twelve more, and gave each of the three boys two apiece. How many had he then left for himself?

11. A boy had seventeen nuts; another gave him three; another seven; another five; and another gave him enough to make his number forty. How many did this last boy give him?

12. Six men bought a horse for seventy dollars. The first gave twenty-three dollars; the second, fifteen; the third, twelve; the fourth, nine; the fifth, seven. How much did the sixth give? and how much did the first give more than he?

13. A man bought a horse for eighty dollars, and paid fifteen dollars for keeping him. He let the horse enough to receive twenty dollars, and then sold him for eighty-three dollars. Did he gain or lose by the bargain, and how much?

14. A man bought a horse for a hundred and twenty dollars; a wagon for fifty dollars; a harness for the same, for twenty-five dollars. He afterwards sold the whole for two hundred dollars. Did he lose or gain, and how much?

SECTION XIV. — *Increase and Decrease by Equal Numbers; or, Multiplication and Division.*

[IN the four lessons that follow, show the aggregate numbers on the Frame, and let the pupils divide them by the eye alone. But, where this does not suffice, the teacher, or one of the class, may occasionally separate them by the fingers. The smaller numbers can be readily separated by the eye, and this should be the chief resort, especially in reviewing.]

1. How many twos in four? How many are twice two, then? How many twos in six? Three twos, then? Two threes? How many twos in eight? Four twos, then? Two fours? How many twos in ten? Five twos, then? Two fives? How many twos in twelve? Six twos, then? Two sixes? How many twos in fourteen? Seven twos, then? Two sevens? How many twos in sixteen? Eight twos, then? Two eights? How many twos in eighteen? Nine twos, then? Two nines? How many twos in twenty? Ten twos, then? Two tens? How many twos in twenty-two? Eleven twos, then? Two elevens? How many twos in twenty-four? Twelve twos, then? Two twelves?

2. How many threes in six? Two threes, then? Three

twos? How many threes in nine? Three threes, then? How many threes in twelve? Four threes, then? Three fours? Threes in fifteen? Five threes, then? Three fives? Threes in eighteen? Six threes, then? Three sixes? Threes in twenty-one? Seven threes, then? Three sevens? Threes in twenty-four? Eight threes, then? Three eights? Threes in twenty-seven? Nine threes, then? Three nines? Threes in thirty? Ten threes, then? Three tens? Threes in thirty-three? Eleven threes, then? Three elevens? Threes in thirty-six? Twelve threes, then? Three twelves?

3. How many fours in eight? Two fours, then? Four twos? Fours in twelve? Three fours, then? Four threes? Fours in sixteen? Four fours, then? Fours in twenty? Five fours, then? Four fives? Fours in twenty-four? Six fours, then? Four sixes? Fours in twenty-eight? Seven fours, then? Four sevens? Fours in thirty-two? Eight fours, then? Four eights? Fours in thirty-six? Nine fours, then? Four nines? Fours in forty? Ten fours, then? Four tens? Fours in forty-four? Eleven fours, then? Four elevens? Fours in forty-eight? Twelve fours, then? Four twelves?

4. How many fives in ten? Two fives, then? Five twos? Fives in fifteen? Three fives, then? Five threes? Fives in twenty? Four fives, then? Five fours? Fives in twenty-five? Five fives, then? Fives in thirty? Six fives, then? Five sixes? Fives in thirty-five? Seven fives, then? Five sevens? Fives in forty? Eight fives, then? Five eights? Fives in forty-five? Nine fives, then? Five nines? Fives in fifty? Ten fives, then? Five tens? Fives in fifty-five? Eleven fives, then? Five elevens? Fives in sixty? Twelve fives, then? Five twelves?

5. Two sixes are how many? Three sixes; another six? Four sixes; another six? Five sixes; another six? Six sixes; another six? Seven sixes; another six? Eight sixes; another six? Nine sixes; another six? Ten sixes; another six? Eleven sixes; another six? Twelve sixes; another six?

6. Two sevens, how many? Three sevens; another seven? Four sevens; another seven? Five sevens; another seven? Six sevens; another seven? Eight sevens; another seven? Nine sevens; another seven? Ten sevens; another seven? Eleven sevens; another seven? Twelve sevens; another seven?

7. Two eights, how many? Three eights; another eight? Four eights; another eight? Five eights; another eight?

Six eights; another eight? Seven eights; another eight? Eight eights; another eight? Nine eights; another eight? Ten eights; another eight? Eleven eights; another eight? Twelve eights; another eight?

8. Two nines, how many? Three nines; another nine? Four nines; another nine? Five nines; another nine? Six nines; another nine? Seven nines; another nine? Eight nines; another nine? Nine nines; another nine? Ten nines; another nine? Eleven nines; another nine? Twelve nines; another nine?

9. Two tens? Three tens? Four tens? Five tens? Six tens? Seven tens? Eight tens? Nine tens? Eleven tens? Twelve tens?

10. Two elevens? Three elevens? Four elevens? Five elevens? Six elevens? Seven elevens? Eight elevens? Nine elevens? Ten elevens? Eleven elevens? Twelve elevens?

11. Two twelves? Three twelves; another twelve? Four twelves; another twelve? Five twelves; another twelve? Six twelves; another twelve? Seven twelves; another twelve? Eight twelves; another twelve? Nine twelves; another twelve? Ten twelves; another twelve? Eleven twelves; another twelve? Twelve twelves; another twelve?

[This section will require more frequent repetition than the others. Omit, in reviewing, the words "another six," "another seven," &c.]

SECTION XV.—*Explanatory.*

NUMBERS are not always expressed in words. What are called *figures*, are frequently used for that purpose. These figures are only nine in number, as may be seen below. They should be well studied, so as to be readily known, wherever they may appear.

1, stands for one.	6, stands for six.
2, " " two.	7, " " seven.
3, " " three.	8, " " eight.
4, " " four.	9, " " nine.
5, " " five.	

These are all the figures that stand for numbers. But how, then, do we manage, when we wish to use a number larger than nine? The same figures are used, only they are put in a different *place*. Every figure becomes ten-fold greater by being removed *one place to the left*. Thus, the figure 1 stands for *one*, when alone, or at the right hand of other figures; for *ten*, when placed the second from the right; and for a *hundred*, or *ten* times *ten*, when it stands the third from the right. Thus, the three figures below,

111

stand for one hundred and eleven (or one-teen): the first figure on the left standing for *one hundred*, the second for *one ten*, the third for a single *one*. It is the same with all the other figures. Thus,

444

stands for *four* hundred and *for*ty- (or four tens) four; and

666

stands for *six* hundred and *six*ty-*six*.

These places for the figures are called *ranks*, or *orders*, and are reckoned *from the right*. Every figure placed in the first order, stands for as many *ones*, or *units*, as it represents; when placed in the second order, for as many *tens*, or *teen;* and for as many hundreds when it stands in the third order. A figure placed in the next order to the left (the fourth order) would stand for so many *thousands*, each of which is equal to *ten* hundred. Thus, in the following number,

4536

the 4 stands for so many thousands, the 5 for hundreds, the 3 for ty, or tens, the 6 for units, or ones. The whole number should be read thus: four thousand, five hundred, and thirty-six.

This is very much like the arrangement of the Frame. [Exemplify on the frame.] A single bead on any of the upper ten wires stands for *one*. Each row of beads stands for *ten*, any one of which is called *teen*, if units be added to it. Each bead on the lower row stands for 100, and the whole row, of course, for ten hundred, which is a *thousand*. Each of these numbers increases tenfold, just as the figures do from the *place* in which they stand.

[Let the following figures now be written *vertically* on the slate or blackboard, and named repeatedly by the class till they are familiar. 7 9 6 3 8 5 1 4 2.]

But it is frequently necessary to write a number in which one or more of the orders is wanting: for example, *two thousand and fifty-four*. Here we must have four *places*, or *orders*, to represent *thousand*, and yet we have only three figures, viz., 2 for two thousand, 5 for fifty, and 4 for four. In all such cases, we use this character, 0, which is called *cipher* or *nothing*, because it stands for nothing. Our number, two thousand and fifty-four, becomes 2054. There are no hundreds, you perceive, and the 0 fills that place. Had it not been put there, the 2 would have stood in the third order, and thus represented 2 hundred instead of two thousand. The cipher, accordingly, is sometimes called *figure of place*, because it is only used to show the *place* of the other figures.

Take notice, however, that a cipher is useless unless it occupies the place of units, or stands between a significant figure and the place of units. Thus, if we wish to write three hundred and seventy-four, the cipher is not wanted, although there are only three figures, because each figure can stand in its proper order, 374, without any cipher. But a cipher must be used in expressing two hundred and five, since we have only two figures, while the hundred is in the third order. Accordingly the number is written 205. For a similar reason, the number three thousand and forty-five must be written with a cipher, 3045.

Write the following numbers in figures on the slate or blackboard, and then read them over without the book:

1. Four hundred and thirty-five.
2. Two thousand, six hundred, and four.
3. Three thousand, and forty-two.
4. Six thousand, three hundred, and seventy-six.
5. Four thousand, four hundred, and forty-four.
6. Two hundred and three.
7. One thousand and twelve.

Sometimes one thousand is considered as *ten hundred*, as in the following:

8. Fifteen hundred and sixty.
9. Eighteen hundred and two.

[Specimen of questions to the class on the above numbers, when they have changed them from words to figures on the blackboard or slate.]

For No. 1. — What does the 4 stand for? [Point to the figures as they are spoken.] Why hundreds? The 5? Why units? The 3? Why *ty*, or tens?

No. 2. — What is the value of the 2? The 6? The 4?

What rank does the cipher occupy? Why? *Ans.* Because there are no ——.

No. 3.—What is the value of the 3? The 4? The 2? What rank does the cipher occupy here? Why?

No. 4.—Why is there no cipher in this number?

No. 5.—What is the value of the first figure on the right? Why? The fourth from the right? Why thousands? The third? Why? How many times is the third greater than the second? The third than the first? The fourth than the second? The fourth than the third? The fourth than the first? How many times is the first contained in the second? In the fourth? In the third? How many times is the second contained in the fourth? In the third? How many times is the third contained in the fourth?

No. 6.—What is the use of the cipher here? Why is there none in the place of thousands? *Ans.* Because the cipher is useless, unless it stands, &c. [Show this principle by an example on the blackboard.] Does the cipher stand for any number? What would this number be, if the cipher were omitted? If another cipher were placed beside the first, thus: [place one] what effect would it produce on the 2? *Ans.* Its value would be ——fold. What effect would be produced on the 3? If a cipher were placed after the 3 [place one], what effect would be produced on the number? Would both the 2 and 3 be increased tenfold?

No. 7.—If another cipher were introduced between the two 1s, what effect would be produced, that is, what figures would change their value? Add a cipher after the 2, and then say What change is thus produced, on each figure severally, and on the whole number?

No. 8.—What effect would a cipher produce on this number, if placed to the left of the 1? To the right of the 1? Between the 5 and 6? After the 6?

No. 9.—What effect would a cipher produce on this number, if placed to the left of the 1? On its right? Beside the other cipher? To the right of the 2?

How many are 10 times 26? How many tens in 2050? How many hundreds in 2500? How many tens in 3700? How many hundreds in 2000? Tens in 540? Tens in 270?

[While proceeding with the following sections, the class should still be exercised in notation and numeration, as above, varied till the subject is perfectly familiar.]

Section XVI. — *Multiplication by Higher Numbers.*

1. How many are 12ty? *Ans.* A hundred and twenty. Why? Because 10ty are a hundred, and 2ty are twenty. How many are 15ty, then? How many are 13ty? 16ty? 18ty? 17ty? 14ty? 19ty?

2. How many are 20ty? *Ans.* Two hundred. Why? Because each of the 10ty make a hundred. How many are 24ty? 27ty? 23ty? 36ty? 11ty? 45ty? 72ty? 69ty? 37ty? 84ty? [Continue and extend similar questions till sufficiently familiar.]

3. How many are 100ty? *Ans.* A thousand. Why? Because *ty* means *tens*, and ten times 100 are a thousand. How many are 160ty? 140ty? 170ty? 240ty? 110ty? 520ty? 370ty? [Continue and extend till familiar.]

4. How many are 124ty? Why? 356ty? Why? 247ty? 563ty? 116ty? 218ty? 311ty? &c.

[In reviewing, these questions should be varied by asking, How many are 10 times 12, 16, 84, 270, &c., in place of 12ty, 16ty, 84ty, 270ty, &c.]

5. How many are 2 times 20? Why? Because, as 2 times 2 are 4, 2 times 2ty are 4ty. How many are 2 times 30? 50? 40? 70? Why? Because, as 2 times 7 are 14, 2 times 7ty must be 14ty. 60? 90? 80?

6. How many are 3 times 20? 40? *Ans.* 12ty or 120. 50? 30? 70? 90? 60? 80?

7. How many are 4 times 20? 50? 30? 60? 40? 90? 70? 80?

8. How many are 5 times 20? 90? 30? 70? 60? 40? 80? 50?

9. How many are 6 times 20? 40? 70? 30? 90? 50? 80? 60?

10. How many are 7 times 20? 80? 60? 40? 50? 30? 90? 70?

11. How many are 8 times 20? 30? 90? 70? 50? 80? 60? 40?

12. How many are 9 times 20? 90? 60? 70? 40? 80? 50? 30?

13. How many are 2 times 13? How do you know? *Ans.* Because 2 times 10 are 20 and 2 times 3 are 6. [In oral arithmetic, the higher order should always be multiplied first,

because the figures are thus taken in their natural order, but chiefly because in practice it is found more easy and convenient.] 2 times 14? Why? [Repeat why after the questions that follow, till the reasoning is perfectly familiar.] 2 times 15? 16? 24? 27? 34? 45? 17? 47? 28? 19? 39?

14. How many are 3 times 13? 15? 14? 16? 19? 17? 21? 18? 24? 37? Why? Because 3 times 30 are 9ty, and 3 times 7 are 2ty one; together 11ty one, or a hundred and eleven. 3 times 54? Why? 72? Why? 87, &c.? 96? 38? 34? 37?

15. How many are 4 times 13? 19? 26? 54? 18? 72? 87? 96? 38? 34? 87? 53? 62? 99? 79? 88? 56? 49?

16. How many are 5 times 15? 13? 17? 26? 22? 23? 32? 47? 73? 31? 54? 27? 85? 96? 74?

17. How many are 6 times 13? 18? 15? 17? 19? 54? 36? 28? 72? 69? 93? 77? 65? 59? 48?

18. How many are 7 times 18? 13? 15? 19? 14? 27? 94? 36? 52? 73? 87? 76? 84? 55? 29?

19. How many are 8 times 13? 19? 16? 14? 17? 94? 85? 22? 73? 87? 54? 45? 95? 17? 57?

20. How many are 9 times 27? 35? 13? 18? 72? 81? 58? 62? 73? 95? 46? 32? 17? 29? 55? 84?

21. How many are two times 126? 2 times 524? 2 times 346? 725? 274? 373? 644? 375? 863? 588? 453?

22. How many are 3 times 132? 144? 365? 427? 629? 863? 275? 529? 246?

23. How many are 4 times 132? 321? 126? 428? 125? 637? 528? 276? 677?

24. How many are 5 times 132? 234? 621? 532? 724? 452? 671? 346? 248?

25. How many are 6 times 132? 342? 254? 524? 362? 241? 526? 728? 126?

26. How many are 7 times 132? 241? 324? 246? 542? 233? 621? 126? 272?

27. How many are 8 times 132? 244? 166? 342? 725? 637? 256? 428? 572?

28. How many are 9 times 132? 214? 326? 148? 522? 615? 926? 328? 218?

SECTION XVII. — *Definitions.*

1. WHEN two or more *unequal* numbers are joined together into one, the process is called *addition*, and the whole number is called the *sum*, or *amount*. Thus, joining 2 and 4 to make 6, or 3, 4, and 5, to make 12, is called *adding* those numbers, and 6 is called the *sum*, or the *amount*, of 2 and 4, and 12 the *sum* or the *amount* of 3, 4, and 5.

2. When two or more *equal* numbers are joined into *one*, the process is called *multiplication*. The number which is to be repeated is called the *multiplicand*, and the number which shows *how many times* the multiplicand is to be repeated is called the *multiplier*, and the increased number, or the multiplicand repeated as often as is required, is called the *product*. Thus, the process 4 times 5 are 20, is *multiplication*. The 4, which shows the *number of times* that 5 is to be taken, is the *multiplier*, 5 is the *multiplicand*, and 20 is the *product*. A more convenient name for the multiplicand and multiplier, as it applies equally to both, is that of *factor*. It is evident that both may always be called by the same name, since 4 times 5 is the same as 5 times 4, a remark applicable to any two numbers whatever. The word *factor*, in this connection, signifies *maker*; *product* signifies the number *made*, or *produced*. *Multiplication*, then, is nothing but a short way of performing *addition*, when the numbers to be added are *equal*. For, to say 4 times 5 are 20, is precisely the same as to say 5 and 5 and 5 and 5 are 20.

3. When one number is to be taken away *once* from another number, the process is called *subtraction*. The number to be diminished is called the *minuend*, the number to be taken away the *subtrahend*, and the number remaining after the subtrahend is taken away is called the *difference* or *remainder*. Thus, if we take 5 from 8, 3 will remain. Here 8 is the *minuend*, or number to be diminished; 5 the *subtrahend*, or number to be subtracted, or taken away; and 3 the *difference*, or *remainder*.

4. When *many* subtractions of the *same number* are to be performed, or when we wish to find *how many times* one number can be taken from another, the process is called *division*. This is, evidently, nothing more than a short way of performing *subtraction*, since it comes to precisely the same thing, whether we find, at once, that 5 is contained in 20 4 times,

which is called *division*, or produced by the slower method called *subtraction*, taking 5 from 20 *as many times* as possible, thus changing the 20 to 15, to 10, to 5, and to 0. The number to be divided is called the *dividend;* the number by which we divide is called the *divisor;* and the result of the division is called the *quotient*. Thus, if it be required to find how many times 4 is contained in 20, 4 is the divisor, 20 the dividend, and 5, the number of times that 4 is contained in 20, is the *quotient*. Sometimes the *divisor* is not contained an exact number of times in the *dividend*, and, consequently, there will be a *remainder* at the close of the operation. Thus, if it be required to find how many times 5 is contained in 22, we find it to be 4 times, and 2 over. The 2 is the remainder, and it forms an undivided part of the dividend.

5. It is evident that the dividend is a product of the divisor and quotient, since, if 4 be contained 5 times in 20, it is plain that 4 times 5 will make 20, and so of any numbers whatever. As the divisor and quotient, then, may be considered factors of the dividend, division may be defined *the process for finding one factor when the product and the other factor are given.* When a remainder occurs, as this remainder is an undivided portion of the dividend, it must be added in if the divisor and quotient are multiplied to reproduce the dividend. Thus, if there be 4 fives in 23, and 3 over, the dividend evidently consists of 3 more than the 4 fives. [Show this on the black-board.]

6. The termination *end*, *ent*, or *and*, in several of these terms, is derived from a Latin word signifying *being*, or *thing*. In this connection it stands for *number*. Hence, multiplic*and* signifies the *number* to be multiplied; minu*end*, the *number* to be diminished; subtrah*end*, the number to be subtracted; divid*end*, the *number* to be divided, and quoti*ent*, the *number* showing *how many* (the *quota*) times the divisor is contained in the dividend. The termination *er*, or *or*, signifies a man, or thing, that *works*, as in the words bak*er*, mill*er*, print*er*, farm*er*, &c. Hence, the multipli*er*, the fact*or*, and the divis*or*, are the numbers by which the *work is performed*, whether in multiplication or division.

7. Signs, or characters, have been invented to express these different processes. Thus, a vertical cross, +, is the sign of addition, and an inclined cross, in the shape of the letter X, ×, is the sign of multiplication, or contracted addition. Thus,

4+5 are 9, and 4×5 are 20. The sign of addition, +, is generally read *plus*, which is a Latin word signifying *more*. The sign of multiplication, ×, is called *multiple*. Thus, 4+5 is read *four plus five*, and 4×5 is read *four multiple five*. A dot is also frequently used as the sign of multiplication. Thus, 4 • 5 is the same as 4×5.

8. A short horizontal line, —, called *minus*, or *less*, is the sign of subtraction. The same sign, with a dot above and below it, ÷, is the sign of contracted subtraction, or division. Thus, 20—5, which is read *twenty less five*, are 15; and 20 ÷5, read *twenty divided by five*, gives only 4. Sometimes a *part* of the sign of division is used in place of the whole. Thus, 20 : 5, or $\frac{20}{5}$, is precisely the same as 20÷5, all three of them signifying 4.

9. Two parallel lines, =, form the sign of equality. It signifies that the numbers placed on each side of it are equal. Thus, 20—5=15, is read twenty *less* five *is equal to* fifteen; and 20÷5=4, is read twenty *divided by* five *is equal to* four.

10. A line drawn over several numbers is called a *vinculum*. It signifies that the numbers thus joined are to be considered as one number. Thus, $\overline{4+5}\times 3$, signifies that the *sum* of 4 and 5, and not 5 alone, is to be multiplied by 3; and $\overline{6-2}$ ÷2 signifies that the difference between 6 and 2 is to be divided by 2. Two parentheses are sometimes used instead of a vinculum. Thus, (4+5)×3 is the same as $\overline{4+5}\times 3$.

[Write the following lines on the blackboard, the first five to be read, the rest to be solved by the pupils.]

16+4=20

16×4, or 16 • 4=64

16—4=12

16÷4, or 16 : 4, or $\frac{16}{4}$=4

18—$\overline{2+7}$, or 18—(2+7)=9

24+8=

24×8, or 24 • 8=

24—8=

24÷8, or 24 : 8, or $\frac{24}{8}$=

24—$\overline{5+3}$, or 24—(5+3)=

[The class should practise similar exercises till they become familiar.]

What is addition? What is the result of addition called? What is multiplication? What is the multiplicand? the multiplier? the factors? the product? What is subtraction? What is the minuend? the subtrahend? the difference, or remainder? What is division? What is the dividend? the divisor? the quotient? What are the factors of the dividend? What is the dividend a product of? What is the difference between addition and multiplication? between subtraction and division? What is the sign of addition? its name? the signs of multiplication? their names? of subtraction? its name? of division? In how many ways can multiplication be expressed by signs? In how many ways can division be expressed by signs? What is the sign of equality? What is the result of addition called? the result of multiplication? of subtraction? of division? In multiplication, what is the number to be repeated called? What is the number called which shows how many times the multiplicand is to be repeated? What is the general name for both these terms? Why may they be properly called by the same name? In subtraction, what is the number to be diminished called? What is the number called which is to be taken away? In division, what is the number to be divided called? the number by which we divide? What do we call what is left? What is a vinculum? What does it signify? What characters are sometimes used in place of a vinculum? What is the precise meaning of the terminations *end*, *ent*, and *and?* What do they signify in arithmetic? What do the terminations *er* and *or* signify? What in arithmetic?

SECTION XVIII. — *Shortened Multiplication, or Multiplication by Easy Numbers.*

1. How many are 10 times 4? 10 times 24? 37? 45? 72? 158? 326? [Write a few such numbers on the blackboard, to be multiplied by 10, thus:

$$24\times10=$$
$$158\times10=, \text{\&c.},$$

and direct the attention of the class to the fact, that the significant figures remain unchanged when multiplied by 10.]

2. How many are 10 times 8? 84? 16? 49? 52? 93? 176? 248?

3. How many are 5 times 8? half of 10 times 8? 5 times 4? half of 10 times 4? [Here direct attention, on the board, to the fact, that a number multiplied by 5 produces the same amount as *half the same number* multiplied by 10; consequently, the easiest way to multiply a number by 5, is to multiply its half by 10. Thus, $72\times5=\frac{72}{2}\times10$, or 36×10.

4. How many are 5 times 16? *Ans.* Half of 16 or 8 ty. 5 times 24? 36? 28? 46? 72? 64? 84? 34? 58? 96? 128? 136? 248? 372?

5. How many are 5 times 17? [Here direct attention, on the board, to the fact, that every *odd* number of fives may be considered as *the next lower even number of fives and one five more.* Thus, $73\times5=\frac{72}{2}\times10+5=36$ ty and 5, or 365; and $27\times5=\frac{26}{2}\times10+5$, or 135.] How many are 5 times 19? 13? 21? 35? 37? 65? 49? 77? 33? 95? 67? 129? 247? 653? 875? 555?

6. How many are 15 times 14? [Show that 15 times any number is 10 times and 5 times that number. Our 14 times 15, then, becomes 14 times 10 and the half of 14 times 10, together $21\times10=210$. Thus, to multiply by 15, it is only necessary to increase the number to be multiplied by its half, and multiply by 10.] How many are 15 times 22? *Ans.* 22 and half of $22=33\times10=330$. How many are 15 times 24? 42? 48? 36? 28? 54? 72? 84? 58? 96? 64? 68? 94?

7. How many are 15 times 17? [Here we have 17×10 and 17×5. By the 5th question above, 17×5 becomes 16×5 and 5. Thus, when an *odd* number is to be multiplied by 15, we add half the next lower even number, multiply by 10, and add 5. Thus, 17×15 becomes $\overline{17+8}\times10+5$, or 255, and 23×15 becomes $\overline{23+11}\times10+5=345$.] How many are 15 times 25? 19? 23? 47? 75? 37? 49? 55? 97? 83? 45? 33? 87? 75? 29? 85? 67? 77? 53? 57? 95?

8. How many are 15 times 34? 67? 26? 57? 74? 128? 39? 156? 159? 234? 562? 325? 628? 473? 654? 637? 429? 579? 777?

9. How many are 20 times 24? [As 20 times any number is twice ten times that number, we have only to double the number to be multiplied by 20, and then multiply it by 10; or, expressed more briefly, multiply twice the number by 10. Thus, $20\times24=2\times24\times10=480$.] How many are 20 times

32? 41? 72? 93? 156? 428? 349? 572? 643? 377? 756? 278? 542? 503? 637?

10. How many are 25 times 8? [As every 4 times 25 makes 100, we have only to find *how many fours* are in any number, to know how many hundreds that number will make when multiplied by 25. Thus, $25\times24=100\times\frac{24}{4}=600$. And $25\times16=100\times\frac{16}{4}=400$. [How many are 25 times 36? 44? 28? 52? 60? 32? 56? 40? 72? 128? 436? 372? 116? 348?]

11. How many are 25 times 37? [Dividing 37 by 4 gives 9 and 1 over; therefore, $37\times25=9$ hundred, and 1 twenty-five, or 925. In the same manner, $38\times25=9$ hundred, and 2 twenty-fives, or 950; and $39\times25=9$ hundred, and 3 twenty-fives, $=975$. Every remainder, then, gives as many twenty-fives as it contains units to be added to the hundreds.] How many are 25 times 17? 15? 22? 19? 47? 54? 63? 95? 86? 74? 125? 237? 355? 178? 323? 218? 346?

12. How many are 25 times 20? 120? 55? 84? 173? 267? 348? 133? 87? 195? 388? 193? 327? 136? 113? 125? 239?

13. How many are 30 times 24 $(3\times24\times10)$? 30 times 45? 76? 255? 327? 54? 96? 238? 126? 272? 49? 78? 232?

14. How many are 35 times 24? $(3\times24+\frac{24}{2}(=84)\times10=840.)$ 35 times 37? $(3\times37+\frac{37}{2}\times10.)$ [Let it always be remembered that the remainder 1, in such cases, is always one 5. Thus, 35 times $37=3\times37(=111)+\frac{37}{2}$(18 and 1 over)$=129\times10+5=1295$.] 35 times $29=\overline{87+14}\times10+5$. 35 times 14? 27? 96? 128? 85? 74? 254? 93? 232? 75?

15. How many are 40 times 24 $(4\times24\times10)$? 40 times 27? 84? 56? 47? 53? 125? 67? 238? 152? 95? 73? 182? 245?

16. How many are 50 times 24 $(24\times50=\frac{24}{2}\times100)$? 50 times 27 $(\frac{27}{2}\times100)$? [In halving odd numbers, the remainder 1 is one 50. Thus $\frac{27}{2}\times100=1300$ and one $50=1350$.] 50 times 36? 48? 57? 73? 94? 85? 29? 132? 173? 178? 127? 185? 142? 155? 187? 143? 172? 189?

17. How many are 45 times 24? $(45=50-\frac{50}{10}.)$ Therefore, 45 times 24 is $50\times24=1200$, minus the tenth of that number $(120)=1080$. 45 times 26? $(1300-130=1170.)$ 45 times 29? $(1450-145=1305.)$ 45 times 36? 93? 45? 27? 39?

96? 78? 124? 104? 156? 118? 97? 138? 125? 187? 152? 175? 126?

18. How many are 55 times 24? ($55=50+\frac{50}{10}$.) Therefore, 55 times 24=50×24=1200, plus the tenth of that number, 120=1320. 55 times 26 (1300+130=1430.) 55 times 34? 75? 82? 53? 72? 96? 28? 29? 73? 125? 146? 132? 165? 184? 173? 105? 115? 123? 136? 145?

19. How many are 60 times 24? (6×24×10.) 60 times 35? 94? 36? 52? 65? 72? 39? 46? 53? 76? 89? 37? 48?

20. How many are 90 times 24? ($24\times100=2400-\frac{2400}{10}=$ 2160.) How many are 90 times 36? 45? 58? 73? 92? 84? 42? 35? 27? 29? 57? 99?

The above are the easy methods of multiplying mentally by every fifth number from 5 to 60 inclusive, and also by 90. The intermediate numbers are managed as follows: Consider 8 and 9 as 10—2 and 10—1; 11* and 12 as 10+1 and 10+2; 13, 14 as 15—2, 15—1; 16, 17 as 15+1 and 15+2; and so of all the others. Thus, 17×24=(15×24)+(2×24); and 23×24=(25×24)—(2×24). Thus, the intermediate are solved like the others, excepting that *once* or *twice* the multiplier has to be added or subtracted. The following table will make this more clear.]

Real factors.	Factors used.	Real factors.	Factors used.
4	5—1	14	15—1
5	5	15	15
6	5+1	16	15+1
7	5+2	17	15+2
8	10—2	18	20—2
9	10—1	19	20—1
10	10	20	20
11	10+1	21	20+1
12	10+2	22	20+2
13	15—2		

From merely glancing the eye down this table, it becomes apparent that no multipliers need be used under 60, except the easy numbers 5, 10, 15, 20, 25, &c., the intermediate factors being rectified by the addition or subtraction, as the case may

* The product of 11 and any number between 10 and 99 inclusive is found by placing their sum between the two figures of the latter factor. Thus 11×34=7 (sum of 3 and 4) between the 3 and 4=374; and 11×45=495. Why? When the sum of the figures exceeds 9, the first figure of course must be increased by 1. Thus, 11×48=528. Why?

be, of once or twice the multiplicand. The same principle may be applied to a variety of other numbers, such as 68 to 72; 78 to 82; 98 to 102; 90; 180; 270; 360, &c., the last four numbers being the same as 100, 200, 300, 400, less one tenth.

21. How many are 8×24? 16×37? 27×24? 36×25? 44×15? 72×30? 64×24? 85×25? 92×27? 76×28? 116×25? 47×32? 94×38? 77×28? 56×49? 49×49? 52×47? 84×27? 43×26? 18×144? 55 (50+$\frac{1}{10}$ of 50) ×86? 32×24? 78×7? 45×45 (50−$\frac{1}{10}$ of 50)? 23×72? 99×28?

SECTION XIX. — *Practical Questions.*

1. A FARMER sold a flock of 300 sheep at 2 dollars a head, and bought 25 cows at 18 dollars each. How much money had he left?

2. What is the cost of 28 bushels of oats at 32 cents a bushel?

3. What cost 27 bushels of corn at 49 cents per bushel?

4. How much must be paid for 15 thousand feet of boards at 18 dollars a thousand; and 6 thousand shingles at 3 dollars a thousand? [16×18. Why?]

5. How much is due to a laborer for working 24 days at 75 cents per day?

6. What cost 18 bushels of corn at 58 cents a bushel?

7. A man had 40 barrels of flour. He sold 16 of them at 6 dollars a barrel, and the rest at 7 dollars a barrel. How much did he get for the whole? [40×6+24.] Why?

8. What is the cost of 14 bureaus at 15 dollars each?

9. What cost 24 bedsteads at 23 dollars each?

10. A man bought 7 barrels of sugar at 13 dollars a barrel, and paid 48 dollars. How much remained due?

11. A farmer had an apple orchard, consisting of 16 rows of trees, and 14 in each row; and an orchard of peaches, of 13 rows and 17 in each. Which orchard had the greater number of trees, and what was the difference?

12. Two trains of cars leave a depot in different directions, one going eastward for 14 hours at the rate of 18 miles an hour, and the other westward for 18 hours at 16 miles an hour.

How far were the trains then apart? [(32×16)+(14×2.)] Why?

13. How much must be paid for 16 tierces of rice at 15 dollars a tierce, and 18 barrels of sugar at 17 dollars a barrel? [(34×15)+(2×18.)] Why?

14. What would be the cost of 23 pounds of black tea at 34 cents a pound, and 27 pounds of green tea at 44 cents a pound? [50×44—230.] Why?

15. What would be the cost of 28 yards of cloth at 84 cents a yard, and 27 yards at 44 cents per yard? [(55×84)—(27×40.)] Why?

16. A man bought 23 bushels of corn at 60 cents a bushel, and 17 bushels of rye at 64 cents a bushel. What was the cost of the whole? [··×··—92.] Why?

17. What must I give for 19 bushels of buckwheat at 45 cents a bushel, and 25 bushels of oats at 43 cents a bushel? [··×··—50.] Why?

18. A farmer bought from a merchant 23 yards of satinet at 38 cents a yard, and paid him 27 bushels of oats at 42 cents a bushel. What is the balance, and to whom must it be paid? [65×4.] Why?

19. What is the difference in value of 24 pounds of tea at 35 cents per pound, and 30 bushels of oats at 40 cents a bushel? [(24 × 5) + (6 × 40.)] Why?

SECTION XX. — *Division of Large Numbers by a Single Digit.*

DIVISION of large numbers mentally by a single digit is generally considered more difficult than any of the other elementary operations. But it may be very much facilitated by some previous practice in the rapid resolution of a dividend, mentally, or by inspection, into separate MULTIPLES of the divisor; that is, into numbers containing the divisor an *exact number of times*, when its respective digits are not so divisible. Thus, neither of the digits in 92 are divisible by 4; but the number can be readily resolved by the eye, or in the mind, into 8ty and 12; and 192, in like manner, is readily perceived to be resolvable into 16ty and 32. Again, 84 requires no resolution for division by 4; but, if 3 be the divisor, it should be considered as 6ty and 24; while, if the divisor be 7, 84 should be considered as 7ty and 14. This process of rapid resolution is useful, even when the dividend is not exactly a multiple of the divisor, and consequently produces a remainder. Thus, for division by 3, 44 may be resolved into 30, 12, and 2; and for 6, 93 may be resolved into 60, 30 and 3.

1. Prepare 72 for division by 3; by 4; by 6.
2. Prepare 54 for division by 2; by 6.
3. Prepare 57 for division by 5; by 4; by 6 [30, 24, 3].
4. Prepare 48 for division by 3; by 5.
5. Prepare 87 for division by 7; by 6; by 5.
6. Prepare 174 for division by 8; by 7; by 6; by 5; by 4; by 3.
7. Prepare each of the following numbers for division by 9: 100; 114; 125 [90, 27, 8]; 136; 153; 142; 171; 281.
8. Prepare each of the following numbers for division by 8: 176; 196; 232; 352; 268; 576; 455.
9. Prepare each of the following numbers for division by 7: 244; 638; 272; 289; 356; 376; 485.
10. Prepare each of the following numbers for division by 6: 532; 236; 748; 255; 192.
11. Prepare each of the following numbers for division by 4: 236; 347; 252; 375; 458; 137.
12. Prepare each of the following numbers for division by 3: 226; 452; 478; 246; 329.
13. Divide the following numbers severally by 9: 192; 353; 468; 736; 228.
14. Divide the following numbers severally by 8: 176; 325; 424; 636; 478; 528.
15. Divide the following numbers severally by 7: 325· 236; 124; 242; 188; 378; 434; 272.
16. Divide the following numbers severally by 6: 532, 312; 636; 234; 152; 444; 228.
17. Divide the following numbers severally by 5: 325; 432; 876; 152; 645; 138.
18. Divide the following numbers severally by 4: 316; 237; 251; 138; 246.
19. Divide the following numbers severally by 3: 213; 231; 316; 415; 632; 174.

SECTION XXI. — *Practical Questions.*

1. A COMPANY of 9 men undertook to build a bridge, for which they were paid 836 dollars. How much would each receive if the money was equally divided among them?

2. How much more would each have received if the company had only consisted of 6 persons instead of 9?

3. How many weeks are there in a year of 365 days?

4. Five brothers sold a piece of property, in which each had an equal interest, for 3215 dollars. How much would each receive?

5. Four equal partners found, at the close of a year's business, that the profits on their sales amounted to 8426, while the expenses amounted to 2000 dollars. How much would be the dividend of each partner?

6. A merchant owned an eighth of a vessel which traded to the West Indies. In one of her voyages the net profits amounted to 3424 dollars. What would be the amount of that merchant's share?

7. Three men went to fish for mackerel, and at the close of the day found they had caught 456 fish. How many would each man have, if they were equally shared?

8. The property of a man dying intestate was found to amount to $3492, of which, by law, one-third goes to his widow, and the rest is divided equally among his 6 children. What will be the share of the widow, and of each child?

9. But, supposing the expenses of settling the estate amount to $450, what would then be the respective shares of the widow, and of each child?

10. A captain and 3 men went out on a fishing excursion, in which 498 fish were taken. The owner of the boat and fishing apparatus was to have one share, the captain two shares, and each of the seamen one share. What would be the respective shares of each person interested?

11. Four men contracted to build 2 bridges. For one of them, the price was to be $1500; and for the other, $1450. The materials, tools, &c., cost them $400. What would be each man's share of the remainder?

12. A farm, on which was a mortgage of $300, was sold for $1400. It belonged to three brothers. How much would be each one's share of the net proceeds?

CHAPTER II.

INCREASE AND DECREASE OF FRACTIONS, OR BROKEN NUMBERS.

SECTION I. — *Fractions of Numbers exceeding Unity.*

Definition. — When any thing, or any number, is divided into *two* equal parts, each of these parts is called the *half* of the thing, or the number. When any thing, or any number, is divided into *three* equal parts, each of these is called the *third* of the thing or number. When divided into *four* equal parts, one of them is called the *fourth* part. In like manner, when a thing or number is divided into *five*, *ten*, *twenty*, or any other number of equal parts, each of these parts is called the *fifth*, *tenth*, *twentieth*, &c., part.

Such numbers as these are called *fractions*, a word which means *broken* into parts. They are represented by figures as follows: $\frac{1}{2}$ signifies one-half; $\frac{1}{3}$ signifies a third; $\frac{1}{4}$ a fourth; $\frac{1}{10}$ a tenth, and so on. When we wish to express more than one of the parts, they are represented in a similar manner. Thus, $\frac{2}{3}$ stands for two-thirds; $\frac{3}{5}$ for three-fifths; $\frac{15}{24}$, fifteen twenty-fourths, &c.

[Here the teacher will write a variety of fractions on the blackboard, to be named by the class till the subject is familiar.]

Fractions always arise from division, and are, therefore, very properly expressed by the same character, namely, by a horizontal line between the dividend and the divisor. — [See Chap. I., Sect. XVI. 8.] Thus, in $\frac{1}{5}$, the figure above the line stands for the *thing*, or *number*; the line itself expresses *division*; and the figure under the line shows into *how many parts the thing or number is to be divided*. Thus, also, the upper figure, which is the dividend, is called the *numerator* (or nummerer), because it shows the *number* of parts contained in the fraction; while the lower figure, which is the divisor, is called the *denominator* (or namer) because it points out the *name* or value of the parts expressed by the numerator. The numerator and denominator are called the *terms* of the fraction. Both terms taken together represent the quotient in the division. Thus, 1 divided by 5, gives for quotient $\frac{1}{5}$.

1. What is a fraction? What is meant by a *half* of a thing, or number? How is it represented? [Answer orally or by the blackboard.] What is a third? two-thirds? How represented? A fifth? three-fifths? an eighteenth? five eighteenths? an eleventh? a thirteenth? twelve thirteenths? Where is the numerator placed? Why so called? Where is the denominator placed? Why so called? What does the horizontal line express? Which is the divisor? the dividend? the quotient?

2. What is the $\frac{1}{2}$ of 2 beads? [Use the frame.] One is what part of two? *Ans.* One is the half of two. Two times one are how many times two? How many halves make a whole one? Three are how many times two? *Ans.* Once 2 and a $\frac{1}{2}$ of two. Four are how many times two? 5? *Ans.* Two times two and a half of 2. Six are how many times 2? 7 how many times 2? 8? 9? 10? 11? 12? 15? 16? 25? 26? 37? 38? 85? 93? 77? 50? 100? 102? 116? 105? 107? 117? 128? 125? 145?

3. One is what part of 3? *Ans.* One is a third of 3. Two is what part of 3? *Ans.* Two is two-thirds of 3. Three times one are how many times 3? How many thirds make a whole one? Five are how many times 3? *Ans.* Once three and two-thirds of 3. What is a third of any thing? [See definition above.] What is two-thirds? Six are how many times 3? 7 how many times 3? 8? 9? 10? 11? 12? 13? 17? 24? 26? 29? 30? 33? 34? 35?

4. How many fourths make a whole one? What part of 4 is 1? *Ans.* One is the fourth part of 4? What part of 4 is 2? 3? What do you understand by $\frac{1}{4}$? $\frac{2}{4}$? $\frac{3}{4}$? 5 are how many times 4? 6 how many times 4? 7? 8? 9? 10? 11? 13? 14? 15? 16? 17? 18? 19? 20?

5. How many fifths make a whole one? One is what part of 5? 2 what part of 5? 3? 4? Six are how many times 5? 7 how many times 5? 8? 9? 10? [Show, by the frame, that every *ten* or *ty*, consists of 2 *fives*.] Eleven how many

times 5? 12? 13? 24? 45? 49? 58? 60? 78? 84? 100? 120? 133? (13ty+3=26×5+3) 94? 86? 74?

6. How many sixes make a whole one? One is what part of 6? Two what part of six? 3? 4? 5? 7 how many times 6? 8 how many times 6? 9? 10? 11? 12? 13? 14? 15? 16? 17? 18? 19? 20? 24? 27? 29? 30? 35? 39? 36? 44? 56? 60? 63? 75? 120? 122? 123? 129? 144? 157? 132? 135? 244?

7. One is what part of 7? 2 are what part? 3? 5? 8 are how many times 7? 9 how many times? 10? 15? 28? 22? 34? 46? 78? 84? 93? 97? 144? 138?

8. Two is what part of 8? 3 what part? 5? 6? 7? 9 are how many times 8? 10 how many times? 11? 13? 15? 16? 19? 22? 25? 43? 37? 34? 47? 54? 72? 83? 84? 93? 96? 120? 135? 144? 147? 139?

9. One is what part of 9? 4 is what part? 5? 6? 10 are how many times 9? 11 how many times? 13? 15? 16? 19? 22? 25? 43? 37? 34? 47? 54? 72? 83? 84? 93? 96? 120? 135? 144? 147? 139?

10. Three is what part of 10? 4 what part? 7? 8? 9? 11 are how many times 10? 12 how many times? 13? 15? 65? 142? 364? 567? 278?

Section II. — *Practical Questions.*

1. If a yard of cloth be worth 3 dollars, and it be cut into 3 equal pieces, what will one of the pieces be worth? that is, what will $\frac{1}{3}$ of a yard be worth? What will $\frac{2}{3}$ of a yard be worth? See Example 3, in last section.

2. If a barrel of fish cost 3 dollars, what part of a barrel can you buy for 1 dollar? How many barrels for 7 dollars? for 9 dollars? 10? 12? 25? 27? 35? 29? 38?

3. If two bushels of wheat will buy a yard of cloth, what part of a yard will 1 bushel buy? How many yards will 4 bushels buy? 5 bushels? 7? 18? 34? 26? 57? 139?

4. If a pound of rice cost 4 cents, what part of a pound will 1 cent buy? 2 cents? 3 cents? How much for 5 cents? 8 cents? 9? 11? 12? 13? 14? 24? 32? 65? 96? 87? 54? 72? 93?

5. If a barrel of flour cost 5 dollars, what part of a barrel

can be bought for 1 dollar? 3 dollars? How many barrels for 10 dollars? For 12 dollars? 15? 18? 24? 27? 39? 96? 152? 274? 385?

6. If a barrel of beef cost 6 dollars, what would $\frac{1}{6}$ of it be worth? $\frac{2}{6}$? $\frac{3}{6}$? $\frac{5}{6}$? How much could be bought for 7 dollars? For 9 dollars? 10? 12? 15? 17? 23? 24? 27? 35? 32? 36? 43? 54? 65? 97? 84? 156? 132?

7. At seven dollars for a barrel of flour, how much could be bought for 1 dollar? For 5 dollars? How much would $\frac{1}{7}$ of a barrel be worth? $\frac{5}{7}$? How much for 8 dollars? 9 dollars? 15? 18? 23? 27? 28? 29? 35? 46? 65? 53? 57? 81? 63? 97? 78? 83?

8. If sugar be 8 cents a pound, how much could be bought for 1 cent? For 3 cents? 4 cents? 5? 7? 9? 11? 12? 15? 17? 18? 19? 22? 35? 36? 37? 38? 45? 54? 72? 63? 87? 65? 56? 66? 77? 49? 59? 37? 26?

9. When wheat is 9 shillings a bushel, how much can be bought for 1 shilling? 2 shillings? 5? 6? 7? 8? 10? 15? 37? 24? 25? 56? 62? 73? 82? 95? 36? 44? 57? 29? 93?

10. If sugar is 10 dollars a hundred weight, how much can be bought for 3 dollars? 7 dollars? 9? 14? 17? 36? 45? 27? 30? 46? 54? 82? 194 (19ty and 4)? 277? 567? 394? 846? 732? 589?

11. If you had 45 dollars, how many barrels of flour could you buy at 5 dollars a barrel? At 9 dollars? At 6? 4? 7? 11? 8? 3?

12. For 59 cents how many pounds of meat can be bought, at 6 cents a pound? At 8 cents? At 7? 5? 9? 11? 10? 13?

13. How many hours would it take you to travel 75 miles, if you go 3 miles an hour? 5 miles? 7? 10? 25? 16? 24? 14? 13? 18? 22?

14. If you had 81 pounds of butter that you wished to put into boxes that would contain 9 pounds each, how many boxes would it take?

15. At 30 dollars for 5 barrels of flour, how much for 1 barrel? For 4 barrels? For 6? 9? 13? 15? 20? 24?

16. At 75 dollars for 15 barrels of flour, how much for 1 barrel? For 7? 9? 24? 32?

17. If 6 horses consume 48 quarts of oats in a day, how many quarts will serve 8 horses the same time? 5 horses? 9? 24? 15? 7? 22?

18. If 3 horses consume 27 quarts of oats a day, how much will 8 horses consume per day? 2 horses? 9? 14? 18? 25? 28? 35? 42?

SECTION III. — *Fractions of Numbers exceeding Unity, continued.*

1. Two times 2 beads and $\frac{1}{2}$ of two beads how many? [Use frame, and repeat the word *beads* in this and the following examples, till the class attain a clear idea of the nature of the numbers, but no longer.] 3 times 2 beads and $\frac{1}{2}$ of 2 beads? 4 times 2 and $\frac{1}{2}$ of 2? 5 times 2 and $\frac{1}{2}$ of 2? 6 times 2 and $\frac{1}{2}$ of 2? 9 times 2 and $\frac{1}{2}$ of 2?

2. Three times 3 [beads, if necessary] and $\frac{1}{3}$ of 3, how many? 4 times 3 and $\frac{1}{3}$ of 3? 5 times 3 and $\frac{2}{3}$ of 3? 6 times 3 and $\frac{1}{3}$ of 3? 7 times 3 and $\frac{1}{3}$ of 3? 8 times 3 and $\frac{1}{3}$ of 3? 9 times 3 and $\frac{2}{3}$ of 3? 17 times 3 and $\frac{1}{3}$ of 3? 19 times 3 and $\frac{1}{3}$ of 3? 25 times 3 and $\frac{1}{3}$ of 3? 72 times 3 and $\frac{1}{3}$ of 3?

3. Three times 4 [beads, if necessary] and $\frac{1}{4}$ of 4, how many? 5 times 4 and $\frac{1}{4}$ of 4? 7 times 4 and $\frac{2}{4}$ of 4? 9 times 4 and $\frac{3}{4}$ of 4? 12 times 4 and $\frac{3}{4}$ of 4? 13 times 4 and $\frac{3}{4}$ of 4? 16 times 4 and $\frac{3}{4}$ of 4? 35 times 4 and $\frac{3}{4}$ of 4? 27 times 4 and $\frac{3}{4}$ of 4?

4. Four times 5 and $\frac{3}{5}$ of 5, how many? 6 times 5 and $\frac{4}{5}$ of 5? 7 times 5 and $\frac{2}{5}$ of 5? 9 times 5 and $\frac{3}{5}$ of 5? 16 times 5 and $\frac{4}{5}$ of 5? 19 times 5 and $\frac{2}{5}$ of 5? 37 times 5 and $\frac{3}{5}$ of 5? 95 times 5 and $\frac{2}{5}$ of 5?

5. Seven times 6 and $\frac{2}{6}$ of 6, how many? 8 times 6 and $\frac{5}{6}$ of 6? 9 times 6 and $\frac{4}{6}$ of 6? 15 times 6 and $\frac{5}{6}$ of 6? 25 times 6 and $\frac{3}{6}$ of 6? 49 times 6 and $\frac{5}{6}$ of 6? 27 times 6 and $\frac{1}{6}$ of 6? 32 times 6 and $\frac{5}{6}$ of 6?

6. Six times 7 and $\frac{4}{7}$ of 7, how many? 7 times 7 and $\frac{3}{7}$ of 7? 9 times 7 and $\frac{5}{7}$ of 7? 8 times 7 and $\frac{6}{7}$ of 7? 12 times 7 and $\frac{2}{7}$ of 7? 14 times 7 and $\frac{3}{7}$ of 7? 19 times 7 and $\frac{6}{7}$ of 7? 43 times 7 and $\frac{6}{7}$ of 7? 84 times 7 and $\frac{2}{7}$ of 7?

7. Five times 8 and $\frac{3}{8}$ of 8, how many? 9 times 8 and $\frac{5}{8}$

of 8? 42 times 8 and $\frac{5}{8}$ of 8? 73 times 8 and $\frac{1}{8}$ of 8? 95 times 8 and $\frac{6}{8}$ of 8? 47 times 8 and $\frac{3}{8}$ of 8? 69 times 8 and $\frac{3}{8}$ of 8?

8. Nine times 9 and $\frac{3}{9}$ of 9, how many? 18 times 9 and $\frac{6}{9}$ of 9? 37 times 9 and $\frac{5}{9}$ of 9? 25 times 9 and $\frac{8}{9}$ of 9? 27 times 9 and $\frac{5}{9}$ of 9? 99 times 9 and $\frac{7}{9}$ of 9? 67 times 9 and $\frac{3}{9}$ of 9?

9. Eight times 10 and $\frac{7}{10}$ of 10, how many? 17 times 10 and $\frac{3}{10}$ of 10? 14 times 15 and $\frac{8}{15}$ of 15? 17 times 25 and $\frac{13}{25}$ of 25? 16 times 34 and $\frac{17}{34}$ of 34? 18 times 23 and $\frac{7}{23}$ of 23? 25 times 19 and $\frac{7}{19}$ of 19? 36 times 24 and $\frac{5}{24}$ of 24?

10. Eleven times 7 and $\frac{5}{7}$ of 7, how many? 18 times 9 and $\frac{6}{9}$ of 9? 37 times 9 and $\frac{5}{9}$ of 9? 15 times 14 and $\frac{5}{14}$ of 14? 16 times 18 and $\frac{3}{18}$ of 18? 35 times 24 and $\frac{9}{24}$ of 24? 27 times 36 and $\frac{1}{36}$ of 36? 28 times 19 and $\frac{3}{19}$ of 19? 27 times 15 and $\frac{4}{15}$ of 15?

Section IV. — *Practical Questions.*

1. If a yard of cloth cost 4 dollars, what will $4\frac{3}{4}$ yards cost? $8\frac{2}{4}$ yards? $7\frac{1}{4}$? $15\frac{3}{4}$? $19\frac{1}{4}$?

2. If sugar cost 6 dollars a hundred weight, what will $8\frac{3}{6}$ hundred weight cost? $5\frac{5}{6}$ hundred weight? $9\frac{1}{6}$? $15\frac{4}{6}$? $24\frac{1}{6}$? $26\frac{2}{6}$?

3. If a man pay at the rate of 7 dollars a week for his board, how much will he have to pay for $7\frac{3}{7}$ weeks? $8\frac{4}{7}$ weeks? $15\frac{5}{7}$? $24\frac{6}{7}$? $49\frac{1}{7}$?

4. If nutmegs are 10 cents an ounce, what must be paid for $6\frac{5}{10}$ ounces? $9\frac{7}{10}$ ounces? $24\frac{3}{10}$? $72\frac{9}{10}$? $85\frac{7}{10}$?

5. At 6 dollars for a barrel of flour, what will $9\frac{2}{6}$ barrels cost? $11\frac{5}{6}$ barrels? $13\frac{1}{6}$? $18\frac{5}{6}$? $15\frac{3}{6}$?

6. If butter is 7 dollars a firkin, how much will $18\frac{3}{7}$ firkins cost? $12\frac{5}{7}$ firkins? $19\frac{6}{7}$? $45\frac{2}{7}$?

7. If a man travel at the rate of 24 miles a day, how far will he travel in $19\frac{5}{24}$ days? In $25\frac{3}{24}$ days?

8. If a pound of currant jelly cost 40 cents, what will be the cost of $16\frac{8}{40}$ pounds? $5\frac{9}{40}$ pounds? $7\frac{6}{40}$ pounds?

9. If a yard of bocking cost 75 cents, what will $3\frac{25}{75}$ yards cost?

SECTION V. — *Fractions of Numbers exceeding Unity, continued.*

1. THREE times 4 are how many times 6? How many times 2?

2. Four times 4 are how many times 2? 8? 5? *Ans.* Three times 5 and $\frac{1}{5}$ of five. How many times 7? *Ans.* Two times 7 and $\frac{2}{7}$ of 7.

3. Three times 5 and $\frac{3}{5}$ of 5 are how many times 7? How many times 4? 6? 8? 14? 9? 17?

4. Six times 9 and $\frac{3}{9}$ of 9 are how many times 12? 15? 8? 4? 26?

5. Nineteen times 6 and $\frac{3}{6}$ of 6 are how many times 4? 7? 9? 20? 35? 17? 18?

6. Eighteen times 4 and $\frac{3}{4}$ of 4 are how many times 12? 14? 8? 5? 8? 16?

7. Twenty-five times 8 and $\frac{5}{8}$ of 8 are how many times 7? 24? 42? 36? 44? 52?

8. Twenty-nine times 7 and $\frac{3}{7}$ of 7 are how many times 9? 14? 25? 30? 57? 34?

9. Forty-nine times 4 and $\frac{1}{4}$ of 4 are how many times 7? 15? 24? 35? 42?

10. Fourteen times 5 and $\frac{3}{5}$ of 5 are how many times 4? 15? 7? 25? 32? 6? 8? 5?

11. Seventeen times 3 and $\frac{2}{3}$ of 3 are how many times 8? 6? 9? 15? 25? 19? 11?

12. Thirteen times 9 and $\frac{4}{9}$ of 9 are how many times 8? 10? 7? 12? 15? 18? 32?

13. Twelve times 7 and $\frac{3}{7}$ of 7 how many times 9? 8? 15? 24? 15? 10? 25?

14. Twenty-four times 8 and $\frac{3}{8}$ of 8 how many times 7? 9? 12? 10? 50? 26?

SECTION VI. — *Practical Questions.*

1. BOUGHT 9 bushels and $\frac{4}{7}$ of a bushel of salt, at 7 shillings a bushel; how many dollars is that, at 6 shillings to the dollar?

2. Bought $8\frac{3}{4}$ yards of cloth at 4 dollars a yard, and paid for it with flour at 5 dollars a barrel; how much flour did it take?

3. A man sold $34\frac{3}{5}$ barrels of flour, at 5 dollars a barrel, and was paid in cloth, at 4 dollars a yard; how many yards did he receive?

4. A merchant exchanged $24\frac{5}{8}$ yards of fine linen, at 8 shillings a yard, for broadcloth, at 18 shillings a yard; how many yards did he receive?

5. A farmer exchanged $6\frac{3}{4}$ cords of wood, at 4 dollars a cord, for flour, at five dollars a barrel; how much flour did he receive?

6. A man bought 16 head of young cattle, at $8\frac{3}{16}$ dollars per head, and paid for them in flour, at 6 dollars a barrel; how much flour had he to pay?

7. If shad cost 8 dollars a barrel, what will $6\frac{5}{8}$ barrels cost? and how much plaster of Paris must be paid for the shad, at 3 dollars a ton?

8. When rice is $3\frac{3}{8}$ cents a pound, and flour 4 cents a pound, how much flour must be paid for 8 pounds of rice?

9. If salmon cost $8\frac{5}{16}$ dollars a barrel, how much flour, at 6 dollars a barrel, will pay for 16 barrels of salmon?

10. A man bought 8 hundred weight of rice, at $3\frac{3}{8}$ dollars per hundred weight, and 3 hundred weight of coffee, at $6\frac{1}{3}$ dollars per hundred weight, and paid for it in cranberries, at 11 dollars per barrel; how many barrels were required?

SECTION VII. — *Fractions of Numbers exceeding Unity, continued.*

1. Two is $\frac{1}{2}$ of what number? 4 is 2 times what number?

2. Three is $\frac{1}{2}$ of what number? $\frac{1}{3}$ of what number? $\frac{1}{6}$? $\frac{1}{16}$? $\frac{1}{25}$? $\frac{1}{32}$? $\frac{1}{49}$?

3. Four is $\frac{1}{5}$ of what number? $\frac{1}{8}$ of what number? $\frac{1}{17}$? $\frac{1}{19}$? $\frac{1}{36}$?

4. Five is $\frac{1}{8}$ of what number? $\frac{1}{9}$? $\frac{1}{56}$? $\frac{1}{38}$? $\frac{1}{24}$?

5. Nine is $\frac{1}{7}$ of what number? $\frac{1}{19}$? $\frac{1}{25}$? $\frac{1}{17}$? $\frac{1}{36}$?

6. Nine is 3 times what number?

7. Twelve is 3 times what number? 4 times? 6 times?

8. Sixteen is 4 times what number? 8 times? 2 times?

9. Twenty-four is 12 times what number? 3 times? 4 times? 2 times? 8 times?

10. Fifteen is 3 times what number? 5 times?

11. If 4 be $\frac{2}{3}$ of some number, what is $\frac{1}{3}$ of that number? If $\frac{1}{3}$ of a number be 2, what is that number? Then 4 is $\frac{2}{3}$ of what?

12. If 6 be $\frac{3}{4}$ of some number, what is $\frac{1}{4}$ of the same number? 2 is $\frac{1}{4}$ of what number? Then 6 is $\frac{3}{4}$ of what?

13. If 8 be $\frac{4}{7}$ of some number, what is $\frac{1}{7}$ of that number? 2 is $\frac{1}{7}$ of what number? Then 8 is $\frac{4}{7}$ of what?

14. 6 is $\frac{3}{8}$ of what number?

15. 8 is $\frac{2}{5}$ of what number?

16. 18 is $\frac{3}{9}$ of what number?

17. 25 is $\frac{5}{9}$ of what number?

18. 32 is $\frac{4}{7}$ of what number? Of how many times ten? *Ans.* Of 5 times 10 and $\frac{6}{10}$ of 10.

19. 56 is $\frac{7}{8}$ of what number? Of how many times 4?

20. 42 is $\frac{6}{8}$ of what number? Of how many times 9? *Ans.* Of 6 times 9 and $\frac{2}{9}$ of 9.

21. 28 is $\frac{4}{9}$ of what number? Of how many times 10? 9? 6?

22. 21 is $\frac{3}{5}$ of how many times 10? 15? 14? 26? 7? 9?

23. 50 is $\frac{10}{7}$ of how many times 12? 19? 6? 8? 4? 5?

24. 42 is $\frac{7}{8}$ of how many times 5? 12? 7? 4?

25. 36 is $\frac{4}{9}$ of how many times 10? 5? 12? 8?

26. 72 is $\frac{8}{5}$ of how many times 12? 7? 9? 6?

27. 18 is $\frac{3}{5}$ of how many times 11? 10? 12? 16? 8?

28. 15 is $\frac{5}{9}$ of how many times 8? 15? 6? 8? 4?

29. 21 is $\frac{3}{11}$ of how many times 10? 12? 6? 8? 5?

30. 45 is $\frac{5}{7}$ of how many times 6? 21? 14? 5? 12?

31. 80 is $\frac{16}{20}$ of how many times 10? 12? 8? 7?

32. 56 is $\frac{8}{11}$ of how many times 20? 10? 5? 6?

33. 60 is $\frac{12}{13}$ of how many times 10? 15? 8? 7?

34. 55 is $\frac{11}{12}$ of how many times 8? 14? 15? 9?

35. 52 is $\frac{4}{9}$ of how many times 12? 7? 20? 30?

36. 12 is $\frac{3}{11}$ of how many times 7? 9? 8? 6? 15?

37. 14 is $\frac{2}{9}$ of how many times 8? 10? 20? 25?

38. 48 is $\frac{6}{7}$ of how many times 11? 5? 9? 4?

39. 32 is $\frac{8}{11}$ of how many times 4? 6? 9? 11?

40. 27 is $\frac{3}{10}$ of how many times 11? 12? 15? 6?

Section VIII. — *Practical Questions.*

1. If $\frac{2}{3}$ of a pound of coffee cost 6 cents, what would $\frac{1}{3}$ of a pound cost? What would 1 pound cost? 2 pounds? How did you find that 2 pounds cost 18 cents?

2. If $\frac{3}{5}$ of a pound of sugar cost 6 cents, what would $\frac{1}{5}$ cost? What would 1 pound cost? 6 pounds? How did you find that 6 pounds cost 60 cents?

3. If $\frac{4}{7}$ of a bushel of wheat cost 36 cents, what cost 1 bushel? 5 bushels? Why?

4. If $\frac{3}{8}$ of a pound of tea cost 24 cents, what cost 6 pounds? Why?

5. If $\frac{4}{9}$ of a box of raisins cost 28 cents, what was the price of the whole box? Why?

6. A man being asked the age of his eldest son, answered, that his youngest son, who was 9 years old, was just $\frac{3}{5}$ of the age of his eldest son. How old was his eldest son? Why?

7. A man sold a cow when dry for 15 dollars, which was just $\frac{3}{5}$ of what he paid for her when fresh. How much did he pay for her?

8. A man bought 12 yards of cloth, and sold it for 54 dollars, which was $\frac{9}{8}$ of what it cost him. What did it cost per yard? and how much did he gain by the whole transaction?

9. There is a pole standing in the water, so that 12 feet of it are below the surface, being $\frac{3}{4}$ of its entire length. How long is the pole? and how long the part above the surface?

10. There is a pole $\frac{3}{4}$ under water and 4 feet out. How long is the pole?

11. A pole is $\frac{2}{5}$ under water and 6 feet out. What is its length?

12. There is a school in which $\frac{2}{9}$ of the pupils learn grammar, $\frac{3}{9}$ learn arithmetic, $\frac{1}{9}$ geography, $\frac{1}{9}$ geometry, and 12 learn to write. How many attend the school? and how many attend to each study?

13. In a certain congregation one Sunday, $\frac{1}{12}$ were in the singers' gallery, $\frac{2}{12}$ in the strangers' gallery, $\frac{4}{12}$ in the wall slips, and 100 in the middle slips. Of how many did the congregation consist? and what was the number of singers?

14. A man being asked the age of his youngest son, answered, that the age of his eldest son was 18 years, which was $\frac{2}{5}$ of his own age, and that his own age was 9 times that of his youngest son. What was the age of his youngest son?

15. A man sold a cow for 21 dollars, which was only $\frac{7}{10}$ of what she cost him. He paid for her with cloth, at 8 dollars a yard. How much cloth did he give for her?

16. A man gave 25 cents for his breakfast, which was only $\frac{5}{8}$ of what he paid for his dinner. What did his dinner cost him? He paid for both meals in buttons, at 6 cents per dozen. How many buttons did it require?

SECTION IX. — *Fractions of Numbers greater than Unity, continued.*

[THERE are three questions in each of the following problems; but the first two are merely leading questions to assist the learner in the solution of the third. To bright pupils they may be unnecessary, and of course should be omitted. With the more dull the section should be repeatedly reviewed, till the third question in each problem can be solved and explained without hesitation, and without the aid of the first and second questions.]

1. $\frac{5}{6}$ of 24 how many? $\frac{5}{6}$ of 24 are $\frac{10}{7}$ of what number? $\frac{5}{6}$ of 24 are $\frac{10}{7}$ of how many times 5? How do you know? Explain the process. *Form of the solution.* — $\frac{1}{6}$ of 24 is 4; therefore, $\frac{5}{6}$ are 20; if 20 are $\frac{10}{7}$, $\frac{1}{7}$ is 2, and $\frac{7}{7}$ 14; which is 2 times 5 and $\frac{4}{5}$ of 5.

2. $\frac{3}{7}$ of 28 how many? $\frac{3}{7}$ of 28 are $\frac{2}{8}$ of what number? $\frac{3}{7}$ of 28 are $\frac{2}{8}$ of how many times 7? Explain the process.

3. $\frac{4}{5}$ of 30 how many? $\frac{4}{5}$ of 30 are $\frac{6}{7}$ of what number? $\frac{4}{5}$ of 30 are $\frac{6}{7}$ of how many times 8? Explain.

4. $\frac{6}{8}$ of 32 how many? $\frac{6}{8}$ of 32 are $\frac{8}{9}$ of what number? $\frac{6}{8}$ of 32 are $\frac{8}{9}$ of how many times 5? Explain.

5. $\frac{4}{9}$ of 36 how many? $\frac{4}{9}$ of 36 are $\frac{8}{10}$ of what number? $\frac{4}{9}$ of 36 are $\frac{8}{10}$ of how many times 6? Explain.

6. $\frac{3}{4}$ of 40 how many? $\frac{3}{4}$ of 40 are $\frac{5}{7}$ of what number? $\frac{3}{4}$ of 40 are $\frac{5}{7}$ of how many times 8? Explain.

7. $\frac{6}{9}$ of 45 how many? $\frac{6}{9}$ of 45 are $\frac{3}{5}$ of what number? $\frac{6}{9}$ of 45 are $\frac{3}{5}$ of how many times 7? Explain.

8. $\frac{5}{6}$ of 48 how many? $\frac{5}{6}$ of 48 are $\frac{10}{7}$ of what number? $\frac{5}{6}$ of 48 are $\frac{10}{7}$ of how many times 3? Explain.

9. $\frac{4}{7}$ of 63 how many? $\frac{4}{7}$ of 63 are $\frac{6}{5}$ of what number? $\frac{4}{7}$ of 63 are $\frac{6}{5}$ of how many times 8? Explain.

10. $\frac{5}{9}$ of 72 how many? $\frac{5}{9}$ of 72 are $\frac{4}{7}$ of what number? $\frac{5}{9}$ of 72 are $\frac{4}{7}$ of how many times 9? Explain.

11. $\frac{4}{5}$ of 15 how many? $\frac{4}{5}$ of 15 are $\frac{6}{10}$ of what number? $\frac{4}{5}$ of 15 are $\frac{6}{10}$ of how many thirds of 21? Explain. *Form of the solution.* — $\frac{1}{5}$ of 15 is 3, and $\frac{4}{5}$ of course 12; if 12 be $\frac{6}{10}$, $\frac{1}{10}$ will be 2, and $\frac{10}{10}$ 20; $\frac{1}{3}$ of 21 is 7, and 20 is two times 7 and $\frac{6}{7}$ of 7.

12. $\frac{4}{3}$ of 18 how many? $\frac{4}{3}$ of 18 are $\frac{8}{9}$ of what number? $\frac{4}{3}$ of 18 are $\frac{8}{9}$ of how many sevenths of 35? Explain.

13. $\frac{6}{7}$ of 21 how many? $\frac{6}{7}$ of 21 are $\frac{2}{3}$ of what number? $\frac{6}{7}$ of 21 are $\frac{2}{3}$ of how many sixths of 24? Explain.

14. $\frac{5}{4}$ of 24 how many? $\frac{5}{4}$ of 24 are $\frac{10}{7}$ of what number? $\frac{5}{4}$ of 24 are $\frac{10}{7}$ of how many fifths of 40? Explain.

15. $\frac{5}{8}$ of 32 how many? $\frac{5}{8}$ of 32 are $\frac{2}{5}$ of what number? $\frac{5}{8}$ of 32 are $\frac{2}{5}$ of how many fifths of 35? Explain.

16. $\frac{4}{7}$ of 63 how many? $\frac{4}{7}$ of 63 are $\frac{6}{8}$ of what number? $\frac{4}{7}$ of 63 are $\frac{6}{8}$ of how many ninths of 45? Explain.

17. $\frac{3}{7}$ of 56 how many? $\frac{3}{7}$ of 56 are $\frac{4}{9}$ of what number? $\frac{3}{7}$ of 56 are $\frac{4}{9}$ of how many fourths of 28? Explain.

18. $\frac{3}{8}$ of 64 how many? $\frac{3}{8}$ of 64 are $\frac{6}{10}$ of what number? $\frac{3}{8}$ of 64 are $\frac{6}{10}$ of how many sixths of 30? Explain.

19. $\frac{2}{8}$ of 72 how many? $\frac{2}{8}$ of 72 are $\frac{3}{10}$ of what number? $\frac{2}{8}$ of 72 are $\frac{3}{10}$ of how many fifths of 40? Explain.

SECTION X. — *Practical Questions.*

1. As 2 boys were counting their money, one said he had 20 cents. The other said, $\frac{4}{5}$ of your money is exactly $\frac{2}{7}$ of mine. How many cents had he? $\frac{4}{5}$ of $20 = \frac{2}{7}$ of what?

2. A merchant kept his silver money in one part of his till, and his cents in the other. Having found 18 dollars in the former, his clerk, who counted the cents, told him that $\frac{3}{4}$ of the dollars were equal in number to $\frac{4}{17}$ of the cents. How many cents were in the drawer? Prove.

3. A man being asked the age of his eldest son, replied, that his youngest son was 6 years old, and that $\frac{2}{3}$ of the age of the youngest was just $\frac{2}{11}$ of that of the eldest. What was the age of the eldest? Prove.

4. A man, being asked how many sheep he had, said, that he kept them in two pastures. In one pasture he had 24; and

that $\frac{3}{4}$ of these were just $\frac{1}{5}$ of what he had in both. How many sheep had he in the second pasture? Prove.

5. Two boys, talking of their ages, one said he was 9 years old. Well, said the other, $\frac{2}{3}$ of your age is exactly $\frac{3}{4}$ of mine. What was his age? Prove.

6. A farmer and his son went out one day to look after his sheep, which were in two separate fields. They counted 35 in one pasture, and the farmer told his boy that $\frac{3}{5}$ of these were just $\frac{7}{18}$ of what were in his other pasture. How many had he in both pastures? Explain the process, or prove.

7. A farmer had sheep in three different pastures. In the first he had 100. In the second he had $\frac{4}{5}$ of $\frac{9}{10}$ of what he had in the first. [How many are $\frac{9}{10}$ of 100? Then how many are $\frac{4}{5}$ of that?] In the third he had $\frac{9}{10}$ of $\frac{5}{4}$ of what he had in the second. How many sheep had he in all? Explain the process.

8. A gentleman had three sons. Being asked the age of the youngest, he replied, that the age of the eldest was 16; the age of the second son was $\frac{2}{3}$ of $\frac{4}{8}$ of that of the eldest; and that the age of the youngest was $\frac{3}{4}$ of that of the second son. What was the difference in age between the eldest and youngest son? Explain the process.

9. A young lady, who was 16 years of age, having asked her female friend how old she was, received the following reply: If $\frac{3}{4}$ of your age were added to your age, the sum would exactly show how old I am. How much older was she than the lady who put the question? Explain the process.

CHAPTER III.

FRACTIONS OF UNITY.

SECTION I. — *First Principles.*

[IN the following exercises, let the teacher write the questions on the black-board, and add the answers as fast as they are announced by the class, in the following form, viz.:

$$2 \times \tfrac{1}{16} =$$
$$3 \times \tfrac{1}{16} =]$$

1. Which number is the numerator of a fraction? Which the denominator? What are both called? How much is 2 times $\frac{1}{16}$? How many times does $\frac{2}{16}$ contain $\frac{1}{16}$? *Ans.* Two times. How much is 3 times $\frac{1}{16}$? How many times does $\frac{3}{16}$ contain $\frac{1}{16}$? How much is 5 times $\frac{1}{16}$? How many times does $\frac{5}{16}$ contain $\frac{1}{16}$? [Point to the black-board.] Will multiplying the numerator of a fraction by any number, then, always make the fraction so many times greater? Remember, then, that *multiplying* the numerator *multiplies* the fraction.

2. Divide $\frac{12}{16}$ by 2; in other words, what is the half of $\frac{12}{16}$? How many times is $\frac{6}{16}$ contained in $\frac{12}{16}$? What is the third part of $\frac{12}{16}$? in other words, divide $\frac{12}{16}$ by 3. How many times is $\frac{4}{16}$ contained in $\frac{12}{16}$? [Point to the black-board.] Will dividing the numerator of a fraction by any number, then, always make the fraction so many times less? Remember, then, that *dividing* the numerator *divides* the fraction.

3. What principle may be drawn from these two exercises?

Ans. — The fraction is { *multiplied* by *multiplying* / *divided* by *dividing* } the numerator.

4. If an apple be cut into two equal parts, what is one of them called? [See Chap. II., Section I.] How shall I write it on the black-board? *Ans.* By drawing a horizontal line, and writing 1 above and 2 below it? [Write it.] If I should divide one of these halves into 2 equal parts, what should one of them be called? [Write it.] Which of these fractions [$\frac{1}{2}$,

$\frac{1}{4}$] is the larger? How many times? Has multiplying the denominator by 2, then [point to the fractions on the black-board], multiplied or divided the fraction? By what number? How many times is $\frac{1}{3}$ greater than $\frac{1}{6}$? Than $\frac{1}{9}$? Than $\frac{1}{12}$? Will *multiplying* the denominator of a fraction by any number, then, always *divide* the fraction by that number? Remember, then, that *multiplying* the denominator *divides* the fraction.

5. [Write $\frac{1}{4}$ on the black-board.] If I divide the denominator 4 by 2, what will the fraction be? [Write it after the answer is received.] Which is the larger fraction? [Write $\frac{1}{9}$.] If I divide the denominator of this fraction by 3, what will be the fraction? [Write it.] Which is the larger? How many times? Will dividing the denominator of a fraction, then, by any number, always multiply the fraction by that number? Remember, then, that *dividing* the denominator *multiplies* the fraction.

6. What principle have you discovered from the last two exercises?

Ans.—The fraction is { *multiplied* by *dividing* / *divided* by *multiplying* } the denominator.

7. [Write $\frac{4}{16}$ on the board.] If this *numerator* [point] be multiplied by 2, or by any other number, what effect will be produced on the fraction? If this *denominator* be multiplied by 2, what will be the effect on the fraction? If a fraction be multiplied by 2, or by any other number, and then be divided by the same number, will the fraction be unchanged, or will it be greater or less than at first? What will be the effect, then, on this or any other fraction, if both terms be multiplied by the same number? Remember, then, that *a fraction is unchanged when both terms are multiplied by the same number.*

8. [Write $\frac{4}{16}$ on the board, as before.] If this *numerator* be divided by 2, or by any other number, what will be the effect on the fraction? If this *denominator* be divided by 2, or by any other number, what will be the effect on the fraction? If a fraction be divided by 2, or by any other number, and then be multiplied by the same number, will the fraction be unchanged, or will it be greater or less than at first? What will be the effect, then, on any fraction, if both terms be divided by the same number? Remember, then, that *a fraction is unchanged, when both terms are divided by the same number.*

9. What principle may be drawn from the last two exercises?

Ans. — The value of a fraction is not changed by { multiplying or dividing } both terms by the same number.

[From this principle it is evident that the same fraction may be represented in literally an infinite variety of forms. For, by continually multiplying both terms of $\frac{1}{2}$ by 2, we have $\frac{2}{4}$, $\frac{4}{8}$, $\frac{8}{16}$, &c., to infinity. And, what is still more remarkable, if the operation is commenced by the use of *odd* numbers as denominators, the fraction $\frac{1}{2}$, or any other, may be represented in an *infinite series* of forms, each series of which may be continued to infinity. From these considerations, the extraordinary fact results, that an infinite number of men, say, as an approximation, every person now alive, and all that ever have existed, or ever will exist, might be employed from the present moment through all eternity, each person writing a series of fractions, all equal to one another, yet no two composed of exactly the same figures. And still further, the time in which this indefinite number of persons might have been thus employed, may be extended through the past eternity as well as through the eternity to come.]

[As the three principles developed above are exceedingly important, it may be well, the more strongly to impress them on the mind of the pupil, to present the subject in another point of view, as follows:]

10. [Write $\frac{8}{16}$ on the black-board.] Which of these two numbers is the dividend? [See p. 67, l. 24.] Which is the divisor? Where is the quotient [or quota, or share]? *Ans.* Both terms, namely, the fraction. If the dividend or thing to be divided, be increased [point to it], will the quotient, or share, or fraction, be increased or diminished? What will be the effect if the dividend be decreased? What will be the effect on the quotient, then, of multiplying the dividend by 2, 3, 4, or any other number? [See the first principle developed in this section.]

11. The divisor shows the number of parts into which the dividend is to be divided: if that be increased, then, will the quotient (or share, or fraction), be thereby increased or diminished? What will be the effect if the divisor be decreased? What will be the effect, then, of multiplying it by 2, 3, or any other number? Of dividing it by any number? [See the second principle developed in this section.]

12. If both dividend and divisor are multiplied by the same number, what will be the effect on the quotient (or fraction)? If both be divided by the same number, what will be the effect? [See the third principle developed in this section.]

13. [Write the following exercises on the board, and the answers, as fast as they are given by the class, as follows: $\frac{1}{2}$ of $1 = \frac{1}{2}$, &c.] What is $\frac{1}{2}$ of 1? $\frac{1}{2}$ of 2? $\frac{1}{2}$ of 3? Of 6? Of 7? What is $\frac{1}{3}$ of 1? Of 2? Of 3? Of 4? What is $\frac{1}{5}$ of 1? Of 2? Of 3? Of 4? What is $\frac{2}{5}$ of 1? Of 2? Of 3? Of 4? $\frac{3}{5}$ of 1? Of 2? Of 3? Of 4? Now look at the board, and say, did you get these answers by adding, subtracting, multiplying, or dividing? What is $\frac{1}{2}$ of $\frac{1}{2}$? $\frac{1}{3}$ of $\frac{1}{2}$? $\frac{2}{3}$ of $\frac{1}{2}$? $\frac{1}{2}$ of $\frac{3}{4}$? What is the operation, then, when the numbers on each side of *of* are both fractions? What, then, does the word *of* imply, when connected with fractions? Remember, then, that *the word of* connected with fractions always *implies multiplication.* What is the divisor in $\frac{6}{1}$? In $\frac{4}{1}$? Perform the division in both these cases? Does the numerator, then, always express the fraction when the denominator is 1? Does the value of an integer, then, remain unchanged when 1 is placed under it as a denominator?

14. What principle is involved in the last exercise?

Ans. A whole number may be expressed fractionally by writing 1 under it as a denominator.

15. *Recapitulation.* — What effect is produced on a fraction by multiplying its numerator? By multiplying its denominator? By dividing its numerator? By dividing its denominator? By multiplying both its terms? By dividing both its terms? What does the word *of* imply when connected with a fraction? How may an integer be expressed fractionally?

16. What are the principles developed in this section?

Ans. 1. If multiplication or division be performed on the numerator, the *same* effect is produced on the fraction.

2. If multiplication or division be performed on the denominator, a *contrary* effect is produced on the fraction.

3. *No change of value* is produced on the fraction when both terms are multiplied or divided by the same number.

4. A whole number may be expressed fractionally by writing 1 under it as denominator.

5. The word *of* connected with a fraction implies multiplication.

Section II.—*Prime Factors, Common Multiples, and Common Divisors.*

Definitions.—I. Numbers may be divided into two classes, namely, Prime numbers and Composite numbers. A *prime number* is a number which can be divided exactly only by itself, or by unity, as 1, 2, 3, 5, 7, 11, 17. A *composite number* is a number which can be measured exactly by a number exceeding unity, or which can be formed by multiplying two or more numbers together, each exceeding unity, as 4 from 2×2; 12 from $2 \times 2 \times 3$; 18 from $2 \times 3 \times 3$. One number is said to be *prime to another* when unity is the only integer by which both can be measured. Thus, 4 and 9 are neither of them *prime numbers*, but they are *prime to each other;* because unity is the only integer which will measure them both.

II. A number greater than unity that will exactly divide two or more numbers, is called their *common divisor;* and the greatest number that will so divide them is called their *greatest common divisor.* Thus, 5 is a *common divisor*, and 10 the *greatest common divisor* of 10 and 50; and 3 is a *common divisor*, and 9 the *greatest common divisor* of 9, 18, and 27.

III. A number that contains another an exact number of times, is a *multiple* of that number. Thus 4, 6, and 8, are each multiples of 2. A number that contains two or more numbers as factors is a *common multiple* of those factors. Thus, 6 is a common multiple of 2 and 3; and 24 is a common multiple of 2, 3, and 4. The *smallest* number that contains two or more numbers as factors, is their *least common multiple.* Thus, 24 is a common multiple of 2 and 3; but it is not their *least* common multiple, for 18 and 12 contain them also; but as no number smaller than 6 contains them, 6 is their least common multiple.

1. What is a prime number? Give examples, and say why they are prime. What is a composite number? Give examples, &c. What is a common divisor of two or more numbers? Give examples, &c. What is the greatest common divisor of two or more numbers? Give examples, &c. What is a multiple of a number? Give examples, &c. What is a common multiple of two or more numbers? Give examples, &c. What is a least common multiple of two or more numbers? Give examples, &c.

2. Is 2 a prime or a composite number? What is an even number? Can any even number except 2, then, be prime? Can the product of two or more odd factors ever be even? Why not? *Ans.* Because every even number has —— for one of its factors. From the answers to these questions, it is plain that the most simple method of resolving a number into its prime factors is to continue halving it as long as it remains *even* (each operation giving 2 for a factor), and then to examine it by the odd numbers, beginning with the smallest. Thus, to resolve 120 into its prime factors, say 2 • [60] = 2 • 2 [30] = 2 • 2 • 2 [15] = 2 • 2 • 2 • 3 • 5. These operations, however, should be performed mentally, unless the number be very large; and, after a little practice, the three 2's may be discovered at once (and the same with other numbers), by a mere inspection of the composite number. The following exercises will render the resolution of composite numbers sufficiently easy, and, after some practice, exceedingly rapid.

3. Is 10 divisible by 2 without remainder? Are 2 tens? 3 tens? Any number of tens? How many tens in 80? In 380? 270? 1250? Are all these numbers, then, divisible by 2? Since any number of *tens*, then, is divisible by 2, by what rank of figures can it be determined whether a number is divisible by 2? What will be the respective remainders, if any, on dividing the following numbers by 2? 154? 379? 1976? 3285? [These numbers, and others, if necessary, to be written on the black-board. They are not to be divided. The pupil should tell at a glance.] What, then, is the *sign* that 2 is a factor in a number? *Ans.* That it is an —— number, or a number ending in —, —, —, —, or —.

4. What will be the remainder if 10 be divided by 9? When 2 tens are divided by 9? 3 tens? 5 tens? 7 tens? What will remain if 100 be divided by 9? 200? 500? Any number of hundreds? When 1000 is divided by 9? 2000? 3000? Any number of thousands? [Write 253674 on the black-board, and other numbers, if necessary, and ask such questions as follow, pointing to the figures.] What will be the remainder when this 2 is divided by 9? The 6? 5? 4? 7? 3? What, then, is the *sign* that 9 is a factor? *Ans.* That the sum of the significant figures is divisible, &c.

5. What is the remainder when 1 is divided by 3? When 2? 3? 4? 5? 6? 7? 8? 9? When 10 is divided by 3? 2 tens? 3 tens? 4 tens? &c. When 1000? 2000?

3000 ? &c. What, then, is the sign that 3 is a factor ? That the sum of the significant, &c.

6. [Write the following and other numbers on the black-board, and ask questions as follows.] 2768, 32562, 5237, 8246. What is the remainder when this number is divided by 2 ? By 3 ? By 9 ? *Note.* — All numbers divisible by 9 are also divisible by 3; but the converse is not always true. For instance, 51, 111, 213, are divisible by 3, but not by 9.

7. Is 100 divisible by 4 ? 200 ? 300 ? Any number of hundreds ? What, then, is the *sign* that 4 is a factor ? *Ans.* That the tens and units are, &c. Is 20 divisible by 4 ? 40 ? 60 ? 80 ? Any even number of tens ? Give another *sign* of 4, then. Unite the two signs. *Ans.* When the units and tens are divisible, if the tens are odd; when the units alone are divisible, if the tens are even.

8. Is 10 divisible by 5 ? Any number of tens ? What, then, is the *sign* that 5 is a factor ? Is 100 divisible by 25 ? What, then, is the *sign* of 25 ?

9. Is 1000 divisible by 8 ? What, then, is the *sign* of 8 ? Is 200 divisible by 8 ? Any even number of hundreds ? Give another *sign* of 8, then. Unite the two signs.

10. What is the *sign* of ten being a factor ? Of 50 ? Of $125=(\frac{1000}{8})$? Of 225 (9×25) ?

11. If an apple be divided into 2 equal parts, and each of these be again divided into 3 equal parts, into how many equal parts will the apple be divided ? Would the number of parts have been precisely the same, had the apple been divided at once into 6 parts ? Divide 24 by 2, and the quotient by 3. Divide 24 by 6, the product of 2 and 3. Is it the same thing, then, in all cases, whether we divide by two numbers separately, or by their product ?

12. Is 3 a factor in 342 ? Is 2 ? Is 6 a factor, then ? Is 2 a factor in 546 ? Is 3 ? Is 6 a factor, then ? What is the sign of 6 ? An *even* number divisible by what ?

13. What is the *sign* of 15 (3×5) ? Of 18 (2×9) ? Of 20 ? *Ans.* An *even* number of —. What is the sign of 24 (3×8) ? Of 75 (3×25) ? Can an even number be prime ? Why ? Can a number ending in 5 be prime ? Why ?

14. [Write 4236981 on the black-board, and frequently exercise the class as follows, on this and other large numbers, till the questions can be answered correctly and rapidly.] What is the remainder, if any, when divided by 2 ? By 3 ?

4? 5? 9? 25? 50? 125? Is it divisible by 6? 15? 18? 20? 24? 75? 225?

☞ Signs to discover the factors 7 and 11 might be developed, but they are too numerous to be of much use in practice. It may be well to mention, however, that 7 is factor in all numbers consisting of 2 or of 3 figures, where the left hand figure, or figures, is double that of the right hand, as 42, 84, 63, 168, 147, &c.; and also if the right hand figure be $\frac{1}{9}$ of the left hand ones, as 91, 182, 273, 364, &c. Eleven is readily discovered in 2 figures, since they must be alike, as 44, 77; it is also a factor in numbers consisting of 3 figures, when the sum of the figures at the right and left is equal to that in the middle, as 473, 374, 286, 385, &c.

[The class may now be directed individually to form on the slate, *a table of numbers*, from 1 to 100, or to 1000, or to 10,000, as the teacher may see fit, *analyzed to their prime factors*. When this is done, let the pupils exchange slates, and each examine his school-mate's work, and mark any errors he may find. Lastly, let the tables be carefully examined by the teacher, to see that no composite numbers are placed among the prime factors, which will probably be the case on the first trial, such as 20=4•5, in place of 20=2•2•5. Writing such a table twice without copy or assistance, will generally make every pupil sufficiently familiar and ready with analysis. If not, the exercise should be repeated till he becomes apt in observing the prime factors in a number, and can declare at once how often they are repeated.]

Examples of Numbers resolved to Prime Factors.

Resolve the following numbers into their prime factors: 1st. 132; 2d. 625; 3d. 488.

Solution 1*st*. 132. Is 2 a factor in 132? How often? Is 3? Then 2•2•3 being factors, 4•3=12. What is the quotient of 132 by 12? Is 11 a prime? Then the prime factors of 132 are 2•2•3•11. 2d. 625. Is 2 a factor in 625? Is 3? Is 5? How often? How many 25s, then, in 625? *Ans.* There being 4 in each hundred, of course there are 25 in 625. Then 25 being twice a factor, gives 5•5•5•5 for the prime factors of 625. 3d. 488. How often is 2 a factor in this number? What is the quotient of 488 by 8? Is 61 prime or composite? Then the prime factors of 488 are

2·2·2·61. Observe, however, that nearly all these steps will become superfluous by practice, and should be dispensed with as soon as possible.

Specimen of the FORM of the Table of Prime Factors.

Nos.		Nos.		Nos.	
1	Prime	11	——	21	
2	——	12		22	
3	——	13		23	
4	2·2	14		24	
5	——	15		25	
6	2·3	16		26	
7	——	17		27	
8	2·2·2	18		28	
9	3·3	19		29	
10	2·5	20		30	
				&c.	

As soon as the members of a class have each formed a table from 1 to at least 100, it may be well to direct their attention to the fact that *every prime* number greater than 3 is either 1 more or 1 less than 6, or one of its multiples. This singular property of numbers may be thus accounted for. Neither 6 nor either of its multiples can be prime, since they must necessarily be *even;* now, if the numbers between any two of its adjacent multiples be examined, take, for instance, 13, 14, **15**, 16, 17, it will be obvious that the middle number cannot be prime, since every multiple of 6 must also be a multiple of 3, and the middle number contains exactly one 3 more; and as the *even* numbers on each side of the middle one cannot of course be prime, it follows that *no number greater than* 3 *can be prime, unless it be* 1 *greater or* 1 *less than* 6, *or than one of its multiples.* Observe, however, that, although primes can be found in no other situation, it does *not* necessarily follow that the converse is true, namely, that *every number in that situation must be prime,* as an inspection of the table will show.

15. Write in a column on the slate the prime numbers between 5 and 97 inclusive, and repeat the exercise daily till they become familiar.

16. What are the prime factors of 14 and 35? What is their greatest common divisor; that is, what prime factors are *common* to both numbers? What are the prime factors of 16 and 12? What is their greatest common divisor? Mention the prime factors of 16 and 18, and say which are common. What is the greatest common divisor of 28 and 42? Of 16 and 36? 18 and 42? 19 and 57? 72 and 30? 20 and 45? 51 and 17? 75 and 125? 39 and 26? 36 and 54? 14 and 63? 15 and 125? 27 and 84? 30 and 81?

17. What is the greatest common divisor of 12, 27, and 51? Of 9, 45, and 54? 63, 18, and 36? 15, 39, and 27? 4, 26, and 38? 14, 49, and 63? 15, 105, and 75? 24, 78, and 42? 85, 34, and 51?

18. What is the greatest common divisor of 4, 32, 12? Of 45, 75, 60? 33, 77, 22? 39, 91, 78? 46, 69, 92? 34, 85, 102? 16, 128, 64? 116, 29, 87?

19. Mention two numbers of which 10 is the common multiple? Of which 15? 22? 26? 34? 35? 42? 46? 51? 52? 106? 112?

20. What is the least common multiple of 2 and 3? Of 3 and 5? 7 and 3? 5 and 11?

21. Is 18 a common multiple of 3 and 2? Its least common multiple? Is 260 a common multiple of 5 and 13? Its least common multiple?

SECTION III. — *Fractional Change of Form.*

1. WHEN a bushel of wheat is divided into 4 equal parts, what is one of them called? How many of these fourths make half a bushel? Is $\frac{2}{4}=\frac{1}{2}$, then? If a bushel was divided into 6 equal parts, how many of them would make half a bushel? Is $\frac{3}{6}=\frac{2}{4}=\frac{1}{2}$? [Show these fractions on black-board.] If divided into 8 parts, how many would make half a bushel? If divided into 10, 12, 16, 18, &c., parts? [Show a number of these fractions, and let the class observe the relation between the two terms of each fraction.] In how many ways could $\frac{1}{2}$ be represented? [Infinite.] Which is easiest understood, the smallest fraction $\frac{1}{2}$, or one of the larger ones?

2. When an article is divided into 3 equal parts, what is one of them called? If divided into 6 equal parts, what

would one of them be called? How many of these last would make $\frac{1}{3}$ of the article? Is $\frac{2}{6}=\frac{1}{3}$, then? [Black-board.] Express $\frac{1}{3}$ in as many different forms as you can. Which is easiest understood, the smallest or one of the larger? Into how many different forms can a fraction be changed without altering its value? [See p. 82, l. 5.]

3. What is the least common multiple of 2 and 3? How, then, may $\frac{1}{2}$ and $\frac{1}{3}$ be changed to equivalent fractions with a least common denominator? How may $\frac{2}{3}$ and $\frac{3}{5}$ be changed to equivalent fractions with a least common denominator? $\frac{3}{4}$ and $\frac{4}{5}$? $\frac{5}{7}$ and $\frac{2}{3}$? $\frac{7}{8}$ and $\frac{3}{5}$? $\frac{7}{9}$ and $\frac{3}{7}$?

4. Change $\frac{3}{4}$ to an equivalent fraction of the same denomination as $\frac{5}{8}$. $\frac{4}{5}$ to same denomination as $\frac{6}{15}$. Make $\frac{1}{2}$, $\frac{3}{4}$ and $\frac{5}{8}$, of same denomination, by a change in the two former? What is the least common multiple of 3, 5, and 25? Change $\frac{2}{3}$, $\frac{3}{5}$ and $\frac{8}{25}$, then, to equivalent fractions with the least common denominator. [Use black-board.] Change $\frac{2}{3}$, $\frac{3}{4}$, and $\frac{5}{15}$, to fractions with least common denominator, commencing by a change in the last fraction. Change $\frac{3}{7}$, $\frac{2}{8}$, $\frac{1}{5}$, $\frac{2}{4}$, and $\frac{4}{20}$, to fractions with least common denominator, commencing by a change in the second and fifth.

5. Change the following sets of fractions to equivalent ones, with the same lowest denominator by division: $\frac{4}{8}$ and $\frac{3}{4}$; $\frac{1}{4}$ and $\frac{8}{16}$; $\frac{5}{20}$ and $\frac{3}{4}$; $\frac{8}{24}$, $\frac{3}{16}$, and $\frac{8}{32}$; $\frac{6}{9}$, $\frac{1}{3}$, and $\frac{16}{24}$; $\frac{4}{10}$, $\frac{3}{5}$, and $\frac{10}{25}$.

6. What is the least common denominator of $\frac{4}{5}$, $\frac{3}{8}$, $\frac{1}{4}$, $\frac{2}{3}$, $\frac{3}{24}$, $\frac{5}{20}$? [Black-board.] How shall $\frac{4}{5}$ be changed to an equivalent fraction of that denomination? How shall $\frac{3}{8}$? $\frac{1}{4}$? $\frac{2}{3}$? $\frac{3}{24}$? $\frac{5}{20}$? What is the least common denominator of $\frac{4}{9}$, $\frac{5}{21}$, $\frac{6}{7}$, $\frac{2}{3}$, $\frac{8}{27}$? How shall $\frac{4}{9}$ be brought to that denomination? $\frac{5}{21}$? $\frac{6}{7}$? $\frac{2}{3}$? $\frac{8}{27}$? What is the least common denominator of $\frac{1}{4}$, $\frac{3}{8}$, $\frac{2}{10}$, $\frac{3}{25}$, $\frac{1}{7}$, $\frac{3}{14}$, $\frac{4}{28}$? How shall $\frac{1}{4}$ be brought to that denomination? How shall $\frac{3}{8}$? $\frac{2}{10}$? $\frac{3}{25}$? $\frac{1}{7}$? $\frac{2}{14}$? $\frac{4}{28}$?

☞ The above exercises should be studied without the aid of the slate. In reciting, the teacher should write the given fractions on the black-board, and call on the pupils to work out the answers mentally.

SECTION IV. — *Addition and Subtraction of Common Fractions.*

Explanation. — Numbers of *different denominations* can neither be added together nor subtracted from each other. Thus, 6 chairs and 3 tables make neither 9 tables nor 9 chairs; and 3 tables cannot be taken from 6 chairs, nor 3 chairs from 6 tables. Their *denomination*, however, can be made the *same*, and then they can be either added or subtracted. Thus, by calling both chairs and tables pieces of furniture, they have the same denomination; and, when added, make 9 pieces of furniture; and the 6 or the 3 may be subtracted from the 9.

The same remark holds good with respect to *abstract* numbers (that is, numbers used without being applied to things), whether they are whole or fractional numbers. Thus 6ty cannot be added to 3 hundred or to 3 units, because they are of different denominations, and would make neither 9 hundred, 9ty, nor 9 units. Neither can 3ty be subtracted from 6 hundred without changing one of the hundred (mentally) to 10ty, and thus we should have 570. This becomes more evident when applied to fractions. $\frac{1}{4}$ can neither be added to, nor subtracted from, $\frac{5}{6}$, while both retain their present form. But $\frac{1}{4}$ can be changed to $\frac{3}{12}$, and $\frac{5}{6}$ to $\frac{10}{12}$, and, as they have now the same denomination (twelfths), they can either be added or subtracted.

But the case is entirely different in regard to multiplication and division. In multiplication, as has already been observed, although it is in reality nothing more than addition, yet one of the factors is *not to be added.* It merely points out the *number of times* that the *other factor is to be taken.* Thus, the 6 chairs and 3 tables may be multiplied by 3, giving 18 chairs and 9 tables, because 3 *times* 6 chairs make 18 chairs, and 3 *times* 3 tables make 9 tables; and so with any other number whatever. Abstract integers may also be multiplied or divided by numbers of different denominations, since it is evident that 6 hundred or 6ty can be taken three times as well as 6 units can; and it is equally evident that either of these digits can be divided by 3 of any denomination whatever. In like manner, $\frac{2}{3}$ and $\frac{3}{4}$, or any other fraction, though of different denominations, may be multiplied or divided by 2, 3, $\frac{1}{2}$, or any other number, since the one is only taking each of these fractions *so many times*, and the other is only finding *how many times* they contain the divisor.

1. Can numbers of different denominations be added together or subtracted from each other? Give me examples to show why. What should be done, then, when it is necessary they should be added or subtracted? Can factors of different denominations be used in multiplying? Give an example to show why. May the divisor and dividend be of different denominations? Give an example to show why.

2. By how many methods can a fraction be changed without altering its value? [See p. 83, l. 36.] Which is more intelligible, a fraction with large or with small terms? Which, then, is the preferable mode of changing the form of a fraction? Can the form of a fraction *always* be changed by division? When *can* division be used? *Ans.* When its terms have a common divisor. Can $\frac{12}{27}$ be changed by division? Can $\frac{14}{52}$? $\frac{6}{8}$? $\frac{3}{9}$? $\frac{5}{8}$? Why? Can $\frac{7}{42}$? $\frac{7}{36}$? $\frac{15}{39}$? Can every fraction be changed in form by multiplication? Why? Because, though every two numbers may not have a common divisor, yet any two, &c.

3. What is the least common multiple of 3 and 15? What, then, is the sum of $\frac{2}{3}$ and $\frac{3}{5}$? What is the least common multiple of 4 and 8? What, then, is the sum of $\frac{3}{4}$ and $\frac{5}{8}$? What is the sum of $\frac{2}{7}$ and $\frac{3}{9}$? Of $\frac{4}{5}$, $\frac{5}{4}$, and $\frac{6}{20}$? Of $\frac{2}{7}$, $\frac{3}{28}$, and $\frac{3}{4}$? Of $\frac{2}{3}$, $\frac{3}{4}$, and $\frac{6}{15}$? Of $\frac{1}{5}$, $\frac{2}{8}$, and $\frac{3}{20}$? [By division and multiplication the least common denominator becomes 20.] Of $\frac{3}{4}$, $\frac{1}{6}$, and $\frac{5}{12}$? Of $\frac{3}{12}$, $\frac{1}{4}$, and $\frac{8}{16}$? [Least common denominator 4.]

4. What is the least common denominator of $\frac{3}{4}$ and $\frac{5}{8}$? What is the difference of these fractions, then? The difference of $\frac{6}{5}$ and $\frac{4}{20}$? Of $\frac{2}{7}$ and $\frac{5}{9}$? Of $\frac{1}{5}$ and $\frac{1}{4}$?

5. How many fourths are there in 1? In 2? In 5? In 9? How many fourths in $1\frac{1}{4}$? In $5\frac{3}{4}$? What, then, is the difference between $6\frac{3}{4}$ and $1\frac{1}{4}$? How many fifths in 1? Eighths in 1? Sixths? Ninths? What, then, is the difference between $4\frac{5}{6}$ and $2\frac{3}{4}$?

SECTION V. — *Contracted Addition and Subtraction of Common Fractions, usually called Multiplication and Division.*

1. IN how many ways can a fraction be multiplied? [See p. 80, l. 16, and p. 81, l. 16.] Name them. By which method is the fraction rendered most intelligible? Which, then, is the preferable mode? Can division be used for multiplying a fraction in all cases?* Why not? By how many methods can a fraction be divided? Which, then, is the preferable mode? Can a fraction be divided or multiplied by division in all cases? Why not?

2. Multiply $\frac{5}{12}$ by 3 by multiplication; by division. Multiply $\frac{3}{21}$ by 7 by multiplication; by division. Can $\frac{4}{17}$ be multiplied by 3 by division? Why not? *Ans.* Because 3 is not a factor in —. Can it be multiplied, then, by multiplication? What is the product? Multiply the following factors by both methods, changing the fraction, when not already so, to its lowest denomination, and observing whether or not the result of the two methods is alike: $\frac{3}{15}$ by 3; $\frac{3}{16}$ by 4; $\frac{5}{18}$ by 3; $\frac{6}{28}$ by 7.

3. Divide $\frac{5}{12}$ by 3 by multiplication. Can it be done by division? Why? Divide $\frac{2}{21}$ by 7 by multiplication. Can it be done by division? Why? Can $\frac{4}{17}$ be divided by 2 by both methods? Why? Divide the following fractions as indicated, changing the quotient, when not already so, to its lowest denomination, and observing whether or not the result of the two methods is alike: $\frac{3}{15}$ by 3; $\frac{8}{16}$ by 4; $\frac{10}{18}$ by 2; $\frac{15}{28}$ by 5.

4. Multiply $\frac{1}{5}$ by $\frac{2}{3}$. [Write it on the black-board.]

Suggestive Questions. — What part of 2 is $\frac{2}{3}$? [See Chap. II., Sect. I., 13.] If $\frac{1}{5}$ be multiplied by 2, then, how many times too large will the product be? If it be 3 times too large, how can it be rectified? [This analysis will be sufficiently clear when exhibited on the black-board, if the teacher write

* A fraction *may* be multiplied and divided in all cases by division; but it becomes complicated when the divisor is not a factor of the dividend. Thus $\frac{3}{5}$ multiplied by four by division becomes $\frac{3}{1\frac{1}{4}}$, and $\frac{7}{9}$ divided by 4 by division becomes $\frac{1\frac{3}{4}}{9}$. It is, therefore, more convenient, in such cases, to multiply and divide by multiplication, which presents the fractions in the more intelligible forms of $\frac{12}{9}$ and $\frac{7}{36}$. When it is said, then, that one number is not divisible by another, all that is meant is that the quotient would be complicated with fractional parts. But, whenever this would not be so, a change of fractional form by division is always preferable.

the answers to the suggestive questions by the class, as follows, especially if the multiplication be only *indicated*, as below, not performed. Thus, $\frac{1}{5}\times\frac{2}{3}$. First step, $\frac{1\cdot2}{5}$; second step, $\frac{1\cdot2}{5\cdot3}=\frac{2}{15}$.] If $\frac{1}{5}$ be multiplied by 2, then, in place of $\frac{2}{3}$, and, because it is 3 times too much, divided by 3, what terms of the two fractions will be multiplied together? Will this be the case, whatever may be the numbers, when one fraction is to be multiplied by another?

5. What principle, then, may be drawn from this exercise?

Ans. To multiply one fraction by another, multiply the — for a new numerator, and the — for a new denominator.

6. Multiply $\frac{2}{7}$ by $\frac{3}{5}$; $\frac{4}{9}$ by $\frac{2}{7}$; $\frac{5}{6}$ by $\frac{1}{8}$; $\frac{3}{4}$ by $\frac{2}{5}$; $\frac{7}{8}$ by $\frac{3}{9}$; repeating, step by step, the above analysis.

7. Multiply $\frac{5}{6}$ by $\frac{3}{10}$. [Write it on black-board.]

Suggestive Questions.—What factor is common to 5 and 10? What factor is common to 3 and 6? What two factors, then, are common to both terms? What, then, will be the product of the two fractions, if the common factors be dropped; that is, if both terms be divided by 15?

8. Mention the products of each of the following pairs of fractions by inspection merely, casting out the factors common to both terms mentally, so as at one step to present each product in its lowest denomination: $\frac{5}{8}\times\frac{4}{20}$; $\frac{7}{8}\times\frac{4}{35}$; $\frac{9}{10}\times\frac{5}{6}$; $\frac{5}{9}\times\frac{6}{15}$; $\frac{3}{8}\times\frac{24}{27}$; $\frac{6}{7}\times\frac{14}{18}$; $\frac{3}{7}\times\frac{7}{8}$; $\frac{2}{5}\times\frac{5}{7}$; $\frac{3}{4}\times\frac{4}{5}$.

9. What addition to the principle developed from the 4th exercise may be drawn from the last two exercises?

Ans. When one or more factors can be found in one or in both of the —— which can also be found in one or in both of the ——, they may be cast out of both before the multiplication, and thus leave the fractional product in its lowest denomination.

10. Divide $\frac{3}{5}$ by $\frac{2}{3}$.

Suggestive Questions. — What part of 2 is $\frac{2}{3}$? If $\frac{3}{5}$ be divided by 2, then, instead of $\frac{2}{3}$, how many times too small will be the quotient? How, then, shall it be rectified? [Let the division be indicated on the black-board as before in Ex. 4, as follows: $\frac{3}{5}\div\frac{2}{3}$. First step $\frac{3}{5\cdot2}$; second step $\frac{3\cdot3}{5\cdot2}$.] If $\frac{3}{5}$, then, be divided by 2 in place of $\frac{2}{3}$, and, because the quotient is 3 times too small, it be multiplied by 3, what terms of the two fractions are found to be multiplied together? [Repeat the above analysis, step by step, in each of the following problems.] What is $\frac{5}{8} \div \frac{4}{9}$? $\frac{7}{8} \div \frac{9}{10}$? $\frac{5}{9} \div \frac{3}{5}$? $\frac{3}{8} \div \frac{24}{27}$? $\frac{6}{7} \div \frac{7}{6}$? $\frac{3}{7} \div \frac{7}{8}$? $\frac{2}{5} \div \frac{5}{7}$? $\frac{3}{4} \div \frac{4}{5}$? In place of multiplying crosswise, would the same result be attained by reversing the divisor, and completing the process by multiplying the two fractions? [Show the process on the black-board in one or two of the above cases, as follows: $\frac{5}{8} \div \frac{4}{9} = \frac{5.9}{8.4}$, by division; $\frac{5}{8} \times \frac{9}{4}$ (the divisor reversed) by multiplication.] As the result, then, is exactly the same, we shall in future pursue the process by *reversing the divisor, and multiplying.* Repeat the above problems by the reversing process.

11. What principle may be drawn from the above exercises?

Ans. To divide one fraction by another, reverse the ——, and proceed as in multiplication.

12. Divide 3 by $\frac{1}{5}$. The most simple method of resolving such questions is to give the integer (3) a fractional form, as $\frac{3}{1}$. But, after a little practice, giving the fractional form becomes a *superfluous step.* Divide 9 by $\frac{4}{7}$; 8 by $\frac{3}{4}$; 14 by $\frac{4}{5}$.

13. How many $\frac{1}{5}$ in $4\frac{1}{5}$? Divide $4\frac{1}{5}$, then, by $\frac{3}{4}$. How many $\frac{1}{7}$ in $6\frac{3}{7}$? Divide $6\frac{3}{7}$ by $\frac{4}{5}$; $3\frac{5}{8}$ by 9. $7\frac{1}{9}$ by $2\frac{1}{8}$. $5\frac{1}{6}$ by $2\frac{5}{9}$.

14. Divide 2 by 7 [$\frac{2}{7}$]. 1 by 5; 3 by 8; 5 by 8; 2 by 6; 1 by 4.

15. What does the word *of* signify when connected with fractions? [See Section I., 13 of this chapter.] How much, then, is $\frac{4}{5}$ of $\frac{3}{8}$? Express $\frac{3}{4}$ of $\frac{7}{8}$ of $\frac{32}{45}$ of $\frac{9}{25}$, in its most simple terms, mentally casting out equal factors. [Black-board.] Express, in their lowest terms, $\frac{7}{9}$ of $\frac{3}{14}$; $\frac{5}{8}$ of $\frac{12}{15}$; $\frac{3}{7}$ of $\frac{21}{27}$; $\frac{6}{54}$ of $\frac{9}{21}$ of $\frac{3}{4}$; $\frac{5}{17}$ of $\frac{51}{65}$.

SECTION VI. — *Practical Questions.*

1. A MERCHANT sold 8 barrels of flour, at $6\frac{3}{8}$ dollars per barrel. How much did they come to? [The pupil should explain the process in all the questions that follow.]

2. A countryman sold $4\frac{3}{4}$ bushels of cranberries to one store-keeper, and $3\frac{3}{8}$ bushels to another, at $3\frac{3}{8}$ dollars a bushel. How much money did he receive?

3. One man bought $2\frac{1}{4}$ bushels of the same cranberries, and another man bought $1\frac{1}{8}$ bushels. How much did the first man pay more than the other?

4. What will 9 bushels of rye come to, at $\frac{11}{16}$ of a dollar per bushel?

5. A man bought 5 yards of cloth for $16\frac{1}{4}$ dollars. How much was that for one yard? What would 3 yards cost at that rate?

6. If 6 men can build a piece of wall in $3\frac{3}{8}$ days, in what time could 1 man build it? In what time could 4 men build it?

7. If 3 horses will eat $16\frac{1}{8}$ tons of hay in a year, how much will 1 horse eat in the same time? How much will 5 horses?

8. If 3 barrels of flour last a family $6\frac{3}{5}$ months, how long will 1 barrel last them? How long will 5 barrels?

9. If 6 yards of cloth cost $13\frac{1}{5}$ dollars, what will 1 yard cost? What will 9 cost?

10. If a man can travel $10\frac{1}{8}$ miles in 3 hours, how much can he travel in 1 hour? In 5 hours?

11. If $2\frac{1}{4}$ bushels of wheat last a family 3 weeks, how much will last them 1 week? 5 weeks?

12. If 5 boxes of raisins cost $11\frac{2}{3}$ dollars, what will be the cost of 1 box? Of 7 boxes?

13. If 3 ounces of silver cost $3\frac{6}{20}$ dollars, what will be the cost of 1 ounce? Of 8 ounces?

14. If $3\frac{3}{7}$ pounds of bread be sufficient for 6 men for a day, how much is that for 1 man? For 5 men?

15. If 9 men receive $11\frac{1}{4}$ dollars for a day's work, how much is that for each man? How much would seven men earn at that rate?

16. If the freight of 9 hogsheads of sugar on a railroad be 16 dollars, what is the freight for 1 hogshead? For 7 hogsheads?

[Repeat the above from Ex. 5 to 16, omitting in each the leading question, "How much for 1," &c. They should be done as follows: 5th $3 \times 5\frac{1}{4}$, 6th $20\frac{1}{4} \div 4$.]

17. There is a pole standing so that $\frac{1}{4}$ of it is in the ground, and $\frac{3}{8}$ of it in the water. How much of it is in the air?

18. A pole is standing so that $\frac{1}{4}$ of it is in the ground, $\frac{1}{8}$ in the water, and 10 feet in the air. How many feet are in the ground, how many in the water, and what is the length of the pole?

19. In a certain school, $\frac{1}{5}$ of the pupils are learning to read, $\frac{2}{5}$ studying arithmetic, $\frac{1}{5}$ studying geography, and the rest, 6 in number, learning grammar. How many are reading? how many studying arithmetic? how many geography? and how many in all?

20. In a congregation $\frac{1}{4}$ were men, $\frac{3}{8}$ women, $\frac{1}{8}$ boys, and 100 girls; how many persons were in the congregation?

CHAPTER IV.

DENOMINATE FRACTIONS, OR FRACTIONS EXPRESSED IN CONCRETE WORDS, NOT IN FIGURES.

SECTION I. — *Change of Form.*

1. TEN cents make a dime, and 10 dimes make a dollar. How many cents, then, make a dollar?

2. Ten dimes make a dollar, and 10 dollars make an eagle. How many dimes, then, in an eagle?

3. Ten mills make a cent, 10 cents make a dime, 10 dimes make a dollar, and 10 dollars an eagle. How many mills, then, in an eagle?

4. Twelve pence make a shilling, and 20 shillings a pound. How many pence, then, in a pound?

5. Four farthings make a penny, 12 pence a shilling. How many farthings, then, in a shilling?

6. Four farthings make a penny, 12 pence a shilling, and 20 shillings a pound. How many farthings, then, in a pound?

7. As 12 pence make a shilling, and 20 shillings a pound, how many pounds and shillings in 540 pence?

8. Sixteen drams make an ounce, and 16 ounces a pound. How many drams in 1 pound? In 3 pounds? In 7 pounds? How many pounds in 512 drams? In 768 drams?

9. Twelve inches make a foot, and 3 feet a yard. How many inches make 5 yards? 15 yards? 25 yards? 14 yards?

10. Three feet make 1 yard, and $5\frac{1}{2}$ yards a rod. How many feet make a rod? 5 rods? 8 rods?

11. Sixteen and a half feet make a rod, and 320 rods a mile. How many feet in a mile?

12. Four quarters make a yard, and 5 quarters make an ell English. How many quarters in 5 yards? How many ells English in 20 quarters? How many ells English, then, in 5 yards?

13. Four quarters make a yard, and 6 quarters an ell French. How many quarters in 12 yards? How many ells French in 48 quarters?

14. Five quarters make an ell English, and 6 quarters an ell French. How many ells French, then, in 12 ells English? Solve this question first by multiplication and division, and then by division and subtraction.

15. Four quarters make a yard, 5 an ell English, and 3 an ell Flemish. How many ells English in 15 yards? and how many ells Flemish in the same?

16. Twenty shillings make a sovereign, and twenty-one shillings make a guinea. How many guineas in 63 sovereigns?

Solve this question first by multiplication and division, and then by division and subtraction.

17. One of the fields of a farm is exactly square, being 40 rods long and 40 wide, making 1600 square rods. Another field is only 25 rods wide, yet it contains the same number of square rods. Is the last named field longer or shorter than the first, and what is its length?

18. Sixty seconds make a minute, and 60 minutes an hour. How many seconds in an hour? In 5 hours? In 25 hours?

19. Twenty-four hours make a day, and 7 days a week. How many hours in a week? In 4 weeks? In 6 weeks? In 14 weeks? In 19 weeks? [14=15—1; 19=20—1.]

20. Two pints make a quart, 8 quarts a peck, 4 pecks a bushel, and 8 bushels a quarter. How many pints in a quarter?

21. Eight quarts make a peck, and 4 pecks a bushel. How many quarts in 75 bushels?

22. Two pints of milk make a quart, and 4 quarts a gallon. How much will a pint cost, at 40 cents a gallon?

CHAPTER V.

MISCELLANEOUS QUESTIONS.

1. A FARMER sent his son to market with 24 bushels of potatoes, which he sold at 25 cents a bushel. He bought 16 pounds of sugar, at $6\frac{1}{4}$ cents a pound, and 11 pounds of coffee, at 9 cents a pound. How much money did he carry home?

[The teacher should require his pupils to prove this and the following questions, and explain the process somewhat as follows: 24 times 25 is $\frac{1}{4}$ of $24\times100=600$; 16 pounds of sugar, at 6 cents, come to 96 cents, and $\frac{1}{4}$ of $16=4$, making 100 cents; and 11 pounds of coffee, at 9 cents, come to 99 cents; together 199 cents, which, taken from 600 cents, leave 401 cents to carry home.]

2. A storekeeper sold 3 dozen of eggs, at 10 cents per dozen; 3 yards of calico, at $12\frac{1}{2}$ cents per yard; and 16 pounds of rice, at 5 cents per pound. What did the whole amount to?

3. A man divided $360 dollars between his two sons and two daughters, giving 2 shares to each son, and 1 share to each daughter. How much did each son and daughter get.*

4. At a certain meeting of 280 persons, there were 4 times as many children and twice as many women as men. How many were there of each class?

5. Two brothers, one of whom was twice as rich as the other, calculated their property, and found that together they had $3600. How much did each possess?

6. The sum of $275 is to be divided among 2 men in such a way that for every $4 that one receives, the other is to receive one. How much does each receive?

* These fractional formulas, or as many as necessary, may be exhibited on the blackboard at the first recitation, but omitted at the review. But neither slate nor paper should be used in any of the exercises in Oral Arithmetic.

7. If a certain number is multiplied by 4, the product is 24. What is the number? By what process is it ascertained?

8. If a certain number is divided by four, the quotient is 24. What is the number? By what process is it ascertained?

9. Ten was added to a certain number, and 8 subtracted from it; it was then multiplied by 6 and divided by 4, when the result was found to be 24. What was the number? Reverse the process. Why?

10. If 4 pounds of brown sugar cost 25 cents, what will 9 pounds cost?

Suggestive Questions. — By what number must the 25 cents be divided to show the price of 1 pound? By what multiplied to show the price of 9 pounds? Place these numbers so as to indicate such operations; that is, place them in a fractional form, thus, $\frac{9}{4}\times25$. Why?

11. If 8 pounds of rice cost 56 cents, what will 15 pounds cost? $\frac{15}{8}\times56$. Why? [See note on preceding page.]

12. If 6 pounds of refined sugar cost 75 cents, what will $12\frac{1}{2}$ pounds cost? $\frac{25}{12}\times75$. Why not $\frac{12}{6}$?

13. A man who employed 11 laborers, found that the wages he paid came to $77 a week. He afterwards hired 4 more laborers at the same rate. How much would the weekly wages then amount to? 15×7. Why?

14. How much is $\frac{1}{3}$ of 105? If, then, the wages for 6 days amount to $105, how much would they amount to for 4 days?

15. If the wages of 15 men amount to $105 for a week of 6 working days, how much would they amount to for 3 weeks and 2 days? $\frac{20}{6}=\frac{10}{3}\times105$. Why?

16. At $105 for 15 laborers for a week of 6 days, how much does each man earn per day?

17. If a man travel 8 days, when the day is 12 hours long, how many hours does he travel?

18. If a man travel 360 miles in 6 days, when they are 12 hours long, how far will he travel in 8 days, when they are 10 hours long? $\frac{10}{9}\times360$. Why?

19. At $33 for 6 barrels of flour, what will 32 barrels cost? 16×11. Why?

20. If wheat rise in price should the 5 cent loaf be made larger or smaller?

21. If a 5 cent loaf weigh 10 ounces when wheat is 80 cents a bushel, how much should it weigh when it is 100 cents per bushel? $\frac{4}{5}\times10$. Why?

22. A ship was provisioned for 8 months, at the rate of 15 ounces per day to each person. What must be the daily allowance for each person, in order that the provision may last 12 months? More or less? and how much? $\frac{2}{3} \times 15$. Why?

23. A retailer one morning weighed his rice, and found that he had exactly 44 pounds. In the evening he again weighed it, and found that he had remaining 3 times as much as he had sold. How much had he sold? $\frac{44}{4}$. Why?

24. The same man weighed his raisins in the morning, and found them to weigh 40 pounds. In the evening there remained 8 pounds more than was sold. How many pounds were sold?

25. Three men purchased jointly a piece of woods, containing 1600 square rods. By agreement the second was to have 200 rods more than the first, and the third 200 rods more than the second. How many square rods was each man to have? $333\frac{1}{3}$, $533\frac{1}{3}$, $733\frac{1}{3}$. Why?

26. There are two numbers whose sum is 96, and their difference 12. What are the numbers? $\frac{96}{2} - 6$. Why?

27. Two merchants agreed to trade as follows: The first was to receive $60 a month for carrying on the business; the remainder was to be shared so that the second was to receive double the amount of the first, exclusive of his monthly pay. They gained 600 dollars the first month. How much would each receive?

Remark. — Interest is an allowance by a borrower for the use of money, or any kind of property lent. Interest is also frequently allowed on unpaid debts. The rate or amount of interest for a given sum differs at different times and in different places. In the state of New York, lawful interest is 7 per cent.; but in most parts of the United States, it is 6 per cent.; that is, 6 for the use of 100 for one year, whether the money be cents, dollars, pounds, or any other denomination. As 100 cents make 1 dollar, 6 cents is the usual interest for 1 dollar for a year. In reckoning interest, 30 days are considered a month, and 360 days a year.

28. At 6 cents for a dollar for 1 year, what is the interest of 5 dollars for the same time? Of $8? Of $12? Of $15? Of $24? Of $100? Of $96? Of $125?*

* $ is an abbreviation of U. S., and is used to denote *dollars*, the currency of the U. S.

29. At 6 cents for a dollar for one year, what would be the interest for 2 years? 4 years? 5 years? 7 years?

30. At 6 per cent. (or 6 for 100) for 1 year, what is the interest of $2 for 2 years? $8 for 2 years? $6 for 3 years?

31. As there are 12 months in 1 year, what is the interest of $1 for 6 months, at 6 per cent.? Interest of $1 for 2 months? Of $6 for 3 months? Of $6 for 8 months? $6 for 9 months? $6 for 11 months? $6 for 1 month? $6 for 7 months?

32. At 6 per cent., what is the interest of $100 for 6 months? For 8 months? 9 months? 1 month? 7 months? 11 months?

33. At 6 per cent., what is the interest of $150 for 1 year? Of $200? Of 250? 226? 500? 750? 326? 530?

34. At 6 per cent., for 12 months, what would be the interest for 2 months, or 60 days? For 30 days, or one month? For 15 days? For 45 days? For 20 days? For 40 days?

35. At 6 per cent. for 12 months, how much for 2 months, or 60 days? How much for 1 day? 2 days? 5 days? 8 days?

36. At 6 per cent. for 12 months, what is the interest of $100 for 18 months? For 16 months? For 8 months?

37. If the interest of $100 for twelve months is $6, how long must that sum be at interest to produce $9? To produce $12? $4? $1?

38. A man was sent on horseback from Boston to Hanover, and went at the rate of 5 miles an hour. After he had been gone 8 hours, another was despatched after him to bring him back. 1. How far was the first ahead when the second started? The second went at the rate of 8 miles an hour. 2. How much would he gain on the first in an hour? 3. In how many hours would he overtake him? 4. How many miles would the first have travelled before he was overtaken? *Ans. to 3d question*, $\frac{40}{3}$. Why?

39. A man sets out on a journey, and travels 30 miles a day. After he had been gone $3\frac{1}{2}$ days, his son sets out after him, and travels 45 miles a day? In how many days would he overtake him?

40. A cistern has two pipes. The first can fill it in 3 hours, and the second in 6 hours? How much of it would each of

them fill in 1 hour? How much would both together? How long would it take them both to fill it?

41. A cistern has 2 pipes. The first can fill it in 3 hours, the second in 4 hours. How much of it can both fill in an hour? In what time can both fill it?

42. A cistern has 3 pipes. The first can fill it in 4 hours, the second in 6, and the third in 8. In what time would all three fill it?

43. A cistern has 3 pipes; 2 to fill, and one to empty it. The first can fill it in 3 hours, the second in 4, the third can empty it in 6 hours? In what time would it be filled if all three were set running?

44. A and B can do a piece of work in 10 days, and with the assistance of C they can do it in 8 days. How long would it take C alone to do it? $\frac{10}{80} - \frac{8}{80} = \frac{1}{40}$. *Ans.* 40 days. Why?

45. A man and his wife consume together a bushel of meal in 2 weeks; but when the wife was alone, it would last her 5 weeks. How much of it did the man consume in 1 week?

46. If a man could build a piece of wall in 5 days, and another in 6 days, how long would it take both to build it?

47. If a staff 4 feet long cast a shadow 6 feet long, how long a pole could cast a shadow 12 feet long at the same time? How long a pole could cast a shadow 15 feet long?

48. A man bought a horse and wagon for 160 dollars. The horse cost $1\frac{1}{2}$ times the price of the wagon. What was the price of the wagon? $\frac{320}{5}$. Why?

49. A boy being asked how many cents he had, replied, if you multiply the number by 5, take 8 from the product, divide the remainder by 9, and add 2 to the quotient, the amount will be 10. How many cents had he?

50. A mason, 12 journeymen, and 4 laborers, receive together $72 wages for a certain time. The mason receives a dollar daily, each journeyman $\frac{1}{2}$ dollar, and each laborer $\frac{1}{4}$ dollar. How many days must they have worked for the money?

51. In an orchard of fruit trees, $\frac{1}{2}$ bore apples, $\frac{1}{4}$ plums, $\frac{1}{8}$ pears, 7 peaches, and 3 cherries. How many trees are there in the whole, and how many of each sort?

52. Sold 16 pounds of butter, at 25 cents a pound, and received in pay sugar, at 8 cents a pound. How many pounds of sugar would I receive?

53. Sold 24 chickens at 50 cents a pair, for 20 yards of delaine. What was the price per yard?

54. If 5 men in 40 days can make 300 pairs of boots, how many can 1 man make in a day? How many could 15 men make in 4 days?

55. A, B, and C, divide $100 among them, so that B should have $8 more than A, and C $9 more than B. How much does each receive?

[In reviewing, omit the leading questions in the three following examples.]

56. If 49 pounds of bread be sufficient for 14 men for 7 days, how much will suffice for 1 man 1 day? How much for 20 men 4 days?

57. If 4 men can reap 48 acres in 12 days, how many acres can 1 man reap in 1 day? How many can 8 reap in 16 days?

58. If $75 be the wages of 5 men for 3 weeks, what will be the wages of 1 man for 1 week? What will be the wages of 8 men for 4 weeks?

59. How many yards of baize, 4 quarters wide, will line 8 yards of camblet, 3 quarters wide?

60. Four mechanics completed a piece of work, for which they received as wages $36, and it was agreed to divide this in proportion to the time they had been respectively employed. Now, the first had worked 3 days, the second 4, the third 5, and the fourth 6 days. What were their respective shares of the wages?

61. Two men purchased some goods, and sold them again at a profit of $24. The first advanced $100, and the second $200. What share of profits should each receive?

62. Two merchants traded together for a year. The first advanced a capital of $3000 for 12 months, and the other a capital of $6000 for 4 months. The profit was $3000. How should it be divided?

63. How many pounds of coffee, at 9 cents a pound, with twice the weight of sugar, at 6 cents a pound, may be purchased for 4 dollars and 20 cents?

64. If 30 men can do a piece of work in 8 days, how many men could do the same piece of work in 12 days?

65. When 20 men can do a piece of work in 12 days, in what time may 30 men do the same work?

All the above questions should be proved by the class.

PART II.

WRITTEN ARITHMETIC.

INTRODUCTION.

ADDRESSED TO TEACHERS.

WRITTEN Arithmetic is the art of calculating by means of *written characters.* As already shown, when treating of Oral Arithmetic, there are only two operations that can be performed with numbers, namely, *Increase* and *Decrease.* To perform these with skill and rapidity, comprises the whole of Arithmetic. Both may be accomplished by numeration. But such a method is entirely too slow for practice in the extensive operations of civilized society. Shortened processes have therefore been invented, chiefly by *omitting superfluous steps,* which have effected wonderful savings both of time and labor. Indeed, so numerous and important are the abbreviations in constant use, that Arithmetic may not inaptly be defined the *art of increasing and decreasing numbers by* SHORTENED *processes.*

In the following exercises it is of vast importance that pupils should avoid the use of *all unnecessary words,* either in speech or thought. For instance, when 4 and 2 are to be added, there is no occasion to think of, far less to *name,* those numbers. In reading, we never think of nor use the *name* nor *sound* of the *letters* which compose the syllables or words. The sound of the *syllable* or *word* itself occurs at the first glance. It will be the same with the numeral characters, if proper care be taken *from the outset;* for it is much more difficult to *reform* a *bad* habit than to *form* a *good* one. The thought of 6, and the writing of it, will be almost, if not quite, simultaneous with the sight of the figures 4 and 2; and, after a little practice, the mind will readily grasp, in the same way, three, four, five, and more figures, at a time.

The same remark applies to all the other processes. There is no occasion, for instance, to say, or even to think of, *four times five make* twenty. When properly trained, the pupil will think of *twenty* the instant he sees the 4 and the 5. The tedious *talk* about *carrying*, also, should be altogether dispensed with. Thus, in the following example,

$$\begin{array}{r} 84 \\ 5 \\ \hline 420 \end{array}$$

in place of saying, or thinking, *five times four are twenty, nothing and carry two; five times eight are forty, and two are forty-two;* the words *twenty, forty-two* (two words out of eighteen), are all that are necessary. The rest are mere hindrances. And not only does such verbiage lengthen the process, but experience has shown that it actually increases the chance of error, by allowing *time* for the *mind to wander* from the subject. By strict attention to this matter, multiplication and subtraction will be performed as fast as the figures can be written, division will be much shortened, and addition need not occupy more than a fourth of the time usually required.

The mode of performing the elementary processes now universally used in our schools, is the same in principle with the obsolete plan of teaching children to read by *spelling* every word. It differs, however, in one important respect. In reading, the spelling process generally lasts but a few months. In arithmetic, unfortunately, it clings to most persons through life. Let it be wholly abjured from the first, then, and the profit and ease both to teacher and pupil will be found to be immense. The spelling process merely "darkeneth by words." Nor is this all. The advantage will be reflected on all other studies. For, by the improved method, it is obvious the mind is kept continually on the alert, and the slovenly, dreamy, mental habits, which are now the bane of our schools, are entirely avoided.

The whole subject of Written Arithmetic may be suitably arranged in five chapters.

I. A DESCRIPTION of the CHARACTERS, or Notation and Numeration.*

II. The shortened processes of Increase and Decrease, applied to Integers and Decimal Fractions. Under the head of Increase come Addition and Multiplication (or addition of equal numbers), including Involution (or multiplication of equal numbers). Under the head of Decrease come Subtraction and Division (subtraction of equal numbers), including Evolution (division of equal numbers).

III. The same processes applied to Common (or Vulgar) and Denominate Fractions.

IV. Practical application of the methods of increase and decrease, promiscuously arranged.

V. The Comparison of Numbers, or Ratio and Proportion.

To these are added a SUPPLEMENT, containing a few subjects more properly belonging to Algebra, though commonly inserted in books of arithmetic, such as Progression by Differences and Progression by Ratios, usually, though improperly, called Arithmetical and Geometrical Progression, Compound Interest, Permutation, &c.

The rapid method of performing the elementary operations necessarily requires considerable practice in Oral Arithmetic. Such pupils, therefore, as have not already studied Part I. of this treatise, should now recite from it once or twice a day, simultaneously with the study of this part of the work. If this is faithfully attended to, *adding in* such numbers as 25, 27, 28, &c., 35, 36, 37, &c., will soon be found as easy as the addition of single digits.

* The art of *writing* the characters is *Notation*; that of *reading* them *Numeration*.

CHAPTER I.

THE NUMERAL CHARACTERS,

OR, ELEMENTARY PRINCIPLES OF INCREASE AND DECREASE.*

THE idea of numbers is one of the first that enters into the human mind. The infant observes his two hands, or the two eyes of his mother. The rudest savage counts his arrows or his game. *Names* for numbers, therefore, are among the first words invented. A few names, of course, answer every purpose for the child or the savage. But, as the child becomes a man, or the savage becomes civilized, new wants call for new numbers, and these of course call for new names, until it becomes impossible to supply a sufficient variety. For no genius could invent, no memory retain, such a multitude of terms, were a distinct word required for each number. Hence mankind has everywhere been compelled to *classify* numbers. Thus, if sixty or a hundred shells were spread out on the sea-shore, or placed in a row, to explain to a group of savages the number of fish contained in a canoe, the collection or row would not give a clear idea of the actual quantity. But, if the shells were arranged in small heaps, each of which should contain an equal number, there could be no such difficulty. Now, such an arrangement has been actually introduced into every community. Nature herself has provided us with a scale, or measure, which is so obvious and simple as to have forced itself into universal use. This is no other than the *ten* fingers. By the aid of this scale any number whatever can be expressed by the aid of a very small number of terms. Nor is this the only advantage of this scale. It requires only *nine* characters, with an additional one to express *zero*, or *nothing*, to represent this wonderful, this infinite variety of numbers.

* If the class of beginners is young, it would be profitable for the teacher to *read* this chapter to them by sections, with illustrations on the black-board and other explanations.

Ten, then, being the universal scale or measure used in calculation, our system is properly called *Decimal* Arithmetic, the word *decimal* meaning *numbered by tens.*

Various kinds of *characters* have been used at different times, and by different nations, for expressing numbers. But the *Roman* and the *Arabic* numerals are the only ones which it is important for the student of Arithmetic to understand.

The *Roman* numerals are chiefly used for dates, chapters, and sections, of books, and the hours on time-pieces. The characters are derived from the alphabet. Their origin is sufficiently evident; and, as a knowledge of this origin will assist the student in recalling them to mind if they should be forgotten, an explanation of it will not be out of place here.

The *ten fingers* present so obvious and convenient a method of numbering, that every people hitherto known, except the Chinese, and an obscure tribe mentioned by Aristotle, has employed them for that purpose. The rude tribes of Africa and America, however, use the fingers of *one* hand only as their scale; that is, they count onward from one to five, as we do from one to ten, and then commence anew. It may justly be affirmed, then, that nature, in forming the human hand, supplied us, at the same time, with the first elements of calculation.

But the Romans not only used the *digits*, or *fingers*, as the foundation of their method of computing; they also derived several of their characters from them. Thus, a finger, represented by I, stood for *one;* two, three, and four fingers, represented by II, III, IIII, stood for *two*, *three*, and *four.* By holding up the hand with all the five fingers extended, a tolerably correct representation of the letter V will appear, formed by the thumb and index finger. V was accordingly chosen as the character for *five.* In like manner, VI (six) is one hand and one finger of the other; VII (seven) a hand and two fingers, &c., while X (ten) represents both hands, considered as two V's, joined by their apices; or it may be formed by holding up both hands, one thumb resting on the other in the form of a cross. C and M, the initial letters of *centum* and *mille*, the Latin words for a hundred and a thousand, represented these numbers. C was originally written thus, ⊏. Its half L, stood for 50. In like manner the half of M, |\, rounded into D, stood for 500.

Such was evidently the origin of the first Roman numerals;

but, as the eye does not readily recognize more than three characters at a glance, a plan has been adopted to obviate that difficulty, and that is, by causing a *smaller* number placed *before* a larger to be subtracted in place of being added. Thus, in place of IIII (four times one), we have IV (five less one); for VIIII (five and four), we have IX (ten less one); for XXXX (four times ten), we have XL (fifty less ten); and for LXXXX (fifty and forty), we have XC (a hundred less ten.)

Besides the characters already enumerated, IↃ is sometimes used for D, and CIↃ for M; and these, in fact, may possibly be the original characters that represented five hundred and a thousand. For, when brought closely together, they greatly resemble the D and the M. But in other respects they are out of rule. For, when Ↄ is annexed to IↃ, it increases the value of the latter tenfold. In like manner, when C is prefixed and Ↄ annexed to CIↃ the last is increased tenfold. Lastly, the value of a character is increased a thousand fold by drawing a horizontal line over it.

TABLE OF ROMAN NUMERALS.

I. one.	CCC. three hundred.
II. two.	CD. four hundred.
III. three.	D. or IↃ. five hundred
IV. four.	DC., or IↃC, six hundred.
V. five.	DCC., or IↃCC., seven hundred.
VI. six.	DCCC., or IↃCCC., eight hundred.
VII. seven.	CM. nine hundred.
VIII. eight.	M., or CIↃ., a thousand.
IX. nine.	MM., or $\overline{\text{II}}$., two thousand.
X. ten.	MMM., or $\overline{\text{III}}$, three thousand.
XX. twenty.	MMMM., or $\overline{\text{IV}}$., four thousand.
XXX. thirty.	IↃↃ., or $\overline{\text{V}}$., five thousand.
XL. forty.	IↃↃM., or $\overline{\text{VI}}$., six thousand.
L. fifty.	IↃↃMM., or $\overline{\text{VII}}$., seven thousand.
LX. sixty.	IↃↃMMM., or $\overline{\text{VIII}}$., eight thousand.
LXX. seventy.	IↃↃMMMM., or $\overline{\text{IX}}$., nine thousand.
LXXX. eighty.	CCIↃↃ., or $\overline{\text{X}}$., ten thousand.
XC. ninety.	CCIↃↃCCIↃↃ., or $\overline{\text{XX}}$., twenty thousand.
C. a hundred.	&c. &c.
CC. two hundred.	

Every one must see what a tedious affair a large calculation would be according to this cumbrous system of notation; nor is it easy to say what our commercial standing, to say nothing of science, would have been to-day had it never been superseded.

Exercises for the Black-board and Slate.

1. Write the following numbers in words, explain each letter separately, and, lastly, read the whole series in connection:

MDCCCLIV; MMMCCLX; $\overline{\text{III}}$XL; XCIV; CMXCIX; $\overline{\text{X}}$IↃCLII; $\overline{\text{X}}$X; XIX; MMDXXXII; $\overline{\text{III}}$DXXV; XCVII; IↃCXXXVIII; $\overline{\text{XXII}}$CDLVI; LVIII; LXXXVII; XLVII; XVI; XIX; XXIV; XXXIX; $\overline{\text{II}}$V; $\overline{\text{X}}$X; XCIX; XXVIII; XXIX; LVI; CCCLXIV; XXV; XVI; IV.

2. Write the following numbers in Roman numerals, read them, and explain each separately.

Eighteen hundred and fifty-four; eighteen hundred and nineteen; twelve thousand two hundred and sixty; three thousand and forty; ninety-nine; fifty-four; ninety-four; forty-six; nine hundred and ninety-nine; ten thousand six hundred and fifty-two; ten thousand and ten; forty-nine; eighteen thousand seven hundred and thirty-six.

Questions to be put by the Teacher.—What does the I represent? *Ans.* A finger. The V? The X? The C? The L? The M? The D? &c., till all the characters in the table are explained, and their origin pointed out.

The *Arabic Numerals*, as they are called, though they are now generally allowed to be of Indian origin, were introduced into Europe by the Arabs nearly a thousand years ago. They are now used by all civilized nations. The Arabian method unites the important advantages of conciseness, simplicity, and precision. Indeed, it is impossible to conceive anything better adapted to the purposes of calculation. A more convenient *scale* than that of *ten* might have been adopted, so as to have allowed of more equal subdivisions without fractional parts (for instance, the scale of eight, of sixteen, or of twelve); but the *principles* of the notation are incapable of improvement.

The number of characters in the Arabic notation is ten. Nine of these represent numbers, and one stands for nothing, by itself, though indispensable to the system. The Arabic characters have also probably originated from the fingers. But they differ from the Roman numerals in this, that some of them consist of vertical, others of horizontal lines, and others again of both. The following are the characters, with their names. Underneath each is placed their supposed original form:

TABLE OF ARABIC NUMERALS.

one	two	three	four	five	six	seven	eight	nine	nought, or cipher.
1	2	3	4	5	6	7	8	9	0
I	=	≡	□	5	6	9	8	ϑ	O

Thus, one is represented by a vertical line, as in the Roman system; two by two horizontal lines; three by three of the same; four, by a square,—that is, two horizontal and two vertical lines; five, by three horizontal and two vertical; six, three horizontal and three vertical; eight (two fours), two squares; seven, two squares, less one vertical; nine, evidently borrowed from the Greek character for nine (ϑ, theta.) The seven is also supposed to be borrowed from the Greek character for that number (ζ, zeta), to which it certainly bears considerable resemblance. Lastly, the nought, or cipher, which does not consist of lines to be counted like the others, but, on the contrary, is entirely round, to show that of itself it has no value. The first nine characters have been rounded to their present form, doubtless, by rapidity in writing.

Formerly, the *ten* Arabic characters were *all* called ciphers, from the Arabic word *sipher*, to enumerate. Hence, arithmetic is often called *ciphering*. The first nine are now called *digits*, a name derived from the Latin word *digitus*, which signifies a *finger*. They are also called *significant* figures, because each of them has a *peculiar value* of its own, and to distinguish them from the *cipher*, which has no value of itself, though it is an exceedingly important figure, as it often modifies the value of all the other figures, as will presently appear.

By means of these *ten* characters, any number can be expressed, however small or great it may be. This is effected by affixing two kinds of value to each of the significant figures, namely, their *primary*, or *simple*, or *absolute* value, and their *secondary* or *local* value. Their *simple* value is always the same. It is expressed by their names as given above. The *local* value differs according to the *place*, or *rank*, which the character occupies as connected with other figures. The word *local* means pertaining to a *place*. For example: 1 means a single unit, or one, when it stands by itself, or when it stands in the *first* rank at the right hand. But when it is placed in the *second* rank from the right, it is ten times greater; that is, it stands for *one ten*. Thus, the figure 1 in 10 or in 16 stands for ten, because it occupies the second rank. But as the figure in the first rank in 10 signifies *nothing*, being only used to place the 1 in the second rank, the two figures together stand for *ten*. In 16, as the figure in the first rank stands for 6, the the two figures together stand for *sixteen*. The same principle holds with all figures. Thus, 24 stands for twenty-four, because the figure 2, being in the second rank, does not stand simply for two, but for two tens, or twenty; and 40 stands for forty, because the 4 occupies the second rank. 44 is forty-four, because the first 4, being in the second rank, is forty; the second 4, being in the first rank, is simply four. Any number, then, as far as ninety-nine, can evidently be expressed with the ten characters. The next higher number is ten tens, or one hundred. This is expressed by placing the figure 1 one place further to the left; that is, in the *third* rank from the right. Thus, in 100 and 124 each of the ones stands for 100, because it is in the third rank. The 2 in the second number counts for twenty, because it is in the second rank, which is the place of tens. Thus, the three figures together, 124, read one hundred and twenty-four. To express thousands, a figure must stand one place still further to the left, because *ten* hundred make *one* thousand. Thus, in the number given below,

a b c d
3333,

there are four 3s, but each has a different value. The first 3 on the right, marked *d*, stands for three ones. The figure in the second rank, marked *c*, is *ten* times greater than the first; that is, it stands for *three tens*, or thirty. The third, marked *b*, is ten times greater than the second, or a *hundred* times (*ten* times *ten* times) greater than the first; that is, it stands for *three hundred*. Lastly, the fourth figure, marked *a*, is *ten* times greater than the third; a hundred times (*ten* times *ten* times) greater than the second; a thousand times (*ten* times *ten* times *ten* times) greater than the first; that is, it stands for *three thousand.* The whole number, then, reads three thousand, three hundred and thirty-three.

The first principle of decimal arithmetic, then, is derived from the tenfold increase of value of the ranks, or places, of figures. It may be expressed as follows:

I. *When figures are placed horizontally, or side by side, every figure is ten times greater than the same figure immediately on its right, and ten times less than the same figure immediately on its left.*

If a cipher were placed to the right of the above four 3's, as below,

a b c d
33330,

the 3 marked *d* would no longer stand for three units, or ones. It would now be three tens, or thirty, because it occupied the *second* rank from the right, which is the place of tens. The 3 marked *c* has also changed its place. It, also, has become ten times greater. It was formerly three tens; it is now three hundred. The same remark applies to the figures marked *b* and *a*. Each is moved one place further to the left, and thus has become tenfold greater. In a word, the whole number has been increased tenfold by having a cipher placed at its right. Again, by removing the cipher, each of the other figures is changed to one rank further to the right, and thus each figure, and consequently the whole number, is decreased tenfold. The object, then, of the cipher is to enable us to place significant figures in their proper rank, and thus show their true local value.

But any other figure, by changing the rank of these 3s, would have changed their value just as effectually as the cipher. If 6 is put in place of the cipher, as below,

a b c d
33336,

each three has its value increased as before, by having its rank changed one place towards the left. The only difference between the two numbers is, that *six* has been *added* to it in the one, *besides* the tenfold increase; whereas *nothing* has been added to the other. Again, by removing the 6, we not only decrease the value of the other figures tenfold by changing their rank, but also diminish the number by six. It is evident that the same observations will hold good if any other significant figure is added or taken away.

The second principle of decimal arithmetic, then, may be expressed as follows:

II. *Every figure becomes tenfold greater by being removed one rank, or place, to the left, and tenfold less by being removed one rank, or place, to the right.*

The following Numeration Table, which teaches us to read the names of those figures that stand for integers, will now be readily understood:

NUMERATION TABLE, No. I.

	of Trillions.			of Billions.			of Millions.			of Thousands.			of Units.		
&c. &c.	Hundreds	Tens or ty	Units	Hundreds	Tens or ty	Units	Hundreds	Tens or ty	Units	Hundreds	Tens or ty	Units	Hundreds	Tens or ty	Units
	3	6	5,	4	2	7,	9	8	4,	2	8	3,	2	4	7,
	5th Period.			4th Period.			3d Period.			2d Period.			1st Period.		

From this table, it appears that each figure, besides its simple name of *one*, *two*, *three*, &c., has two other names. For instance, the first figure on the left of the table is three *hundreds* of *trillions*, or more simply three *hundred trillions;* the second is six *tens* of *trillions*, or six*ty* trillions (the final syllable *ty* signifying tens), and so forth. The term *units* is always omitted. Hence the third figure is *not* read five *units* of trillions, but simply *five trillions;* and the figure on the right of the table is not read seven *units* of *units*, but simply *seven*. The whole series of figures is read thus: three hundred and sixty-five trillions, four hundred and twenty-seven billions, nine hundred and eighty-four millions, two hundred and eighty-three thousand, two hundred and forty-seven. Higher numbers than these are rarely required. It may be proper to mention, however, that the same principles of nomenclature can be continued to infinity, the classes or periods being named quadrillions, quintillions, sextillions, septillions, &c., to each of which, as before, are assigned three ranks or places, namely, units, tens, hundreds.

It will be observed that the figures in the table are divided by commas into periods, or classes, of three orders of figures each, commencing at the right. This should always be done when a series of figures *exceeds four* in number, for otherwise they cannot so *easily be read.* These periods, it may be noticed, are named Units, Thousands, Millions, Billions, Trillions, &c. The orders, or ranks, are the same in every period, namely, Units, Tens, Hundreds.*

* This is the French mode of separating numbers into periods. Its simplicity has led to its universal use in this country. By the English mode, formerly used here, each period has six figures, and is read as follows:

		Hundreds of thousands	Tens of thousands	Thousands	Hundreds	Tens	Units
4th Period.	of Trillions.	4	7	4	2	1	3,
3d Period.	of Billions.	7	2	5	6	3	8,
2d Period.	of Millions.	1	2	3	4	5	6,
1st Period.	of Units.	7	8	9	1	2	3,

It ought here to be carefully noted that the word *unit*, besides forming the name of the first period, and of the first order or rank of each period, may be applied to figures of any order whatever. Thus a single unit of the first order is expressed by 1

A unit of the second order by 1 and 0 ; thus . . 10

A unit of the third order by 1 and two 0s ; thus . 100

A unit of the fourth order by 1 and three 0s ; thus 1000

and so on for the units of higher orders. But, when units are named simply, without expressing any particular order, units of the *first* order are always meant.

As it is evident from the above table and explanation that there cannot be more than *nine* different numbers of any one denomination, since an addition of one more to the nine enlarges the number to *ten*, and thus carries it into the next higher rank, we thence have the third principle of decimal arithmetic ; namely :

III. *Ten units of any one rank make one unit of the next rank to the left; and one unit of any one rank makes ten units of the next rank to the right.*

Exercises for the Black-board or Slate.

1. Divide 44444444 into periods of threes by commas, commencing at the right.

2. What is the general name of the first period on the right ? Of the second ? Of the third ?

3. Repeat the name of the orders in the first period ? *Ans.* Units, tens, hundreds. Repeat those of the third ; of the second ; of the fourth. Are they the same in every period ?

4. What name is never expressed ?

5. What is the first figure on the left called ? *Ans.* Tens of millions. How many millions does that figure stand for ? What is the second figure on the right called ? The fourth on the right ? The third on the left ?

6. How many times is the second figure on the right greater than the first ? [Point to the figures on the black-board.] The third than the first ? The fourth than the second ? The fourth than the first ? The fourth than the third ? &c.

7. How many times is the first on the right contained in

the second? The first in the third? The second in the fifth? &c.

8. Point off and write in words the following figures: 72358, 700, 1245, 604267, 8956238, 284563002, 123456, 7924502, 3824507266.

9. Increase the first two of the above nine numbers tenfold; the second two a hundred fold; the third two a thousand fold; and the remaining three ten thousand fold. In other words, multiply them by 10, by 100, 1000, 10,000.

[The above exercises should be repeated and varied by the teacher till the subject becomes perfectly familiar to the class.]

But the right-hand figure does not always represent units. Sometimes it becomes necessary to use or speak of a number *less* than *one*. Thus, with respect to the money of the United States, the dollar is considered the unit. But a sum *less* than a dollar frequently enters into a calculation, — cents, for instance, which are hundredths of a dollar; or dimes, which are tenths. These parts of a unit of any kind are called *fractions*, a word signifying *broken into parts*. Or, let us suppose an apple to be cut into ten equal parts. One or more of these fractions or tenths of an apple may enter into a calculation. When this is the case, these tenths, as the smallest part of the number, would occupy the rank on the right. But some mark would then be necessary to show *which rank was occupied by the units*. The character used for this purpose is a reversed comma, called a separatrix,* placed to the right of the rank of units. When a separatrix is used, *any number of figures or ciphers can be added to the right of a number without changing the value of the other figures*. Thus, if we take the four 3s again, marked with letters as before [see p. 113], and put a separatrix after the 3 marked *d*, to show that it stands in the place of units, we can add as many figures as we choose on either hand without changing the value of any of the 3s, for this simple reason, that their *distance from the rank of units is unchanged* by that addition. For example:

1st, *abcd* 3333	2d, *abcd* 3333‘	3d, *abcd* 3333‘465	4th, *abcd* 2343333‘465

* Some writers use a dot, others a comma, for a separatrix. Both are wrong. For, as both these characters are used with figures for other purposes, they thus give rise to much uncertainty and perplexity. For instance, if the *dot* be used as separatrix, there is no means of ascertain-

In the first example, we have the four figures exactly the same as before, the units' rank occupying the right, no separatrix being necessary to designate it. In the second example, we have still the same figures occupying the same stations, with a separatrix at the right of the units. But this is unnecessary. For, as the units occupy the rank on the right, they are sufficiently designated by their local situation. The third example exhibits the same characters as the second, with the addition of three others, 465. Here the separatrix is essential. Without it, the 5 would occupy the place of units, and each of the 3s would be a thousand times greater than in the first and second examples. In the fourth example, three additional figures occur on the left, which of course do not change the value of the others, as has been already sufficiently shown above.

This explanation brings out the fourth principle of decimal arithmetic, as follows:

IV.— *When there is a separatrix, the units' place is immediately on its left; when there is none, the right hand figure represents the units.*

The fractions, of which two examples are given above, are called *Decimal Fractions*, or simply *Decimals*, meaning numbers *broken* into tenths, or tenths of tenths (hundredths), or tenths of tenths of tenths (thousandths), &c. The value of these fractions depends on the same principle as that of integers or whole numbers; that is, each figure is ten times greater than the same figure on its right, and is only one-tenth of the value of the same figure on its left. The manner of reading them may be learned from the following table:

NUMERATION TABLE, No. II.

Thousands	Hundreds	Tens	Units	Tenths	Hundredths	Thousandths
4	4	4	4	‘4	4	4

ing whether 6·5 means six and five tenths, or six times five. If the *comma* be used, 65,231 may either signify sixty-five thousand two hundred and thirty-one, or sixty-five and two hundred and thirty-one thousandths. By the use of the inverted comma, all uncertainty disappears.

From this table it appears that the same names, with the addition of *th*, are used for the numbers tenfold, a hundred fold, a thousand fold, &c., *less* than units, as for those tenfold, a hundred fold, a thousand fold, &c., *greater* than units. Thus, the figure to the left of units is named *tens;* that to the right *tenths.* The second to the left *hundreds;* the second to the right *hundredths*, and so on. Observe, however, that the three figures to the right of the units (the fractions) may either be read four tenths, four hundredths, and four thousandths, or four hundred and forty-four thousandths; or four thousand four hundred and forty tens of thousandths; and so on in an infinite variety of expressions. Indeed, this remark may be applied to any number, whether integral or fractional. Take, for instance, the number 538. The usual expression for this is five hundred and thirty-eight. But it might be considered as five hundred and three tens and eight; or fifty-three tens and eight; or five thousand three hundred and eighty tenths; or fifty-three thousand eight hundred hundredths, &c., without end.

It is also plain from the last table that, by changing the place of the separatrix, the value of every figure is changed; being *increased* tenfold, a hundred fold, &c., by removing it one, two, &c., places to the *right;* and *decreased* tenfold, a hundred fold, &c., by removing it one, two, &c., places to the *left.* And this will evidently be the case whatever may be the figures employed. For instance, in the number 42'56, which reads forty-two and fifty-six hundredths, if the separatrix be removed one place to the right, we shall have 425'6, which reads four hundred and twenty-five and six tenths. If it be now removed two places to the left, we shall have 4'256, which reads four and two hundred and fifty-six thousandths. And lastly, by removing the separatrix altogether, the number becomes 4256, four thousand two hundred and fifty-six. Thus, it appears that the principles regulating the notation and numeration of decimal fractions are precisely the same as in whole numbers, as exemplified in *first and second principles*, which see (pp. 114 and 115).

Observe, however, that should there be so many decimal places as to require division into periods for the sake of easy reading, the period adjoining the units should only consist of *two* figures, the rank of units of course being wanting in fractions, as 436,427'38,945. Observe, also, in forming any number, whole or fractional, into periods, always to *begin* with units, whether proceeding to the right or to the left, or both ways. Division of fractions into periods, however, will rarely be necessary.

Exercises for the Black-board or Slate.

1. Point off and write in words the following numbers: 54326‘48; 8043‘805; 2769‘0072; 5214‘3724.

2. Point off and write in words the same figures, increasing the first tenfold, the second a hundred fold, the third and fourth a thousand fold, by changing the place of the separatrix; in other words, multiply them by 10, 100 and 1000.

3. Use the figures in the first exercise once more, decreasing them tenfold, &c.; that is, dividing them by 10, 100, and 1000, by changing the place of the separatrix.

4. Increase ‘092 tenfold; that is, multiply it by 10, and then decrease it 100 fold; that is, divide it by 100.

5. Mention separately the effect that would be produced on each of the following numbers by a removal of the separatrix: 25‘07; 38‘206; ‘525; 92‘3.

6. Mention separately the effect that would be produced on each of the following numbers by placing a separatrix after the first figure on the left: 2346; 18; 398; 27945.

7. Mention what numbers are superfluous in the four numbers that follow, and why: 600; 006; ‘006; ‘600.

8. Name the value of the 4 and of the 6 in the following number: 4060. If the 4 were removed, would the value of the 6 be changed? Why? If the 6 were removed, would the value of the 4 be changed? Why? If the cipher between the two significant figures were removed, what change, if any, would it effect upon the 4? Why? Upon the 6? Why? If the cipher occupying the place of units were removed, what effect would be produced upon the number? Why?

9. Fifty-two millions six thousand and twenty. How many figures are necessary to represent this number? [The number is not to be written in figures?] How many of them are significant figures?

10. Express in words 637 in four different ways. *Ans.* Six hundred and thirty-seven; six hundred thirty and seven; sixty-three tens and seven; six thousand three hundred and seventy tenths.

11. Express in words the following numbers, each in six different ways: 5326; 9478; 124,679; 38,472; 5'4; 7360; 9024; 573'7.

The fifth principle of decimal arithmetic will now be understood without further explanation, namely,

V. *The cipher is superfluous, except where it intervenes between a significant figure and the place of units.*

Exercises, continued.

12. Increase 245 tenfold, or, which is the same thing, multiply it by 10.

13. Decrease 2540 tenfold, or, which is the same thing, divide it by 10.

14. Multiply 3'532 by 100.

15. Divide 453'2 by 100.

16. Multiply 1768 by 10, and then divide it by 100, removing superfluous figures, if any.

17. Multiply 17'68 by 1000.

18. Divide 17680 by 100, removing superfluities, if any.

19. Write down in figures, in separate lines, the following numbers, first determining in your own mind how many figures are necessary for each number, and how many of them are significant:

Thirty-six.

Two hundred and seven.

Two hundred and seventy.

Two hundred and seven tenths.

Two hundred and seven hundredths.

Twenty-five thousand nine hundred and twenty-six.

Twenty-five thousand and nine hundred and twenty-six thousandths.

Three thousand and four and eighteen hundredths.

Four million six thousand and thirty-seven.

Three hundredths.

Fifty-nine thousandths.

Nine millions and nine thousandths.

Thirty-four billions thirty-seven thousand and fifty.

Eleven thousand eleven hundred and eleven (a puzzle.)

Four millions four hundred thousand and forty and four hundredths.

[Let the above exercises be repeated and varied, till each pupil can perform them correctly with rapidity and ease.]

THE FIRST PRINCIPLES OF DECIMAL ARITHMETIC.

[*To be committed to memory by the Pupil.*]

When figures are written horizontally, or side by side,

I. *Every figure is ten times greater than the same figure immediately on its right, and ten times less than the same figure immediately on its left.*

II. *Every figure becomes tenfold greater by being removed one rank or place toward the left; tenfold less by being removed one rank or place toward the right.*

III. *Ten units of any one place make one unit of the next place to the left; and one unit of any one place makes ten units of the next place to the right.*

IV. *When there is a separatrix, the place of units is immediately on its left; when there is none, the right hand figure occupies the unit's place.*

V. *The cipher is superfluous, unless it occupies the place of units, or intervenes between a significant figure and the place of units.*

Questions to be put by the Teacher.—Which are the two most important kinds of numerals? In what cases are the Roman numerals chiefly used? What is their probable origin? What does the capital I represent? The V? The X? What is the origin of the C and M? Of the L and D? On what principle are the Roman numerals horizontally arranged? *Ans.* When a smaller one, &c. What is the probable origin of the Arabic numerals? Of how many characters does the Arabic notation consist? Why are the first nine called *significant* figures? Why called digits? What is the tenth character called? What is its use? What is meant by the *simple* value of a figure? By its *local* value? Why is our system of computation called *Decimal* Arithmetic? What change is made in the value of a figure by removing it one place or rank to the left? One place to the right? Three places to the left? Two to the right? How is a number increased tenfold, a hundred fold, &c., or multiplied by 10,

100, 1000, &c.? How is a number decreased tenfold, a hundred fold, &c., or divided by 10, 100, 1000, &c.? What is the meaning of the word *unit?* Does the word apply to more than one rank of figures? What is the use of the reversed comma, or separatrix? What is the objection to the use of the comma or the period as a separatrix? Where is the place of units when there is a separatrix? When there is none? How many ranks or orders of figures make a period? Why should large numbers be separated into periods? Repeat the names of the first five periods. By what character should the periods be separated? Are the names of the orders the same or different in different periods? Repeat their names. What does the word fraction signify? On what principle does the notation of decimal fractions proceed? *Ans.* On the same principle as that of whole numbers. [See p. 119.] What *is* the principle? In dividing a number into periods, with what rank should we always commence? What is the first principle of arithmetic? The second? The third? The fourth? The fifth? [This, and indeed every chapter, should be studied till the pupil can answer all the questions at the close correctly and without hesitation.]

CHAPTER II.

THE SHORTENED PROCESSES OF INCREASE AND DECREASE OF INTEGERS AND DECIMAL FRACTIONS.

SECTION I.—*Addition.*

[THE following exercises should be transferred to the slate, and practised till the *sum* of each set of two, three, or more figures can be rapidly read off without spelling; (that is, without naming the individual figures), horizontally* as well as vertically; irregularly as well as regularly, without taking into view the *local* value of the figures. Every figure down to

* Every pupil should learn to add horizontally as well as vertically. In ledgers of country merchants, &c., much of this kind of work is necessary.

the range of stars, * * * * *, should be considered as of the denomination of units, the sign of addition, +, being omitted, except in the first range of exercises, merely to save room. The class should be daily exercised on the black-board with these or similar combinations. For an explanation of the signs used in Addition see Oral Arithmetic, p. 57.

1. What is the sum of

1+5	1+3	1+9	1+7	2+5	4+2	3+2	5+9	3+1
2+1	4+1	6+1	8+1	2+2	2+6	2+7	2+2	3+3

38	39	30	48	49	40	58	59	50	69	67	67
53	63	73	54	64	74	55	65	75	66	77	89

89	94	79	89	67	45	16	76	85	72	95	69
88	80	27	58	43	25	84	95	52	36	42	53

2. What is the sum of [take three figures at once, first horizontally, then vertically]

123	538	281	817	405	142	539	293	518	752
456	246	523	352	263	874	218	536	364	134
324	143	162	461	847	635	406	921	246	947

389	283	719	613	328	152	357	731	916	372
527	346	334	724	896	398	416	128	237	415
634	812	486	176	510	276	938	654	948	926

3. What is the sum of [four figures, first horizontally, then vertically. At first these may be thrown together in pairs; afterwards all four should be caught at a glance, just as we do the letters of a word.]

3267	3948	7913	2896	5276	2468	8276	1928
9418	1234	3546	3754	3894	1357	1358	6734
7826	5678	2687	2137	1728	9123	9425	3695
8512	9123	1504	9862	9264	4567	1072	2739

1235	4372.	2864	2816	4321	3142	9817	5942
6347	1893	9351	3743	1672	6417	3246	7637
9165	4715	4234	5214	3514	2738	7234	8424
2738	2637	6712	3746	6328	9516	1832	3056

[In the first range of exercises that follow, two figures in each *column* form 10. In the second range, it takes three figures. In the first four exercises of each range the figures that form the combination are *together*. In the others, they are not. The object of such an exercise is to make the pupil quick at observing the *tens*. The figures should be transferred to the slate, and studied till the class is prepared to read the *sums* on the blackboard as rapidly as words.]

4. What is the sum of [horizontally as well as vertically]

2512	4792	1741	1934	2531	1568	2541	7512
3282	6318	2235	9176	3427	5725	6138	1262
6375	1241	3624	2483	4362	4213	3421	3598
4735	5618	7486	1514	8579	5342	4972	4321
5283	2365	2756	7354	1362	2314	1485	3514
1314	6251	1613	1613	7637	5478	2312	6789
3252	2514	2143	2143	4352	2233	9674	4334
6544	7387	5768	5768	5416	6563	7313	3262

5. In any series of three figures that regularly increase or decrease by the common difference 1, such as 4, 5, 6; or 7, 6, 5; what will be the effect if the common difference, 1, be taken from the largest figure and added to the smallest? Will the same equality be produced by such an operation when the common difference is 2, 3, or any other number, such as 2, 5, 8; or 9, 6, 3; &c. [Show this principle on the black-board, thus: 4, 5, 6 become 5, 5, 5, by carrying 1 from the 6 to the 4; and 5, 7, 9, become 7, 7, 7, by carrying two from the 9 to the 5. The same thing occurs, of course, when the numbers decrease regularly. Thus, 9, 8, 7, become 8, 8, 8; and 15, 12, 9, become 12, 12, 12.

[The following exercises are designed to aid the pupil in rapidly discovering numbers with a common difference, three of which, of course, are equal to *three times the mean number*. Towards the beginning, the series are placed in regular order; afterwards they are arranged irregularly, many having the fourth figure interposed. In the last four exercises, the figure out of the regular series in each column is either 3, 6, or 9, which has the effect of increasing each figure *in* the series, by 1, 2, or 3. Thus, 5, 6, 7, and $3=3\times7$, and not 3×6; because each of the figures in the series, namely, 5, 6, 7, is increased 1 by the 3; while 5, 6, 7, and $6=3\times8$, each of the series being increased 2 by the 6; and, for a like reason, 5, 6,

7, and 9=3×9. In these exercises the numbers should be added vertically only.

5982	2947	8479	4738	3234	1394	2134	1252
3458	6171	7568	8293	5456	4676	5466	7393
2347	7282	6657	6174	7678	7958	8798	2934
1236	8393	2925	7382	9128	8213	1251	3418

2392	1237	1948	6323	3693	4921	2139	4741
4748	2982	4663	5578	5471	5843	5451	3693
1857	3454	7351	7466	6582	6765	8763	5832
6966	5676	8242	1857	7693	3969	3675	6929

* * * * * *

Carrying the Tens.

[Each figure should have its *local*, as well as its *simple*, value in the exercises that follow. In order that the pupil may thoroughly realize this, he should read the first exercise as follows: two thousand six hundred and fifty-four. Seven thousand nine hundred, &c. As some of the exercises contain decimal fractions, it will be proper occasionally to remind the pupil that numbers of *different denominations*, and consequently belonging to *different ranks*, cannot be added together.

6. Add togther the following numbers, placed vertically for the sake of convenience:

The long process.		The shortened process.
2654		2654
7972		7972
3814		3814
2483		2483
14	sum of the thousands.	16923 sum total.
27	sum of the hundreds.	
21	sum of the tens.	
13	sum of the units.	
16923	sum of the whole.	

Addition by these two processes amounts to precisely the same thing. The only difference is this: in the longer process the sum of each different rank is set down separately, and then added into one whole; whereas, in the shortened process, the units only of each rank are set down, and the *tens carried* and added in as units to the figures of the next rank to the left, agreeably to the third principle of arithmetic, p. 117. The operation of adding may be commenced in any of the ranks when the long process is used [let the pupil try this, by commencing separately with each of the four ranks], but in the shortened process, it is necessary to commence with the first rank on the right, because *tens* of each rank are to be added in as *units* of the next rank to the left.

PROOF. *Suggestive Questions.*—If columns of figures be correctly added downwards as well as upwards, will their sums be the same, or will they be different? Will they mutually prove each other, then? If one or more columns of figures be added together, and then all the lines in the column or columns be again added *except the lower one*,* what will be the difference between these two sums? If the second sum be now added to the line that was cut off, will this third sum be the same as that of the first, if the first has been added correctly? Here, then, are two methods, either of which will prove whether the work has been performed properly. If numbers that are placed vertically may be proved by adding them downwards as well as upwards, how may numbers added horizontally be proved on the same principle?

7. Arrange vertically and add the following numbers:

1st. 246+3582+72+9873+855+2144+3792+53.

2d. 4321+2153+3946+2604+4098+24452+13246+6944+8175+4924+5678.

3d. 75630+76042+4942+3294+6757+4275+8641+1975+4132+3609+9063+7429.

4th. 21539+172+184+64577+73722+35392+9077+1814+6137+1691+6105+26+3284.

5th. 1619+1610+3612+95471+63300+14713+832+9468+3215+403+123+8678+9136+7924+8706+321.

6th. 152·13+326·78+49·237+1736·46+5897+13068+9·415+72·5+842·19+335·86+973·437+8642+54·32.

7th. 7172·12+553·14+241·177+877·35+927·13+5679·12+684·24+68+539+28+135+9232+465·12+8472+8·579.

* It is usual to omit the *upper* line in this method of proof, but such a plan is more liable to error, since the figures to be added will recur nearly in the same order. Should the teacher prefer to have the upper line cut off, the other figures should be added downwards

8th. 3157·13+9711·82+76131·31+2854·32+1646·78+532·964+72·341+555·66+8404·26+9373·28+1357·246+8891·372.

9th. 141·32+37·48+96·23+45·67+89·13+579·24+12·45+670·86+26·26+3333·34+975·342+872·48+94·26+38·12.

10th. 1234+5678+9123+4567+8912+5790+2040+6735+9813+4276+1358+9342+88761+3456+72352+4638+1926.

11th. 3825+9638+1326+5431+9425+3873+4284+7965+9123+4476+1358+9123+4782+91234+48245+5796+4312.

12th. 94·37+282·3+496·01+182·94+529·87+651·32+8259·73+5687·74+5035·69+6798·56+6073·87+9087·26+8709·56.

13th. 5687·23+8235·26+1829·4+8708+1324+5639·24+5824·37+8308·12+2358·09+9263·57+9123·24+3281.

14th. 917·23+2872·46+538·7+1316·28+4827+325·16+3243+9127+4832·44+9123·45+1055+6249+5542·46+326·4.

15th. 3408·26+1357·95+2186·37+9345·35+8421·38+1796·24+3875+9394·32+8218·77+54134·28+3276·45+9137·42+376.

8. Add the following numbers horizontally, and then find the amount of their respective sums vertically.

37·28+49·56—32·06+56·28+72·54+62·47 . =
24·36+8·45+94·7+37·4+28·84+97·28+3·5 =
1·5+24·6+·375+9·42+87·43+215·64 . . =
·52+·125+·25+·375+2·25+1·125+3·64 . =
28+572+3·75+4·375+28·49+57·32 . . . =
30+9·02+5·76+·28+94+3·72+8·56 . . =
25+50+3·25+9·125+8·26+9·47+5·4 . . =
32·04+5·67+95·25+7·84+9·375+1·236 . =
5·84+3·57+4·92+65+37·5+6·25 . . . . =
8·27+4·328+59·25+·06+32·5+5·42 . . . =
23·85+29·47+86·32+89·45+27·88 . . . =
81·32+64·39+54·72+38·43+127+55 . . =
24·375+88·84+37·25+17·94+13+85·92 . =

Total

14·57+93·29+6·25+4·375+19·25+16 . . =
39·06+48·27+33·29+74·08+9·5+2·37 . . =
12·42+7·93+5·34+16·22+37·18+27·19. . =
40·87+78·19+28·99+92·42+70·36+12·48 . =
28·56+91·23+45·67+91·35+79·24+6·25 . =
68·98+76·54+41·28+20·16+70·95+4·26 . =
69·09+26·34+20·93+32·15+81·93+26·86 . =
25·86+39·06+8·27+30·33+74·08+9·51 . =
5·12+42·07+93·53+40·87+78·19+128·99 . =
92·42+70·36+12·48+28·56+91·23+45·67 . =
91·35+79·24+73·86+32·89+45·27+24·12 . =

Total

24·37+84·88+37·25+17·94+13·85+92·14 . =
39·06+48·27+33·29+39·06+48·27+33·19 . =
74·08+95·12+42·07+93·53+40·87+21·3 . =
78·19+28·19+92·42+70·36+12·48+28·56 . =
11·23+46·67+19·35+24,79+68·98+76·54 . =
41·28+20·16+70·95+43·26+20·95+16·34 . =
20·93+32·15+18·93+26·86+37·21+43·84 . =
13·72+26·58+12·34+16·78+29·35+19·24 . =
38·26+23·87+29·73+12·54+27·3+26·54 . =
37·38+17·53+12·46+34·2+18·92+15·24 . =
16·92+27·23+18·24+42·32+4·64+16·9 . . =
48·37+52·78+19·23+47·38+52·4+5 . . . =
26·43+21·87+27·94+13·28+47·36+29·5 . =
7·123+56·56+38·25+13·24+17·96+12 . . =

Total

Grand Total

[Two, three, or more exercises like the above may now be formed from the figures in Example 7, and these may again be used by taking the figures in a backward order, thus changing 32·45 to 54·23, and so forth. Addition should be practised till the pupil can run up a column correctly with the utmost ease and rapidity. It would be well if the classes should occasionally practise the addition of columns the whole length of the slate, until he has finished his course of arithmetic. No operation is so often called for in practical business as the summation of numbers.

Specimen of different methods of adding Large Amounts.

1494
3871
2636
9467
8538
7372
2768
4937
3883
1679
5789
9561

1. *Reading by three figures at once.*—First step. *Nineteen* and eighteen are *thirty-seven*, and seventeen are *fifty-four*, and eleven are *sixty-five*; carry six to twenty-one are *twenty-seven*, and seventeen are *forty-four*, and sixteen are *sixty*, and nineteen are *seventy-nine;* carry seven to eighteen are *twenty-five*, and twenty-four are *forty-nine*, and twelve are *sixty-one*, and eighteen are *seventy-nine;* carry seven to fifteen are *twenty-two*, and nine are *thirty-one*, and twenty-four are *fifty-five*, and six are *sixty-one.* As soon as this step has been practised with various long columns of figures till it can be performed with ease and rapidity, the pupil may proceed to the SECOND STEP, which consists in omitting all the words in the first step *except those in italics.* The process of adding the above will then require only the following words: Nineteen, thirty-seven, fifty-four, sixty-five; twenty-seven, forty-four, sixty, seventy-nine; twenty-five, forty-nine, sixty-one, seventy-nine; twenty-two, thirty-one, fifty-five, sixty-one.

2. *Reading by four figures at once.*—This method is the same as the last, except that *four* figures are read at the same time, in place of *three.*

3. *Additional Abbreviations.*—*a.* Never stop at *ty;* that is, if, in the summation of a column, you come to six*ty*, seven*ty*, or

any other exact number of tens, let the eye catch the amount of two, three, or four figures *more*, while you are pronouncing the word mentally, so that you may take sixty-*six*, or sixty-*nine*, in place of simply sixty, &c. Thus, if in summing up a column you come to the number 50, and, while that number is passing through your mind, you see 4 and 5 as the next two figures above, then you just add 9 to your 50, without even mentally repeating the word fifty. Thus, in running up a column, the words forty, fifty, sixty, &c., never occur alone, but always in combination with one, two, or three, &c., additional numbers.

b. Select the tens as much as practicable; that is, if you have, say thirty-*four*, and see a *six* above in the column, even though it may not be adjoining, call your number *forty;* and, while mentally pronouncing that word, add in two or three more figures, so as, not even in such a case as this, to *stop at ty.* By careful practice of this method, an intelligent pupil will soon be able to read off *six*, *eight*, or even sometimes *ten* figures at once.

c. Let the eye glance up one column *while you are writing* the units of the preceding one.

d. Dispense with words altogether. That the mind can call up the idea of the *sum* of three, four, or more figures, without thinking of the names of the individual numbers, will be evident to any one who will give the experiment a fair trial. We open a book, and possibly the first word that meets the eye is one of many letters and syllables, such as *incomprehensibility*, and instantly the *idea* strikes the mind, without its taking cognizance of any of the nineteen letters or eight syllables. The mind seizes it as a whole, without special regard to its individual parts. Such a power as this may be acquired with numeral characters as well as with letters, and the saving of time will be found to be beyond all calculation. Nor is this all. The mental discipline thus acquired will be of incalculable value in every other study.

Questions to be put by the teacher, before the pupil commences the next Section, and to be repeated from time to time till they are answered without hesitation.—What is addition? See p. 56. What is the result of addition called? What is the sign of addition? What is it called? See p. 57. What is the sign of equality? In what order can figures be placed most conveniently for addition? Should they always be placed in this order? See p. 124. Why? Where should we begin to add? Why? Why are the tens of one column carried as units to the next column to the left? What is an integer? See p. 15. What is a fraction? What is a decimal fraction? Are decimal fractions added differently from integers? Are the tens carried from the column of tenths to that of units in the same manner as from one rank of integers to the next? Why? To what rank does the right hand column belong, in a number consisting wholly of integers? Is this the case when a number consists partly or wholly of decimal fractions? How, then, is the local value of the figures ascertained? Explain the two methods of proving addition.

Practical Exercises.

1. By a census taken in the year 1841, the population of the British Islands was as follows: England, 14,995,138; Wales, 911,603; Scotland, 2,620,184; Ireland, 8,175,124; army, navy, &c., 193,469; islands in the British seas, 124,040. What is the total population of these islands?

Ans. 27,019,558.

2. In the same year, the number of emigrants that left the United Kingdom of Great Britain was as follows: to their North American colonies, 38,164; to the United States, 45,-017; to the Australian colonies and New Zealand, 32,625; to all other places, 2786. What was the whole number of emigrants from the kingdom that year? *Ans.* 118,592.

3. The value of the principal articles of manufacture in Great Britain for that year is estimated as follows: cotton, £35,000,000;* woollen, £24,000,000; iron and hardware, £20,000,000; watches, jewelry, &c., £3,000,000; leather, £13,500,000; linen, £8,000,000; silk, £10,000,000; glass and earthenware, £4,250,000; paper, £2,000,000; hats, £2,-000,000. What was the gross amount in pounds sterling?

Ans. £121,750,000.

4. The white population of the United States, according to the census of 1840, classed according to ages, was as follows: *Males*, under five years of age, 1,270,790; of five and under ten, 1,024,072; of ten and under fifteen, 879,499; of fifteen and under twenty, 756,022; of twenty and under thirty, 1,322,440; of thirty and under forty, 866,431; of forty and

* This character, £, stands for pounds in money.

under fifty, 536,568; of fifty and under sixty, 314,505; of sixty and under seventy, 174,226; of seventy and under eighty, 80,051; of eighty and under ninety, 21,679; of ninety and under a hundred, 2,507; of a hundred and upwards, 476. *Females*, under five, 1,203,349; of five and under ten, 986,921; of ten and under fifteen, 836,588; of fifteen and under twenty, 792,168; of twenty and under thirty, 1,253,395; of thirty and under forty, 779,097; of forty and under fifty, 502,143; of fifty and under sixty, 304,810; of sixty and under seventy, 173,299; of seventy and under eighty, 80,562; of eighty and under ninety, 23,964; of ninety and under a hundred, 3,231; of a hundred and upwards, 315. What was the number of free white males? Of free white females? Total of free whites? *Ans. to the last question*, 14,189,108.

5. The number of free people of color in the United States, according to the census of 1840, was as follows: *Males*, under ten years of age, 56,323; of ten and under twenty-four, 52,799; of twenty-four and under thirty-six, 35,308; of thirty-six and under fifty-five, 28,258; of fifty-five and under a hundred, 13,493; of a hundred and upwards, 286. *Females*, under ten, 55,069; of ten and under twenty-four, 56,562; of twenty-four and under thirty-six, 41,673; of thirty-six and under fifty-five, 30,385; of fifty-five and under a hundred, 15,728; of a hundred and upwards, 361. What was the number of free males of color? Of free females of color? Of free persons of color?

Ans. to the last question, 386,245.

6. The number of slaves in the United States, according to the census of 1840, was as follows: *Males*, under ten years of age, 422,599; of ten and under twenty-four, 391,131; of twenty-four and under thirty-six, 235,373; of thirty-six and under fifty-five, 145,264; of fifty-five and under a hundred, 51,288; of a hundred and upwards, 753. *Females*, under ten, 421,470; of ten and under twenty-four, 390,075; of twenty-four and under thirty-six, 239,787; of thirty-six and under fifty-five, 139,201; of fifty-five and under a hundred, 49,692; of a hundred and upwards, 580. What was the number of male slaves? Of female slaves? Of slaves of both sexes?

Ans. to the last question, 2,487,213.

7. What was the whole population of the United States by the census of 1840, including 6100 persons of the navy of the United States, who were not reckoned in either of the last three enumerations? *Ans.* 17,068,666.

8. The following table exhibits the territorial area of the states of Europe and of North America. Find the aggregate, which will be correct if it amounts to 11,658,480.

AREA OF EUROPE.

	Square miles.
Russian Empire, including Poland	2,000,000
Austrian Empire, including Lombardy, &c.	257,368
France, including Corsica	203,736
Great Britain and Ireland, British Islands, and Malta	120,500
Prussia	107,921
Spain	182,270
Turkish Empire	10,585
Sweden and Norway	91,164
Belgium	13,214
Portugal	36,510
Holland	13,598
Denmark	21,856
Germany	90,620
Italy	100,953
Swiss Confederation	14,950
Greece	17,900
Ionian Islands	999
Cracow	488
Andorre	200

AREA OF NORTH AMERICA.

United States			3,306,865
British America,	New Britain	2,598,837	
	Upper and Lower Canada	346,860	
	N. Scotia, N. Brunswick, C. Breton, &c.	104,701	
			3,050,398
Mexico			1,038,834
Central America			203,551
Russian America			394,000
Danish America (Greenland)			380,000

Aggregate

9. The following is a statement of the white population of each of the states and territories of the United States and of the District of Columbia, according to the census of 1850, followed by tables of the free colored and slave population for the same period. Find the sum of each of the columns of males and females, which will be correct if they agree with the several given aggregates. Then take the sum of the aggregates, which will be correct if it amounts to 23,190,998.

STATES, &c.	Males.	Females.	Aggregate.
Maine	296,745	285,068	581,813
New Hampshire .	155,960	161,496	317,456
Vermont . . .	159,658	153,744	313,402
Massachusetts . .	484,093	501,357	985,450
Rhode Island . .	70,340	73,535	143,875
Connecticut . .	179,884	183,215	363,099
New York . . .	1,544,489	1,503,836	3,048,325
New Jersey . .	233,452	232,057	465,509
Pennsylvania . .	1,142,734	1,115,426	2,258,160
Delaware . . .	35,746	35,423	71,169
Maryland . . .	211,187	206,756	417,943
Dist. of Columbia	18,494	19,447	37,941
Virginia . . .	451,300	443,500	894,800
North Carolina .	273,025	280,003	553,028
South Carolina .	137,747	136,816	274,563
Georgia	266,233	255,339	521,572
Florida	25,705	21,498	47,203
Alabama . . .	219,483	207,031	426,514
Mississippi . . .	156,287	139,431	295,718
Louisiana . . .	141,243	114,248	255,491
Texas	84,839	69,195	154,034
Arkansas . . .	85,874	76,315	162,189
Tennessee . . .	382,235	374,601	756,836
Kentucky . . .	392,804	368,609	761,413
Missouri . . .	312,987	279,017	592,004
Illinois	445,544	400,490	846,034
Indiana	506,178	470,976	977,154
Ohio	1,004,117	950,933	1,955,050
Michigan . . .	208,465	186,606	395,071
Wisconsin . . .	164,351	140,405	304,756
Iowa	100,887	90,994	191,881
California . . .	84,708	6,627	91,335

TERRITORIES.			
Minnesota . . .	3,695	2,343	6,038
Oregon	8,138	4,949	13,087
Utah	6,020	5,310	11,330
New Mexico . .	31,725	29,800	61,525
Aggregate, . .			19,552,768

FREE COLORED POPULATION.

	Males.	Females.
Under 10	59,125	59,748
10 and under 15 . .	26,061	26,247
15 and under 20 . .	20,395	23,399
20 and under 30 . .	35,782	41,765
30 and under 40 . .	26,153	29,072
40 and under 50 . .	18,199	19,741
50 and under 60 . .	11,771	12,572
60 and under 70 . .	6,671	7,362
70 and under 80 . .	2,878	3,438
80 and under 90 . .	1,106	1,512
90 and under 100 . .	319	549
100 and upwards . . .	114	229
Age unknown	150	137
Total		
Aggregate		434,495

SLAVE POPULATION.

	Males.	Females.
Under 10	506,251	513,331
10 and under 15 . .	221,480	214,712
15 and under 20 . .	176,169	181,113
20 and under 30 . .	289,595	282,615
30 and under 40 . .	175,300	178,355
40 and under 50 . .	109,152	110,780
50 and under 60 . .	65,254	61,762
60 and under 70 . .	38,102	36,569
70 and under 80 . .	13,166	13,688
80 and under 90 . .	4,378	4,740
90 and under 100 . .	1,211	1,473
100 and upwards . .	606	819
Age unknown . . .	1,581	1,533
Total		
Aggregate		3,203,735

10. Benjamin Franklin was born in the year 1706. He died when he was 84 years of age. In what year did he die?

11. Three merchants entered into partnership. The first advanced $5500 dollars towards the capital; the second advanced $1000 more than the first did; and the third advanced $1500 more than the second. What was the whole amount of their capital? *Ans.* $20,000.

12. The first man was created 4004 years before Christ. How long is it from his creation to the present year?

13. There are two numbers, of which the smaller is 4520, and the difference between them 540. What is the greater number, and what is their sum?

Ans. to the last question, 9580.

14. A man left by will to his widow $5000, and to an adopted daughter $2000. The rest of his estate, after the payment of his debts, he directed to be equally divided among his four sons. The debts amounted to $3426, and each son received $1550. What was the value of the whole property? *Ans.* $16,626.

15. Two brothers set out on a journey in different directions. The one travelled 167 miles, the other 134. How many miles were they then apart?

16. A man sold a house and lot for $6254, which was $1746 less than they cost. What was their cost?

Section II.—*Subtraction.*

[For an explanation of the terms and signs used in subtraction, see p. 56, 3; 58, 8.]

Exercises for the Slate and Black-board.

1. Name the difference between each of the following pairs of figures:

2 1 6 1 7 1 8 1 2 2 4 2 8 2 7 2
1 3 1 4 1 5 1 9 2 3 2 6 2 5 2 9

4 3 8 3 5 3 5 4 6 4 7 5 8 5 9 6
3 6 3 7 3 9 4 8 4 9 4 6 5 7 5 6

7 6 8 9 7 7 4 9 6 8 2 9 3 4 2 9
6 9 6 7 8 7 9 5 7 2 5 9 6 5 8 9

12	5	17	15	8	11	3	11	7	13	13	6
9	13	9	6	14	7	12	4	15	9	4	10
13	5	10	5	18	8	16	7	11	9	12	7
8	12	3	14	9	15	9	10	8	14	4	16
17	9	10	13	6	11	11	8	16	7	11	8
8	15	5	6	14	9	3	12	8	12	5	11
10	7	11	2	6	14	2	5	13	12	5	8
4	14	6	14	15	5	12	14	3	4	16	12

10—9 12—6 13—5 17—8 14—9 19—4 16—7

7—3 15—9 14—8 19—3 17—7 21—5 14—8

[These figures may be studied on the book, or transferred to the slate for that purpose. The pupil should continue to practise them till he can recite the differences rapidly from the black-board, taken regularly as well as in irregular order, *without naming* the subtrahend or minuend. Pupils should be accustomed to name the differences of numbers placed horizontally, and also with the *smaller* number above as well as below. The former is required in balancing accounts; the latter frequently occurs in long calculations.]

2. What is the difference between two heaps of apples, one of which contains 12, the other 16? If 10 more apples be added to each heap, will their difference be changed, or will it remain the same? Will the difference be unchanged by adding 20, 30, 40, or any other number to each? Will the difference between *any* two numbers whatever be changed by adding an equal number to each? May not the following, then, be considered the sixth principle of Arithmetic?

VI. *If equal numbers be added to unequal numbers, their difference remains unchanged.*

3. What is the difference between 10 and 6? If this difference be *added* to the *smaller* number, to what will it be equal? If the difference be *taken* from the larger, to what will it then be equal? Will the same principle hold in *any* two numbers? [Give examples with other numbers on the black-board, when necessary.] May not the following, then, be considered the seventh principle of Arithmetic?

VII. (1.) *If the difference between two numbers be added to the smaller, their sum is equal to the greater.* (2.) *If the difference be taken from the greater, the remainder will be equal to the smaller.*

EXEMPLIFICATION OF SUBTRACTION,

Where some of the figures in the subtrahend are greater than those of the same rank in the minuend.

[*For the Black-board.*]

From 52364878 Minuend.
Take 21436294 Subtrahend.

Leaves 30928584 Difference, or Remainder.

Subtrahend+Difference=52364878=Minuend. Proof No. 1.

Minuend—Difference =21436294=Subtrah'd. Proof No. 2.

4. Commencing at the right, for a reason that the student will presently discover: Four from 8, how many? Can 9 be taken from 7? Adding 10 to the second rank of minuend [see Principle VI. above] 9 from 17? Then adding 10 to second rank of subtrahend, also, will change the 2 in third rank to what? Three from 8, then? Can 6 be taken from 4? Adding 10 to fourth rank of minuend, 6 from 14? Adding 10 to fourth rank of subtrahend, also, how many from 6? Can 4 be taken from 3? Adding 10 to sixth rank of subtrahend, what does the 1 become? And so forth. If the figures in the minuend were always greater than the corresponding ones in the subtrahend, would it be of any consequence where the process of subtraction commenced? [Give an example on the black-board.] Why, then, is it generally necessary to commence on the right?

Proof 1. By the seventh principle of arithmetic, to what is the sum of the subtrahend and the difference or remainder equal? Proof 2. If the difference be subtracted from the minuend, to what will the remainder be equal?

Exercises for the Slate or Black-board.

1. 2763850'26 Minuend.
648273'18 Subtrahend.

Remainder.

Proof No. 1.

Proof No. 2.

2. 826043'251 Subtrahend.
963561'37 Minuend.

Remainder.

Proof No. 2.

Proof No. 1.

3. 6420589'37 Debtor.

Proof.

4570634'49 Creditor.
Balance.

Proof.

4. 9247385'4 Minuend.
13746'8 Subtrahend.

Remainder.

Proof 1.

5. 1234567890'12 Minuend.
813721437'5 Subtrahend.

Remainder.

Proof 2.

6. 643464'38 Debtor.
Balance.

Proof.

2365720'83 Creditor.

Proof.

7. 1638296'42 Minuend.
300512'76 Subtrahend.

Remainder.

Proof 2.

Proof 1.

8. 524698'73 Subtrahend.
1293463'5 Minuend.

Remainder.

Proof 1.

Proof 2.

9. 63745896·731—42638938·9.
10. 5372849·2358—789632·15723.
11. 9716452—3856947·123.
12. 36894726—14239879.

[All the above can be changed to new exercises, by making a slight change in the left hand figures, and substituting the subtrahend for the minuend. If necessary, others can be added by the teacher, or, still better, by the pupil. Subtraction should be practised till the class can perform it as rapidly as the remainder can be written. But this can only be done by *reading without spelling;* that is, by thinking of or writing the difference between numbers *without naming* those numbers either orally or mentally; and in like manner increasing them by ten when necessary, without mentioning that circumstance. Thus, in the following example, all the words are superfluous except the three words in italic, namely, four eight, two:

$$\begin{array}{r} 527 \\ 243 \\ \hline 284 \\ \hline \end{array}$$

Three from seven leaves *four;* four from two, add ten, leaves *eight;* one to two makes three from five leaves *two.*

Subtraction by Addition.

Definition.—The *complement* of a number is the difference between that number and 1 of the next higher rank or order; that is, a number and its complement amount to 1 of the next higher rank or order of figures. Thus, 8 is the complement of 2, because 2 and 8 together make 10. For the same reason 2 is the complement of 8; 30, also, is the complement of 70, and 70 of 30, because together they make 100, 1 of the next higher rank. Thus, also, 28 and 72 are mutually complements. The complement of 0, of course, is 10.

1. What is the complement of 6? Of 5? 3? 7? 4? 2? 9? 1? 6? 8? 20? 50? 70? 40? 60?

2. What are the complements of 2222 and of 73480.

No. 1.	No. 2.
2222 Number.	73480 Number.
7778 Complement.	26520 Complement.
10000 Sum.	100000 Sum.

3. Why does the first 2 on the right of No. 1, require 8 as its complement, while all the others require only 7? [Add the complement and see.]

No. 1. Subtract'n by adding complement.	No. 2. Subtract'n in usual way.
826492 Minuend.	826492
243546 Subtrahend.	243546
826492 Minuend.	582946
756454 Compl. of Subt.	
Dropping 1,—582946 True remainder.	

☞ It is proper to remark here, that in the example "No. 1. Subtraction by adding complement," the third and fourth lines are altogether superfluous, being placed there merely to exemplify. All that is necessary in such operations is to add (without having it written) the complement of the subtrahend to the minuend, dropping 1 of the next higher rank of figures than the highest of the subtrahend.

4. What is the minuend? The subtrahend? The difference, or remainder? By how much is the minuend greater than the remainder? *Ans.* By the s———. If, in subtraction, then, you announce the minuend as the answer in place of the remainder, as in the above example No. 1, how much too large will your answer be? [See answer to last question but one.] Now, as your answer is too much by the amount of the subtrahend, how much too large will it be if you add to it the complement of the subtrahend; that is, how much does a number and its complement amount to? If the answer, then, be too great by the sum of the subtrahend and its complement, how can it be rectified? *Ans.* By dropping 1 of the ——. May not, then, the following be considered as the eighth principle of arithmetic?

VIII. *The difference of two numbers may be obtained by adding to the larger the complement of the smaller, and diminishing this sum by* 1 *of the next higher rank of figures than the highest of the smaller.*

5. Read off the difference of each of the following pairs of numbers, by adding, without writing down or naming, the complement of the smaller to the larger, and diminishing the amount by 1 of the next higher rank of figures than is contained in the smaller, and repeat similar exercises on the black-board and slate till it can be done rapidly.

4	16	5	15	6	17	3	5	7	6	17	15	18	7
6	4	9	9	8	8	9	1	14	3	18	2	9	9

8—2 17—3 14—5 9—0 18—7 25—4 7—2 8—4

[All the exercises in subtraction given above may now be performed, by adding the complement, in place of taking the subtrahend from the minuend.]

[On the same principle, the sum of two or more numbers may be taken from a minuend, whether it consist of one or more numbers, without finding their sum or sums, as in the following examples :]

6. From 46589 take the sum of 2976, 3582, 176, and 24, by addition.

46589 Minuend.
2976
3582
176 } Subtrahend.
24
39831 Difference.

Proof =Sum of difference and subtrahend.

Solution. — 6, 4, 8, 4 (comps.) and 9=31; carrying 3 to 7, 2, 1, 2 (comps.), 8=23; dropping 1 gives 1 to 8, 4, 0 (comps.), 5=18; dropping one, we have 6, 7 (comps.), 6=19; dropping 2 (why 2 ?) from 1 and 4=3.

7. From 7962, take the sum of 5143, 236, 728, 97, 4, and 8, by addition, and prove by subtraction in the usual manner.

8. From 549728, take the sum of 72, 3146, 458, 6, 93, and 872, by addition, and prove by subtraction.

9. From 82493 take 725, 4193, and 6127, and prove by subtraction.

10. From 7248·63 take the sum of 24·5, 784·26, 3158, and 2·34, and prove by subtraction.

11. From 946·783 take the sum of 71·375, 42·6, and 84-·07, and prove.

12. From 8148 take the sum of 7·05, 3·56, 92·4, and 145·3, and prove.

13. Find the difference between the sum of 7643 and 5234, and the sum of 6431 and 978, without finding those sums.

7643 5234	Minuend.
6431 978	Subtrahend.
5468	Difference.

The process reads thus: 2+9 (comps. of 8 and 1) +4+3 =18; carry 1 to 2+6 (comps. of 7 and 3) +3+4=16; carry 1 to 0+5 (comps. of 9 and 4) +2+6=14; set down the 4 and drop the 1 (Why?); 3 (comp. of 6) +5+7=15; drop the 1.

14. From the sum of 5682 and 39476 take the sum of 2158 and 3426, and prove. [Here the 2 ones will be dropped from the rank of tens of thousands.]

15. From the sum of 3678, 5237, 4286, take the sum of 12 and 5213. [Here *one* will be dropped from the rank of hundreds, and *one* from that of tens of thousands.] Prove by subtraction.

16. From the sum of 2·76, 3854, 913·2, take the sum of 346, 2·71, and 1234. How many ones must be dropped in this operation, and from what ranks of figures? Prove by subtraction.

17. From 9876 take the sum of ·71, 360·8, 2·15, 42·76. How many ones must be dropped in this operation, and from what ranks of figures? Prove.

18. From the sum of 8721, 345, and 26·38, take 145. How many ones must be dropped, and whence?

19. From the sum of 526, 3927, and 44, take the sum of 1234 and 600.

When subtraction is performed by addition of the complements, it should be proved by subtraction performed in the usual way. Exercises of this sort afford admirable means for the development of the judgment and imagination, as well as of the memory. The judgment is employed in deciding what number is to be dropped, and when; and the imagination is exercised in calling to mind one figure by the sight of another. Additional exercises may be given by the teacher when necessary; that is, when the pupil or class has not acquired a facility by those given above. It would be still better, however, if the pupil were accustomed to form them for himself.

Questions by the teacher.—What is subtraction? See p. 56, 3. What is the greater number called? The smaller? The result? What is the sign of subtraction? Its name? What is the most convenient mode of arranging the figures? How many modes should be used? Name them. Should subtraction be commenced at the right or left? Why? In what case is it of no consequence where it is commenced?* Will the difference be changed if the same number be added to the minuend and subtrahend? How, then, should we proceed when a figure in the subtrahend is greater than the one of corresponding rank in the minuend? If the subtrahend and difference be given, how may the minuend be found? What, then, is the first mode of proving subtraction? When the minuend and difference are given, how may the subtrahend be found? What, then, is the second mode of proving subtraction? How may the sum of two or more given numbers be subtracted from another number, or from the sum of two or more given numbers, at one operation, without finding either of the sums?

☞ Addition and subtraction may be performed simultaneously by a much easier method than the above, but it does not afford such excellent mental discipline. As, however, it may be preferred for practical business, it is proper to present it here, as follows:

Find the difference between the sum of 2556 and 3798, and the sum of 1324 and 2796, by writing the complement of the subtrahend, and affixing to each complement a hyphen (–), to

* We may commence at the left in every case, provided we take notice as we proceed whether the adjoining figures on the right require 10 to be added, and act accordingly. Exercises so performed would afford excellent intellectual discipline.

show that *one* of the next higher denomination is to be omitted for each number in the subtrahend.

Minuend,	2556	
	3798	
Subtrahend,	–8676	Complement of 1324.
	–7204	Complement of 2796.
Difference,	2234	

Perform all the above exercises from 1 to 19, *writing* the complements of the numbers to be subtracted in place of the actual numbers.

Practical Exercises.

1. Washington died in 1799, at the age of 67 years. In what year was he born?

2. There are two adjoining farms, one of which was sold for $5820, and the other for $376 less. What was the cost of both? [This, and each of the five following exercises, should be performed at one operation by means of the complement.] *Ans.* $11,264.

3. A lady went a shopping with $24 in her purse. She paid $6 for a bonnet, $3 for two pairs of shoes, $5 for a piece of sheeting, and $3 for marketing. How much had she left? *Ans.* $7.

4. A farmer, who was in the habit of settling annually with his creditors, set out for that purpose on New Year's day with $150 in his pocket. He paid his blacksmith $20, his tailor $28, his shoemaker $25, his saddler $30, and his storekeeper $43. After making these payments, how much had he left? *Ans.* $4.

5. A farm, including the stock of cattle, sheep, horses, and hogs, was valued at $8000. The cattle were considered to be worth $240, the sheep $175, the horses $150, and the hogs $75. What was the value of the land? *Ans.* $7360.

6. A merchant sent his clerk to collect some accounts, and directed him to take a purse of silver with him, in case change should be wanted. The clerk collected from one person $25, from another $140, from another $256, and from another $67.

When he came back, he found he had exactly $500. How much had he in the purse when he left home? *Ans.* $12.

7. Two men set out on a journey, travelling in the same direction, and at the end of a week one of them had travelled 200 miles, and the other 240 miles. How far were they then apart?

8. A merchant sold a ship for $8000, which was $1500 more than he paid for it. How much did he pay for it?

9. The value of the gold coined in the mint of the United States in 1831, was $714,270; in 1832, $798,435; in 1833, $978,550; in 1834, $3,954,270. How much more was coined in 1834 than in the other three years taken together?

Ans. $1,463,015.

10. The following is a statement of the revenue of the government of the United States from the year 1837 to 1842, inclusive. The revenue is comprised in two classes, namely, receipts from customs, and from the sales of lands and miscellaneous sources. The latter column is left blank, to be filled by the pupil by horizontal subtraction. If the work be correctly performed, the total amount of the receipts from customs and from the sales of lands, &c., will agree with that of the aggregate of receipts.

Years.	Customs.	Sales of Lands and Miscellaneous.	Aggregate of Receipts.
1837	11,165,970		18,029,528
1838	16,155,455		19,369,639
1839	23,136,397		30,397,515
1840	13,496,834		16,991,191
1841	14,481,998		15,952,293
1842	18,176,721		19,611,599
Total			

11. In 1790, the first census under the constitution of the United States was taken by act of Congress, and it has been followed by similar enumerations every ten years. The following table shows the total population at these several periods. The column of free colored persons is left blank, to be filled by the pupil. The necessary addition and subtraction should be performed at one operation, by aid of the complements of the several numbers. Prove by horizontal addition of the totals.

Years.	Whites.	Free Colored.	Slaves.	Total.
1790	3,172,464		697,897	3,929,827
1800	4,304,489		893,041	5,305,925
1810	5,862,004		1,191,364	7,239,814
1820	7,861,937		1,538,038	9,638,181
1830	10,537,378		2,009,043	12,866.021
1840	14,195,695		2,487,455	17,069,458
1850	19,553,068		3,204,313	23,191,874
Total				

12. The following is a statement of the commerce of the United States, from the year 1831 to 1842, inclusive. Complete the table by finding the difference in value between the exports and imports of each year, and balance the statement by finding the total difference, which may be considered the cost of the freight, and the amount of the profits of the commerce. The computation will be correct, if the balance of the second and third columns is the same as that of the fourth and fifth. Why?

[All the subtractions in the following table should be performed horizontally, as a suitable exercise to prepare the pupil for balancing books and accounts.]

Years.	Value of Exports.	Value of Imports.	Excess of Exports.	Excess of Imports.
1831	$ 81,310,583	$103,191,124		
1832	87,176,943	101,029,266		
1833	90,140,433	108,118,311		
1834	104,336,973	126,521,332		
1835	121,693,577	149,895,742		
1836	128,663,040	189,980,035		
1837	117,419,376	140,989,217		
1838	108,486,616	113,717,404		
1839	121,028,416	162,092,132		
1840	132,085,946	107,141,519		
1841	121,851,803	127,946,177		
1842	104,691,534	100,162,087		
Total				
Bal.				
Aggr.				

13. The following is a statement of the population of some of the largest cities and towns of the United States, by the census of 1840 and that of 1850. The column of *increase in*

ten years is left blank, to be filled by the student by horizontal subtraction. When that is done, add the three columns, and if the *difference* between the sums of the first two columns agrees with the sum of the third, the work is correct.

Cities.	1840.	1850.	Increase in 10 years.
Portland, Me. . . .	15,218	20,815	
Boston, Mass. . . .	93,383	136,881	
Providence, R. I. .	23,171	41,513	
New Haven, Conn. .	12,960	20,345	
New York, N. Y. .	312,710	515,547	
Brooklyn, N. Y. . .	36,233	96,838	
Albany, N. Y. . .	33,721	50,763	
Buffalo, N. Y. . .	18,213	42,261	
Rochester, N. Y. . .	20,191	36,403	
Troy, N. Y. . . .	19,334	28,785	
Newark, N. J. . .	17,290	38,894	
Philadelphia, Pa. .	220,423	340,045	
Pittsburgh, Pa. . .	21,115	46,601	
Baltimore, Md. . .	102,313	169,054	
Washington, D. C. .	23,364	40,001	
Richmond, Va. . .	20,153	27,570	
Charleston, S. C. .	29,261	42,985	
Savannah, Ga. . .	11,214	15,312	
Mobile, Ala. . . .	12,672	20,515	
Louisville, Ky. . .	21,210	43,194	
Cincinnati, Ohio . .	46,338	115,436	
St. Louis, Mo. . .	16,469	77,860	
New Orleans, La. .	102,193	116,375	
Milwaukee, Wis. . .	1,700	29,061	
Cambridge, Mass. .	8,409	15,215	
Roxbury, Mass. . .	9,089	18,364	
Worcester, Mass. .	7,497	17,049	
Poughkeepsie, N. Y.	10,006	13,944	
Syracuse, N. Y. . .	6,500	22,271	
Newburg, N. Y. . .	6,000	11,415	
Reading, Pa. . . .	8,410	15,743	
Wilmington, Del. .	8,367	13,979	
Cleveland, Ohio . .	6,071	17,034	
Columbus, Ohio . .	6,048	17,882	
Detroit, Mich. . .	9,102	21,019	

SECTION III. — *Multiplication.*

[For an explanation of the signs and terms used in Multiplication, see p. 56, 2; and 57, 7.]

[It is usual, in treatises on arithmetic, to present the pupil with a multiplication-table, and require him to commit it to memory. Fortunately this drudgery is wholly unnecessary where Oral Arithmetic has been properly attended to. Should the teacher, however, still think it requisite, the table should be a mere skeleton, as below, to be filled up by the pupil from his own mind. In mathematics, nothing should rest on authority. Neither book nor teacher should furnish ought but definitions, graduated exercises, and suggestive questions.]

1	2	3	4	5	6	7	8	9	10	11	12
2											
3											
4											
5											
6											
7											
8											
9											
10											
11											
12											

Exercises for the Slate and Black-board.

1. Name the product of each of the following pairs of factors, without naming the factors, to be repeated as a daily exercise till it can be done correctly, as rapidly as the words can be spoken.

4	7	5	9	2	9	8	5	1	9	5	4	6	7	2	3
5	3	8	6	8	4	3	2	4	7	8	7	2	7	2	2

4	6	3	9	2	6	8	3	6	9	6	3	4	2	4	8
4	4	3	3	7	4	8	5	7	9	3	4	2	9	8	6

2. Multiply 42579638 by 2, 3, 4, 5, 6, 7, 8, and 9, severally, and prove by addition.

Exemplification for the Black-board.

Where the multiplier consists of one figure only.

```
                42579638 Multiplicand, or 1st factor.
                       6 Multiplier, or 2d factor.
               ---------
               255477828 Product.
               ---------
               { 85159276 1st factor taken 2 times.
Proof by addition, { 127738914 1st factor taken 3 times.
               { 255477828 1st factor taken 6 times.
               ---------
```

Solution. Suggestive Questions.—How many are 6 times 8? 40 of first rank make how many of 2d rank? How many are 6 times 3? How many are 18+4 from the first rank? 20 of second rank = how many of third? How many are 6 times 6+2 from second rank? 30 of third rank = how many of fourth? How many are 6 times 9+3 from third rank? and so on till all the figures are taken 6 times; that is, multiplied by 6.

[The student can hardly be cautioned too frequently to *avoid unnecessary words.* All that are requisite in the above example are (if any are necessary at all) forty-eight, twenty-two, thirty-eight, fifty-seven, forty-seven, thirty-four, fifteen, twenty-five. The student should also be engaged in writing one figure while multiplying the adjoining one; for multiplication, with one figure as factor, should proceed as fast as the figures in the product can be written.]

Proof by Addition. How was the first line of proof found by addition? *Ans.* By adding the first factor to itself. How was the second found? The third? By these three lines the product of the multiplicand by any significant figure could be found. How could four times the first factor be found? (3+1.) How could five times be found? (2+3.) How seven times? Eight times? Nine times?

3. Can 200 be taken 4 times as well as 2? Can 2000, or any number, be taken 4 times? Can 4, 40, or 400, be taken 2000 times? Is it necessary, then, in multiplication, (as in addition and subtraction) that the factors should be of the same denomination or order?

[In multiplication, it is usual to place the smaller factor at the right hand under the larger factor; but, as this is frequently impracticable in long calculations, and as the process is precisely the same, however placed, the student should never be hampered by any peculiar form, but, on the contrary, accustom himself to place the smaller factor in every possible situation,—above, below, at the right, left, or middle; and also in a horizontal position both before and after the larger, from which, in this case, it should be separated by the sign × or •. The latter will be necessary in the bills of parcels below. The student should also be cautioned against the awkward and unnecessary practice of ascertaining the various products on a separate part of his slate.]

4. Multiply 58369247 by each of the significant figures except 1, as a separate exercise, and prove by addition.
5. Multiply 30517284 by each significant figure, and prove.
6. Multiply 73261543 by each figure, and prove.
7. Multiply 85063472 by each figure, and prove.
8. Multiply 14759062 by each figure, and prove.

Exemplifications for the Black-board.

When ciphers occur on the right hand of either factor, or of both factors, or when decimal fractions occur in either or both factors.

9. Multiply 24500 by 50.
10. Multiply 3‘76 by ‘4.
11. Multiply ‘24 by ‘4.

9th.	10th.	11th.
24500	3‘76	‘24
50	‘4	‘4
1225	1504	96

Solution by Suggestive Questions.—9th. The ciphers being neglected in both factors, what is 5 times 245? But, as we have multiplied by 5 in place of 50, how many times is the product too small? How can this be rectified? See p. 115, l. 8. But as 245 has been multiplied, instead of 245 hundred, how many times is the product still too small? How can this be rectified? When ciphers at the right hand of factors are neglected, then, how many ciphers are requisite at the right of the product to rectify the process? *Ans.* As many as there are in — —.

10th. As the separatrix has been neglected in the smaller factor, thus making it 4 instead of 4 tenths, is the product too large or too small? How many times? How can this be rectified? See p. 115, l. 8. Rectify it, then. As the separatrix in the larger factor has also been neglected, thus making it 376 in place of 3'76, is the product too large or too small on this account? How many times? How can this be rectified? Rectify it, then. Now, how many decimal places in the larger factor? In the smaller? In both? In the corrected product? Is the number in the product the same as in both factors? Had the number of fractional places in either or in both factors been less, what would have been the effect on the product? What would have been the effect had there been more? Must the number of fractional places in the product, then, be always equal to the number in both factors?

11th. But in this case only two figures occur, and yet three decimal places are wanting. What character is used when necessary to show the true place of figures? See p. 114, l. 24. Should it be placed to the right or left of the significant figures? See Oral Arithmetic, Chap. I., Sect. XV., p. 52, l. 11.

From these three elucidations, then, may not the following be considered the ninth fundamental principle of arithmetic?

IX. *In multiplication, the number of ciphers at the right of a product, or the number of decimal places which it contains, must always be made equal to the number of either in both factors.*

12. Multiply 28700 by 40; by 600; by 90; by 300; as separate examples.

13. Multiply 25'7 by '6. How many decimal places should be in the product?

14. Multiply 3'58 by '8, previously determining the number of decimal places in this and in the examples below.

15. Multiply 46·42 by ·005.
16. Multiply 2·347 by 9.
17. Multiply 2·347 by ·009. How many decimals?
18. Multiply 2·347 by ·00001.
19. Multiply 5·19 by 9.

Exemplification for the Black-board.

Where the multiplier is greater than 10 *and less than* 20.

20. Multiply 35264 by 14.

The long method.

```
 35264
    14
 ------
141056 Product of multiplicand by 4.
35264  Product of multiplicand by 1 and by 10 by position.
 ------
493696 Product of multiplicand by 14.
 ------
```

The short method.

```
 35264
    14
------
493696
------
```

Suggestive Questions. The Long Method.—How is the multiplicand multiplied by 10 by position? See second Principle, p. 115. To which rank of the product by 4 is the first rank of the multiplicand added? To which rank of the product by 4 is the second rank of the multiplicand added? To which is the third? &c. Could not these figures be added in without writing them over, and thus save two lines, or two thirds of the work? Let us try.

The Short Method. $4\times4=16$; $4\times6=25$ (24 and 1 carried)$+4$ (right hand figure of multiplicand) $=29$; $4\times2=10$ $(8+2$ carried$)+6=16$; $4\times5=21$ $(20+1)+2=23$; $4\times3=14$ $(12+2)+5=19$; $1+3=4$. [This short method only differs from the long method by adding in the right hand figure of the multiplicand *as the work advances*, instead of adding it in *after the completion* of the product of the multiplicand by the units' figure of the multiplier, as will readily be seen by a comparison of the two methods above. Where the class is young, or dull, perhaps it might be proper to postpone the short process till the review. Great care should be taken, as usual, that the pupil does not use too many words; all that are necessary in the above exemplification of the short process are, sixteen; twenty-five twenty-nine; ten, sixteen; twenty-one, twenty-three; fourteen, nineteen; four. After a little practice, four out of these ten words might be dispensed with. [Which four ?]

21. Multiply 23542 by 11 to 19 severally, and prove by the long method; or multiply by the long method, and prove by addition.

22. Multiply as above 4536249 by the numbers from 11 to 19 severally, and prove.

23. Multiply as above 49560·34 by 11 to 19 severally, and prove.

24. Multiply as above 7638·05 by 11 to 19 severally, and prove.

25. Multiply as above 697842·09 by 1·1 to 1·9 severally, and prove.

26. Multiply 3268·534 as above by ·11 to ·19 severally, and prove.

27. Multiply 2343241, by short method, by 21 to 29 severally, and prove.

[This and the following exercises by the short method differ in nothing from the preceding, save that each figure of the multiplicand must be doubled, or trebled, or quadrupled, &c., according to the number of tens in the multiplier, before it is added to the product of the figure to its left. Every example should be proved by addition or by the long method.]

28. Multiply 7286159 by 21 to 29 severally, and prove.
29. Multiply 124932 by 31 to 39 severally, and prove.
30. Multiply 3946072 by 41 to 49 severally, and prove.
31. Multiply 2312412 by 51 to 59 severally, and prove.
32. Multiply 65749 by 61 to 69 severally, and prove.
33. Multiply 98357 by 71 to 79 severally, and prove.
34. Multiply 85679 by 81 to 89 severally, and prove.
35. Multiply 142312 by 91 to 99 severally, and prove.
36. Multiply 35241 by 324.

[This and the following exercises differ in nothing from the preceding, save that three or more products, in place of two, are added in mentally; that is, without being written down.]

Exemplifications for the Black-board.

The long method.

	35241	First factor.	
	324	Second factor.	
Partial products,	140964	by 4	= 4
	70482	by 2 and by 10 by position	= 20
	105723	by 3 and by 100 by position	=300
Total,	11418084	by 324.	324

The short method.

35241 First factor.
324 Second factor.

11418084 Product.

Suggestive Questions. The Long Method.—How is the second partial product multiplied by 10 by position? See *Second Principle*, p. 115. How is the third partial product multiplied by 100 by position? Of what does the units' figure of the total product consist? *Ans.* Of the product of the units of both factors. Of what two figures does the second rank of the total product consist; that is, of the product of 4 by what, and of 1 by what? [Point to the 4 and 1 as they are mentioned.] Of what three figures does the third rank of the total product consist; that is, of the product of 2 by what, of 4 by what, and of 1 by what? [Point to these figures.] Of what three figures does the fourth rank consist? Of what three the fifth? Of what three the sixth? Of what does the seventh and eighth consist? Why could not these products be added in mentally, that is, without writing them during the progress of the work, and thus save nearly three fourths of the figures, as below?

The Short Method.—(4×1) 4; (4×4) 16+(2×1) 2=18; (4×2) 9 (1 carried)+(2×4) 8+(3×1) 3=20; (4×5) 22 (2 carried) + (2×2) 4+ (3×4) 12=38; (4×3)=15+ (2×5) 10+(3×2) 6=31; (2×3) 9+(3×5) 15=24; (3×3) 11. All the *words* necessary are the following: and very many even of these may be omitted after some practice. *Four;* sixteen, *eighteen;* nine, seventeen, *twenty;* twenty-two, twenty-six, *thirty-eight;* fifteen, twenty-five, *thirty-one;* nine, *twenty-*

four; eleven. Those in italics are the only ones really essential.

Exemplification of the Proof by Addition with large numbers.

35241	First factor.	
324	Second factor.	
11418084	Product of 35241 by 324.	
70482	First factor added to itself ×10 by position	= 20
105723	Sum of 1st factor and 4th line ×100 by pos.	=300
140964	Sum of 1st line and 5th line	= 4
11418084	Proof.	324

With larger figures in the multiplier.

32541	First factor.	
978	Second factor.	
31825098	Product of 32541 by 978.	
65082	First factor added to itself. . .	prod. by 2.
130164	The last line added to itself. . .	prod. by 4.
260328	The last line added to itself.	prod. by 8
227787	Sum of 1st, 4th, and 5th lines ×10 by position	= 70
292869	Sum of 4th and 7th lines ×100 by position	=900
31825098	Proof.	978

37. Multiply 213·54 severally by 2·34, by 32·4, and by 423, and prove by the long method, or by addition.

38. Multiply 3521·4 severally by ·451, by 32·5, and by 3·53, and prove.

39. Multiply 765·324 severally by 56·2, by 7·24, and by ·258, and prove.

40. Multiply 9815462 severally by 374, by 865, and by 914, and prove.

41. Multiply 521432 severally by 1324, by 2413, and by 4132, and prove.

[Though the student may be allowed at first to practise by

the long method, yet he ought not to pass on to *Division* till he can use the short method with ease and rapidity.]

Practical Exercises.

1. If 25 men can do a piece of work in 25 days, how long will it take 1 man to do it ?

2. If 16 men can do a piece of work in 14 days, how long will it take 1 man to do it ?

3. Two men set out from the same place, travelling in opposite directions; one at the rate of 42 miles, the other at the rate of 36 miles a day. How far would they be apart at the end of 5 days ?

4. Two men set out from the same place, going in the same direction; the one in railroad cars at the rate of 300 miles a day, the other in a wagon, at the rate of 38 miles a day. How far would they be apart at the end of 3 days ?

5. Two men set out at the same time, but in contrary directions, to travel round a large circular course; the one at the rate of 3, the other at the rate of 5 miles an hour, and after 3 hours' travel they meet each other. How many miles was the circumference of the course ?

6. A carpenter was employed on a building for 25 days, at $1'25 per day. He received at different times $20. How much remained due ?

[The following bills of parcels should be transferred to the slate, and the multiplication be performed horizontally. Where the price is given in cents, as 100 make a dollar, the whole number of cents divided by 100 will give the amount in dollars and cents.]

Boston, Oct. 4, 1853.

7. Mr. James Scott,

Bought of Wm. Smith,

24 lbs. of coffee, at 7 cents . . . .
216 lbs. of sugar, at 6 cents . . . .
5 lbs. of tea, at 65 cents. . . . .
250 lbs. of rice, at 4 cents . . . .
14 lbs. of starch, at 10 cents. . . . . . .
6 gallons of molasses, at 36 cents . . . .
3 gallons of lamp oil, at 94 cents . . . .
175 lbs. of raisins, at 8 cents . . . .

$48'27

Boston, Sept. 15, 1854.

8. Miss Jane Roberts,

Bought of John Smith,

9 yards mousseline de laine,	at 45 cents	. . .	
8 do. do.	at 34 cents	. . .	
32 do. cotton cloth,	at 7 cents	. . .	
6 do. linen,	at 65 cents	. . .	
9 do. calico,	at 15 cents	. . .	
6 do. do.	at 13 cents	. . .	
5 do. gingham,	at 24 cents	. . .	
2 pairs of gloves,	at 58 cents	. . .	
			$17·40

Received payment, John Smith.

Pittsford, Vt., Oct. 19, 1854.

9. Mr. John Fox,

Bought of Henry Sawyer,

3 thousand ½-inch boards,	at $6	. . . .	
4 do. ¾-inch do.	at $7	. . . .	
7 do. inch do.	at $12	. . . .	
2 do. 2-inch plank	at $18	. . . .	
4 do. 3 by 4,	at $8	. . . .	
5 do. shingles,	at $3	. . . .	
Charged in account,			$213

Henry Sawyer.

10. Mr. John Brown,

1854. In account with Clement & Norton, Dr.

Jan. 18.	To 68 gallons of molasses, at 31 cents . .	
	To 425 lbs. brown sugar, at 6 cents . .	
Feb. 21.	To 83 lbs. old hyson tea, at 54 cents . .	
	To 75 lbs. coffee, at 7 cents	

Cr.

Jan. 15.	By 72 bushels corn, at 65 cents	
	By 18 bushels rye, at 75 cents	
	By 32 bushels buckwheat, at 45 cents . .	
Mar. 4.	By cash to balance	21·95

Errors excepted,

Clement & Norton.

11. Mr. Jacob Jones,

Bought of Henry Wheaton,

24 yards broadcloth, at $2'75 . . . . .
36 do. do., at $2'90 . . . . .
48 do. cambric, at 14 cents . . . . .
24 barrels flour, at $6'25 . . . .
27 firkins butter, at $6'75 . . . .
34 barrels pork, at $11'25 . . . .

$891'87

Albany, N. Y., Jan. 4, 1849.

12. Make a bill, like that in Ex. 10, of the following statement: On the 28th of June, 1849, William Jenkins bought of S. Talbot & Co., 34 gallons of molasses, at 36 cents per gallon; 26 bushels of salt, at 85 cents per bushel; 14 pounds of tea, at 42 cents per pound; and paid 36 bushels of corn, at 58 cents per bushel; and 45 bushels of oats, at 42 cents per bushel. The balance, $0'44, was paid in money. By whom was it paid?

13. Make a bill and receipt of the following statement: Mr. A. Williams bought of Samuel Roberts the following articles: A quarter of lamb, weight 7 pounds, at 8 cents a pound; a fillet of veal, weight 9 pounds, at 7 cents a pound; a quarter of mutton, 16 pounds, at 6 cents a pound; a pig, weight 12 pounds, at 10 cents a pound; 2 bunches of celery, 8 cents a bunch; and a bushel of turnips, for 35 cents.

Amount, $3'86.

14. William Hudson sold the following articles to Robert Benson. Make a bill and receipt for them. 325 bushels of corn, at 65 cents a bushel; 73 bushels of wheat, at $1,25; 150 bushels of oats, at 40 cents; and 115 bushels of rye, at 72 cents. Amount, $445'30.

15. Make a bill, like that in Ex. 10, of the following statement: Samuel Brown bought of William Roberts, of Philadelphia, Dec. 31st, 1853, 24 lbs. of tea, at 45 cents per pound; 16 bushels of salt, at 45 cents per bushel; 25 yards of cotton cloth, at 13 cents per yard; and 54 lbs. coffee, at 10 cents per pound. He paid a hog, weighing 425 pounds, at 8 cents per pound. The balance, $7'35, was paid in money. By whom was it paid?

INVOLUTION,

Or Multiplication by Two or more Equal Factors.

INVOLUTION teaches the method of finding the powers of numbers.

DEFINITIONS.

1. A *square* is a figure with four equal sides, and four equal angles.* Squares are employed for the measurement of *surfaces*, or of any thing of which only two dimensions (length and breadth) are considered. In measuring surfaces, the square is the form to which all others are reduced. Thus, painters' work is estimated by the number of square feet covered by the paint; a sail, by the number of square yards it contains; a field, by its contents in square rods. A great mistake is frequently made by using the terms square miles, square yards, &c., for miles square, yards square, &c. For, although 1 square mile is the same as 1 mile square, yet, with all other numbers the result is very different. Thus, 16 square miles are only equal to 4 miles square, as is evident from an inspection of the figure, in which each of the sixteen small squares may represent *one* square mile, or square yard, or square foot, and the whole *sixteen square miles* form but *four miles square*. A slight examination of the figure will show that a square surface, or superficies, is measured by multiplying the length by the breadth.

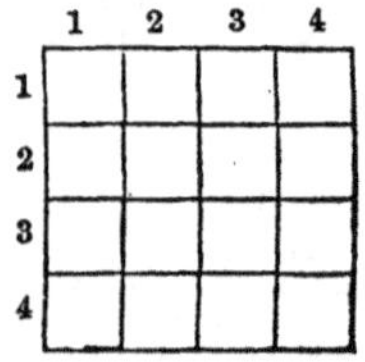

2. If a number be once multiplied by itself, the product is called the *square*, or *second power*, of that number. Thus, $4 \times 4 = 16$; 16 is the square, or second power of 4.

3. A *cube* is a solid body, with six equal square sides, and consequently of three dimensions,—length, breadth, and depth,

* An *angle* is the space comprised between two straight lines which meet in a point, as at A and B; or the quantity by which two straight lines departing from a point diverge from each other. The lines containing the angle are called its sides, or legs. The size of an angle has no reference to the length of its sides. Thus, the angle at A is much greater than the angle at B.

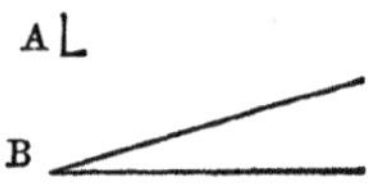

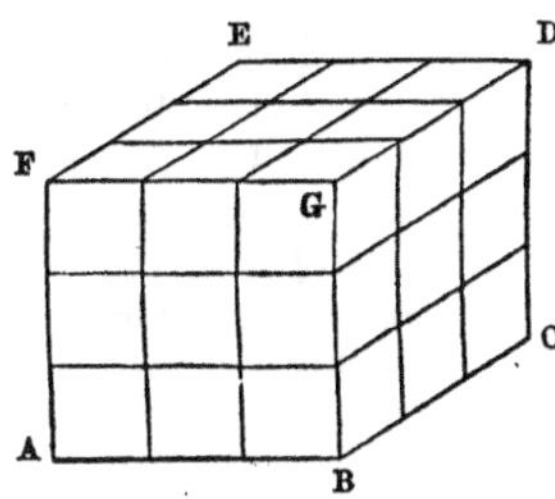

and containing equal angles, as A, B, C, D, E, F. If the length or breadth of one of its sides be twice multiplied by itself, the last product will give the number of *cubic* or *solid* feet, &c. that the figure contains. Thus, if the length of one of the sides of the adjacent *cubic* figure be 3 inches, the contents, $3\times3\times3$, will be 27 cubic inches. Geometry teaches us that all similar solid bodies are to each other as the cubes of their like dimensions. Thus, if one ball, or globe, have a diameter of 2 inches, and another have a diameter of 4 inches, the latter will be 8 times the size of the former, since $4\times4\times4=64$, while $2\times2\times2=8$. For the same reason, a cube whose sides are 3 feet long will be 27 times as large as one of 1 foot long, since the cube of 1 $(1\times1\times1)$ is 1, while the cube of 3 $(3\times3\times3)$ is 27.

4. If the product of a number multiplied by itself be again multiplied by the same number, the last product is called the *cube*, or *third power*, of that number. Thus, $4\times4\times4=64$; 64 is the cube, or third power of 4.

5. The term *power* designates the product arising from multiplying a number a certain number of times, and the number so multiplied is called the *root*. The powers are distinguished from each other by the *number of times* that the *root* is used as factor. Thus, 16 is the *second power* of 4, because it contains that factor *twice ;* 64 is the *third power* of 4, because it contains the factor 4 *three* times; and 256 is the *fourth power* of 4, because it contains 4 as a factor *four times ;* and so on.

6. A power is sometimes denoted by a number placed at the right of the root, and a little above it, which is called the *index*, or *exponent* of the *power*. Thus, 4^3 signifies the *third power*, or *cube* of 4; and 4^5 the *fifth power* of 4.

7. The root of a number is designated by a small figure placed in what is called the radical sign. Thus, $\sqrt[2]{4}$ signifies the square root of 4; and $\sqrt[3]{16}$ the cube root of 16. The sign and number are called the exponent of the root. The figure is generally omitted in the radical sign for the square root. Thus, $\sqrt[2]{9}$ and $\sqrt{9}$ both signify the same number, the square root of 9, namely 3. Roots and powers are also frequently denoted by numbers placed in a fractional form, the numerator expressing

the power, and the denominator the root. Thus $4^{\frac{1}{2}}$ signifies the second or *square root* of the *first power* of 4, which in this case is simply the *square* root of 4, since the first power of a number is the number itself; and $8^{\frac{2}{3}}$ signifies the third, or *cube root* of the *second power* of 8. Thus $8^{\frac{2}{3}}=\sqrt[3]{8\times8}=\sqrt[3]{64}=4$.

8. Numbers whose roots can be exactly found, are called *perfect* powers, and their roots *rational* numbers. Numbers whose roots cannot be exactly expressed in numbers, are called *imperfect* powers, and the approximation to their roots are called *surds*, or *irrational* numbers. Thus, 1, 4, 9, are *perfect* powers, because they have exact roots, namely, 1, 2, 3. But 2, 5, 10, are *imperfect* powers, and their roots, $\sqrt{2}$, $\sqrt{5}$, $\sqrt{10}$, are *surds*, because they cannot be exactly expressed in numbers.

Exercises for the Black-board or Slate.

1. What is the square, or second power of 16?
2. What is the cube, or third power of 12?
3. What is the cube of 3·6?
4. What is the numerical value of $4^2\times3^4$? Of $4^2\times2^2$?

Suggestive Questions. — How much is 4^2? How much 3^4? How much, then, is $4^2\times3^4$?

5. What is the sum of the squares of the prime numbers (see Oral Arithmetic, p. 84) between 1 and 10 inclusive? Of the cubes of the composite numbers between 1 and 10 inclusive? *Ans.* 88; 2521.

6. What is the difference between 2^2 and 2^3? Between 3^2 and 3^3? Between 4^2 and 4^3? *Ans.* 4; 18; 48.

7. Find the square and the cube of each of the digits, arrange them in tabular form as follows, and commit them to memory.

Roots,	1	2	3	4	5	6	7	8	9
Square, or 2d power,									
Cube, or 3d power,									

Exemplification for the Black-board.

8. Involve 24 to the second and third power; in other words, find the square and cube of 24. Perform this in three ways; 1. Keep the units and tens separate throughout, and merely

indicate, without performing the multiplication. 2. Perform the multiplication, but keep the products of the different ranks separate. 3. Involve the number in the usual manner.

	No. 1.	No. 2.	No. 3.
24=	20+4	20+ 4	24
24=	20+4	20+ 4	24
1st Product by the units,	$20 \cdot 4+4^2$	80+16	96
1st Product by the tens,	$20^2+20 \cdot 4$	400+ 80	48
SQUARE of 24,	20^2+twice $20 \cdot 4+4^2$	400+160+16	576
24=	20 + 4	20+ 4	24
2d Product by units,	$20^2 \cdot 4+$twice $20 \cdot 4^2+4^3$	1600+640+64	2304
2d Product by tens,	20^3+twice $20^2 \cdot 4+$ $20 \cdot 4^2$	8000+3200+320	1152
CUBE of 24,	20^3+thrice $20^2 \cdot 4+$thrice $20 \cdot 4^2+4^3$	8000+4800+960+64	13824

Suggestive Questions. — What is the product by the units in No. 1? What does the dot between the 20 and 4 signify? See p. 58, l. 5. What does the character between the two fours signify? The small 2 after the last 4? Read the product by the units in No. 1. [Here direct the attention of the class to this product in Nos. 2 and 3, and show it to be the same.] What is the product by the tens in No. 1? What is the product by the tens in No. 3; 48 or 480? Why? [Position.] What is the sum of the products by the units and by the tens in No. 1? Read it. Why is this called the *square* of 24? Whence comes the *twice* $20 \cdot 4$? [Addition.] Read the 2d product by the units in No. 1, and explain how it is produced; thus, 4 times $4^2=4^3$, &c. Do the same with the second product by the tens. Read it over. In No. 3, is this product 1152, or 11520? Why so? Read the *sum* of the second product by the units and by the tens in No. 1. Why is it called the *cube* of 24? Examine the *cube* in the three processes, and see if they agree. What does the *square* of 24 contain besides the squares of the units and of the tens? *Ans.* Twice their ——. Would that be so, whatever was the number of the tens and units? To what does 20^2 of the fifth line of No. 1 correspond in the same line in No. 2? To what does 4^2 correspond? What does the *cube* of 24 contain besides the cubes of the tens and of the units? *Ans.* Three times the square of the — multiplied by the —, and three times the square of the — multiplied by the —. Would this be so, whatever was the number of the tens and units? Has not, then, the following been developed as the tenth principle of arithmetic? namely,

X. 1. *The* SQUARE *of any number of tens and units is equal to the squares of the tens and of the units taken separately, plus twice the product of the tens and units.* 2. *The* CUBE *of any number of tens and units is equal to the cubes of the tens and units taken separately, plus three times the square of the tens multiplied by the units, and three times the square of the units multiplied by the tens.*

9. Involve 18 to the third power, as in process No. 1 of the last exercise, and repeat the 10th principle from the process that will stand before you on the slate.

10. Involve 65 to the third power, as in the last exercise, and repeat as above.

Questions by the teacher.—What is multiplication? See p. 56, 2. What is the multiplicand? The multiplier? The product? What are the factors? How may multiplication be proved? Should there be ciphers on the right of either factor, or of both factors, will the product be correct, or too small, or too large, if these ciphers be neglected in the multiplication? How, then, can this product be rectified? If decimal fractions occur in either or in both factors, will the product be correct, or too large, or too small, if the separatrix be neglected in the multiplication? How may it be rectified? What is involution? What is the square, or second power of a number? What is a cube, or third power of a number? What is a power of a root? What is the index, or exponent of a root? What is the index, or exponent of a power?

SECTION IV.—*Division.*

[For an explanation of the terms and characters used in division, See p. 56, 4, and 58, 8.]

Exercises for the Slate and Black-board.

1. Name the quotients of the following numbers [to be repeated as a daily exercise till the quotients can be given correctly at a glance, without naming the divisors or dividends]:

4÷2, 8÷2, 6÷2, 12÷2, 18÷2, 10÷2, 14÷2, 16÷2, 9÷3, 18÷3, 12÷3, 6÷3, 24÷3, 16÷4, 25÷5, 18÷3, 20÷5, 27 ÷3, 24÷6, 12÷6, 6÷6, 15÷3, 21 : 3, 21 : 7, 14 : 7, 10 : 5, 30 : 5, 24 : 8, 8 : 8, 18 : 9, 32 : 8, 54 : 9, 28 : 7, 64 : 8, 49 : 7, 36 : 6, 48 : 6, 63 : 7, 72 : 8, 56 : 7, 81 : 9, 32 : 4, 40 : 5, 35 : 5, 36 : 4, 45 : 5.

2. Name the quotients and remainders of the following numbers, in this manner, namely, $\frac{11}{4}$, $\frac{17}{3}$; Two, three; five, two.

$\frac{18}{5}$ $\frac{17}{3}$ $\frac{26}{8}$ $\frac{30}{4}$ $\frac{25}{8}$ $\frac{17}{9}$ $\frac{23}{9}$ $\frac{37}{5}$ $\frac{29}{8}$ $\frac{27}{4}$ $\frac{66}{8}$ $\frac{74}{9}$ $\frac{27}{8}$

$\frac{35}{8}$ $\frac{12}{8}$ $\frac{13}{3}$ $\frac{17}{4}$ $\frac{18}{4}$ $\frac{18}{5}$ $\frac{22}{3}$ $\frac{22}{4}$ $\frac{22}{8}$ $\frac{19}{3}$ $\frac{19}{4}$ $\frac{19}{5}$ $\frac{23}{6}$

$\frac{34}{5}$ $\frac{34}{9}$ $\frac{34}{4}$ $\frac{34}{7}$ $\frac{68}{8}$ $\frac{57}{6}$ $\frac{38}{9}$ $\frac{37}{4}$ $\frac{52}{6}$ $\frac{13}{2}$ $\frac{13}{5}$ $\frac{13}{3}$ $\frac{15}{4}$

$\frac{17}{9}$ $\frac{37}{7}$ $\frac{65}{7}$ $\frac{76}{9}$ $\frac{15}{7}$ $\frac{22}{5}$ $\frac{22}{7}$ $\frac{23}{4}$ $\frac{16}{9}$ $\frac{19}{9}$ $\frac{17}{7}$ $\frac{34}{9}$ $\frac{27}{6}$

3. Name the quotients, and remainders where they occur, of the following numbers:

$\frac{18}{4}$ $\frac{34}{5}$ $\frac{24}{6}$ $\frac{25}{5}$ $\frac{32}{6}$ $\frac{31}{7}$ $\frac{24}{5}$ $\frac{29}{7}$ $\frac{39}{9}$ $\frac{21}{3}$ $\frac{18}{4}$ $\frac{14}{2}$ $\frac{25}{4}$

$\frac{37}{7}$ $\frac{49}{7}$ $\frac{62}{8}$ $\frac{56}{4}$ $\frac{28}{6}$ $\frac{35}{7}$

Exemplification for the Black-board.

Short Division; that is, where the divisor does not exceed 12.

4. Divide 63543 by 4.

```
Divisor,  4)63543 Dividend.
Quotient,   15885 " 3 undivided remainder.
Divisor,        4
            -----
Proof,      63543
            -----
```

Suggestive Questions.— How many fours in 6 of the fifth rank? The 2 that remain of the fifth rank make how many of the fourth rank? How many fours in 23 of the fourth rank, then? The 3 that remain of the fourth rank make how many of the third rank? How many fours in 35 of the third rank, then? How many of the second rank are the 3 that remain of the third rank? How many fours in 34 of the second rank, then? How many of the first rank are the 2 that remain of the second rank? How many fours in 23, then? Our quotient, then, appears to be 15885 and 3 remainder. Is the remainder carried to the right or to the left in division, as above? Which way are numbers carried in all other operations; that is, in addition, subtraction, and multiplication? Why, then, do we commence at the left in division, and at the right in all other operations?

Proof.— In division, what terms are factors of the dividend? See p. 57, 5. What terms, then, multiplied together, will reproduce the dividend, if the work be correctly done? Is the remainder a part of the quotient? Is it also a part of the dividend? Why must it be added in when the dividend is reproduced by multiplying the divisor and quotient?

5. Divide 264852 by each of the digits severally from 2 to 9, also by 11 and 12, proving each problem by multiplication, as above.

6. Divide 65382432 by each digit separately from 2 to 9, also by 11 and 12, and prove by multiplication.

7. Divide 97862432 by each digit separately from 2 to 9, also by 11 and 12, and prove by multiplication.

Exemplification for the Black-board.

Long Division; that is, where the divisor exceeds 12.

8. Divide 64235 by 24. [Place the three methods together on the black-board.]

a. The Long Method.

Dividend,	64235	(24 Divisor.
1st partial product,	48000	2000 ⎫
		600 ⎬ Partial Quotients.
1st remainder,	16235	70 ⎪
2d partial product,	14400	6 ⎭
2d remainder,	1835	$2676\frac{11}{24}$ Total Quotient.
3d partial product,	1680	24 Divisor.
3d remainder,	155	64235 Proof 1, viz. divisor × quot. [+remainder.
4th partial product,	144	
Undivided rem'r,	11	
Proof 2,	64235	Sum of products and last remainder.

b. Contracted Method, by omitting unnecessary ciphers.

Dividend,	64235	(24 Divisor.
1st partial product,	48	$2676\frac{11}{24}$ Quotient.
1st remainder,	162	64235 Proof 1, viz., divisor × quo- [tient + remainder.
2d partial product,	144	
2d remainder,	183	
3d partial product,	168	
3d remainder,	155	
4th partial product,	144	
Undivided remainder,	11	
Proof 2,	64235	Sum of products and last remainder.

c. Abridged Method, by performing the Subtraction mentally.

```
                       Dividend,  64235(24 Divisor.
Partial dividends formed of ( 162    2676 11/24 Quotient.
  remainders and one figure <  183   ———
  from general dividend,    (   155 64235 Proof.
                                  11 undivided remainder.
```

Solution.— The *contracted method* differs only in one respect from the exercises already performed by the pupil in division, and that is, by *writing down* each partial product before subtracting it, instead of performing the subtraction *mentally.* In the *abridged method*, resort is again had to mental subtraction. Take notice, however, that it is frequently necessary to add *more than one ten* to the minuend and subtrahend. Thus, in dividing the first partial dividend, we have $6\times4=24$, which cannot be taken from 2; adding $30=32$ leaves 8; $6\times2=12$, and adding 3 (namely 30 of the rank on the right)$=15$ from $16=1$. In dividing the second partial dividend, we have $7\times4=28$, which cannot be taken from 3; adding $30=33$, leaves 5; then $7\times2=14$, adding 3 (30 of the rank on the right)$=17$ from $18=1$, and so on, adding always as many tens as may be necessary. The *Long Method* is merely given to show the reasons for the different steps in the *Contracted* and *Abridged Methods.*

After the class have studied the exemplifications in the book, they should be written on the black-board; the first in full with all the explanations; the second with the figures only; of the third, merely the divisor and dividend. The divisor is placed at the right, not only to save room, but to bring together the factors of the dividend for the proof. But the mind of the pupil should not be shackled by confining him to particular forms. It will frequently be found convenient in practice to be able to place the terms in division in various positions, as below. That he may not be cramped by forms, let him practise with those different positions in school.

```
          Divisor,                     Divisor,
or Divisor)Dividend(Quotient           Dividend(or Divisor.
                                                Quotient.
```

Questions for the Contracted Method.— What is the upper number on the left? [Point to the dividend.] What is the upper number on the right? What is the first number under the dividend? Of what numbers is it the product? To what rank does it belong by position? [Point, if necessary, to the corresponding number in the *Long Method.*] What is the next number below? Whence comes the 2 on the right? What is the next number below? [Always pointing.] It is 144 of what rank by position? [Here point again, if necessary, to Long Method.] What is the next number? Whence the 3? What is the rank of this number by position? What is the next number below? It is 168 of what rank by position? What is the next number? Whence the last 5? It is 155 of what rank by position? What is the next number? Of what rank, and why? What is the next number? An undivided remainder from what? Why is it placed in the quotient with the divisor written under it? To show that it has not been —. Will the product of the divisor and the integers in the quotient be the exact dividend? What number is necessary to complete the dividend?

[Let the teacher now write out and explain on the black-board the *Abridged Method*, as given above, and then call on one or more of the class to perform similar operations, and give similar explanations on the board.]

It may be proper to observe here that, when the divisor consists of many figures, the pupil at first may not readily ascertain how many times it is contained in the partial dividend. To obviate this difficulty, all the figures in the divisor, except the first [or the first and second], may be neglected for the moment, taking care, however, to make proper allowance for them, especially for the second [or third] figure. Thus, supposing the question was how many times 356 is contained in 932, the 56 may be neglected, provided it be remembered that the 3 (the first figure) represents for the moment more than $3\frac{1}{2}$, consequently can only be contained *twice*, and not *three times*, in 932. Again, supposing the first two figures of the divisor to be 48 or 49, it should be recollected that the first figure is in fact nearer to 5 than 4. But, in spite of the utmost care of a learner, a wrong figure will occasionally appear in the quotient. For division must AT FIRST be merely *a series of suppositions and trials.* Happily a correction is easily made on the slate. If, then, the remainder proves to be equal to, or greater than the divisor, the quotient figure must be too small, since the remainder shows that the divisor is contained in the partial dividend a *greater* number of times. On the contrary, if the product of the quotient figure and the divisor be greater than the partial dividend, it is equally evident that the quotient figure is too large. In either case the correct number must be substituted. Beginners may be materially assisted by forming a table of products of the divisor with each separate digit except the first, as in the following example, according to the contracted method.

```
                          Dividend.
235 • 2=  470             59469805(235 Divisor factor.
    • 3=  705             470       253063 Quotient factor.
    • 4=  940             ----     ----------
    • 5=1175              1246     59469805 Proof.
    • 6=1410              1175
    • 7=1645              ----
    • 8=1880               719
    • 9=2115               705
                           ---
                           1480
                           1410
                           ----
                             705
                             705
                             ---
```

But aids of this sort should be as seldom used as possible, and when used, should be soon discontinued.

Sometimes it happens, after the figure has been brought down from the dividend to the remainder, that the number is still too small to contain the divisor, as in the following example.

```
Dividend, 80520(264 Divisor.
           132   3   Part of the quotient.
```

Here 264 was found to be contained 3 times in 805 with a remainder of 13. Bringing down the 2 from the dividend, the number 132 is found to be too small to contain the divisor. As this shows that there are no tens in the quotient, a cipher should be put in the place of tens. For, in fact, 264 goes *no* times in 132, and leaves the same number (132) as a remainder. Bringing down, therefore, the last figure of the dividend (the 0), we ask how many times 264 is contained in 1320, and finding it to be 5 times, the 5 is placed as usual in the quotient, making it 305.

9. Divide 3628497 by 37, by the abridged method, and prove.

10. Divide 1284634 by 96 by the same.

11. Divide 47389256 by 375; also by 284 and by 763.

12. Divide 83245796 by 2458; also by 372 and by 815.

13. Divide 6529374 by 3275; also by 4762.

14. Divide 5138267 by 789; by 426; and by 738.

15. Perform the same operations with the figures of each dividend reversed; and again, with the figures of each divisor reversed, the dividend carried to three decimal places when necessary.

Remark. — The division can be continued as far as may be thought proper, if there be a remainder after the integers in the dividend are exhausted, by adding ciphers to it, provided a separatrix be placed in the quotient after the *units* resulting from the division of the integers. Why?

The method of *Long Division* is commonly used where the divisor exceeds 12; but, if the class has been properly trained in Oral Arithmetic, it will be found advantageous to use *Short* Division in all cases where the divisor does not exceed two, or even three figures. In practice, it will be found sufficiently easy, and will essentially aid the power of abstraction.

16. Divide 23569248 by every number containing two significant figures between 12 and 100, by short division, and prove.

17. Divide 62543965, and also 34902054 severally by the same, and prove.

18. Divide each of the above two numbers severally by 126, 135, 224, 364, and 452, and prove.

Exemplifications for the Black-board,

Where ciphers occur in the Divisor and Dividend, or in the Divisor alone.

19. Divide 568400 by 2600; and 673428 by 2400.

No. 1. Dividend, 5684|00(26|00 Divisor.
48 $218\frac{1600}{2600}$ Quotient.
224 ———
Remainder, 1600 568400 Proof.

No. 2. Dividend, 6734|28(24|00 Divisor.
193 $280\frac{1428}{2400}$ Quotient.
Remainder, 1428 ———
673428 Proof.

Suggestive Questions.—If a number should be divided into 4 equal parts, and each of these into 3 equal parts, into how many equal parts would the number be divided? If a number should be divided into 4 equal parts, and each of these again into 100 equal parts, into how many equal parts would the number be divided? 4 and 100 are factors of what number? Is the same result obtained, then, when a number is divided by *factors taken separately*, as when divided by *their product*? Into how many equal parts is dividend No. 1 divided by cutting off two figures at the right? See Principle 2, p. 123. What, then, is the quotient of 568400 by 100? What factor remains to divide 5684? What is the quotient when it is divided by 26? To what ranks of the original dividend does the undivided remainder 16 belong? Is it 16, then, or 1600?——What is the quotient of dividend No. 2 by 100? What is the remainder? What other factor should divide 6734? What is the quotient of that division? What is the remainder? To what ranks of the original dividend does this undivided remainder belong? Is it 14, then, or 1400? Is the 28 also an undivided remainder? To what ranks do the 2 and the 8 belong? What, then, is the *whole* undivided remainder of the original dividend?

20. Divide 14260 by 530; 726500 by 670; 257600 by 3400; and 8276000 by 270, and prove.

21. Divide 265023 by 610; 806284 by 7300; and 52648 by 70, and prove.

Exemplifications for the Black-board,

Where Decimal Fractions occur in the Dividend or Divisor, or in both.

22. Divide 2'4 by 6; 2'4 by '6; '24 by '6; 24 by '6; and '24 by 6.

No. 1.	No. 2.	No. 3.	No. 4.	No. 5.
6)2'4	'6)2'4	'6)'24	'6)24	6)'24
'4	4	'4	40	'04

Suggestive Questions.—How many decimal places in the dividend, or product, of No. 1? How many in the divisor factor? As the number of decimal places in the factors, then, must correspond with the number in the product (See *Multiplication*, pp. 153, 154), how many decimal places are required in the quotient factor? How many are required in No. 2, then? Why? How many in No. 3? Why? How many decimal places are there in the divisor factor of No. 4? How many ought there to be in the dividend product, then? How shall that be supplied without altering its value? See p. 118, l. 27. How many decimal places are there in the dividend product of No. 5? Are there any in the divisor factor? How many should there be in the quotient factor, then? How shall the second one be supplied? Is there more than one decimal place in this number, ‘40? Where should the cipher be placed, then, in quotient factor of No. 4?

23. Divide 325‘76 by 23‘4; 589‘42 by ‘72; 68945 by 7‘32; 89728 by ‘8; and 56‘238 by 62, and prove by multiplication.

Practical Exercises.

1. If a piece of work can be accomplished by 1 man in 36 days, how long should 4 men be about it? Prove.

2. If 5 men can do a piece of work in 4 days, how many days should 1 man take to do it? How many days should 4 men? Prove.

3. If 12 barrels of flour cost $72, what will 1 barrel cost? What will 5 cost? Prove.

4. If 5 pounds of brown sugar cost 35 cents, what will 12 pounds cost? [In this and several exercises that follow, the leading question, what will 1 cost, weigh, &c., is omitted, but can readily be supplied by the teacher, if it is found necessary, or, still better, by the pupil himself.]

5. If $420 be paid for 60 acres of land, what will be the cost of 45 acres at the same rate? Prove.

6. If 32 yards of cotton cloth cost $2‘56, what will be the cost of 5 yards at the same rate?

7. If $60 gain $3‘60 interest in one year, what will $100 gain in the same time? Prove.

8. If 6 yards of broadcloth cost $18, what will 15 yards cost? Prove.

9. A man dying, left a widow and 6 children, without a will. In such cases the law directs that the widow shall receive one third of the property for life, and that the remainder shall be equally divided among the children. The estate was valued as follows: a farm, at $6000; a yoke of oxen, $90; 3 horses, at $75 dollars each; 16 cows, at $24 each, 8 young cattle, at $10 each; 100 sheep, at $2‘50 each; farming tools, $100; household furniture, $300; grain and provisions, $62. What was the share of the widow, and of each child?

EVOLUTION.*

Or Division into Two or more Equal Factors.

EXTRACTION OF THE SQUARE ROOT, OR DIVISION INTO TWO EQUAL FACTORS.

a. Pointing off squares into periods of two figures.

Exemplifications for the Black-board.

1. Write in columns, on the slate or black-board, as follows, the squares of ‘01, ‘1, 1, 10, 100, 1000, being the smallest significant figure, and the squares of ‘09, ‘9, 9, 90, 900, 9000, the greatest significant figure. Write, also, a line of ciphers, and point them off into periods of two figures, as under:

TABLE OF SQUARES.

Square of ‘01	=	Square of ‘09	=
of ‘1	=	of ‘9	=
of 1	=	of 9	=
of 10	=	of 90	=
of 100	=	of 900	=
of 1000	=	of 9000	=

PERIODS.

5th 4th 3d 2d 1st

0000000000‘0000

* Evolution is generally placed near the close of the book in treatises on arithmetic. But, as it is strictly an elementary process, and a mere branch of division, it comes much more appropriately here. And, if the preceding part of the book has been thoroughly mastered, the pupil will find no difficulty in extracting either the square or the cube root. Teachers who dislike the arrangement, however, can easily omit *evolution* until the review of the whole book.

Suggestive Questions. — Of how many figures does the square of any number of units consist? See squares of 1 and 9 above. Which period, then, will the square of units occupy? How many ciphers are there in the square of any number of tens? See the square of 10 and of 90 above. Which period, then, will be occupied by the significant figures of the square of tens? Which period will be occupied by those of the hundreds? Of the thousands? &c. Which period will be occupied by those of the tenths? *Ans.* The period to the right of that of the —. Which period by those of the hundredths? In which period, then, should you look for the root of the units? The root of the thousands? Of the tenths? Of the hundreds? Of the hundredths? Why, then, do we divide numbers whose roots are sought into periods of two figures?

b. To find the Square Root, when it consists of tens and units.

2. What is the square root of 2916? Prove.

		Root.	
	$\dot{2}9\dot{1}6$(5	Tens.	
	50^2=2500	4 Units.	
		$\overline{54} \cdot 54 = 2916$	Proof by [invol'n.
Divisor,	$2 \cdot 50 = 100$)416		
See 10th principle, 1, p. 166.	$2 \cdot 50 \cdot 4 =$ 400		
	4^2 16		
	000=Diff. between 2916 and 50^2+twice $50 \cdot 4 + 4^2$.		

Suggestive Questions. — What is the greatest square in 29? What is its root? Is this root 5 units or 5 tens? Deducting, then, the square of 50 from the given square, as above, what must be found in the remainder? See the exemplification in Involution, No. 1, SQUARE of 24, p. 165, or see 10th principle, p. 166. Which of these numbers is now known? *Ans.* Twice the ——. When a product [416] and one of its factors [twice 50] is known, how can the other factor be found? See p. 57, 5. What should the remainder [416] contain, besides twice the product of the tens and units of the root? Is 2916 an exact square, then?

3. Find the square root of 1296, and prove.
4. Find the square root of 2916, and prove.
5. What is the value of $\sqrt{625}$? Prove.
6. What is the value of $\sqrt{2025}$? Prove.

c. To find the Square Root, when it consists of an integer of more than two figures, &c.

7. Find the square root of 105625, and prove.

```
                              ·  ·  ·
                              105625(3  Root of tens  } of
                       30² =  900     2 Root of units } tens.
                              ----    --
        Divisor 2·30=60)156           32  Root of tens.
See 10th prin. 1. { 2·30·2=    120      5 Root of units.
                  { 2²  . . =    4    ---
                               ---    325·325=105625 Pr.
        Divisor 2·320=640)3225
By 10th prin. 1. { 2·320·5=  3200
                 { 5²     =    25
                              ----
                              ....
```

[The method of finding a square root of a larger number of figures does not differ from that of the preceding case, except in slight changes in the *numeration* of the ranks of the root. See Numeration, p. 117, l. 1. For instance, the 3 and 2 found at the beginning of the process, are at first considered as the root of *tens and units* OF TENS; but, as soon as they are found, they are taken together as the *tens of the root;* and bringing down to the remainder [32] the first period [25], making 3225, we proceed to develop the root of the true units [5] as before. When there is a remainder after all the periods have been used, it shows that the number whose root is sought is an *imperfect power*. But we may find a number as near as desirable to the root by annexing periods of ciphers to the remainder, and thence developing roots for the tenths, hundredths, &c., always remembering, however, that all periods for finding square roots, after the first, consist of two figures; and also that we must always *begin* to mark off both ways at the place of units. Should a remainder occur at the close of the whole process, it must of course be added in, if the work is proved by involution. The following example exhibits a case of this nature :

8. Find the square root of 79238, and prove.

. . . Root.

79238(2 Tens } of tens.

$20^2=400$ 8 Units } of tens.

Divisor, $2\cdot 20=40$)392 28 Tens.

By 10th prin. 1. { $2\cdot 20\cdot 8=$ 320 1 Units.

{ $8^2=$ 64 281 Units.

Divisor $2\cdot 280=560$)838 ·4 Tenths.

By 10th prin. 1. { $2\cdot 280\cdot 1=$ 560 281·4 Approximate root.

{ $1^2=$ 1

Divisor $2\cdot 2810=5620$)27700

By 10th prin. 1. { $2\cdot 2810\cdot 4=$ 22480

{ $4^2=$ 16

Remainder, 52·04

Proof $281{\cdot}4^2+52{\cdot}04=79238$.

9. Find the square root of 61504, and prove.

10. Find the square root of 43264, and prove.

11. Find the square root of 61928, to two places of decimals, and prove.

12. Find the square root of 363729·61, and prove.

13. Find the square root of 432·64, and prove.

14. Find the square root of 92165·4, and prove.

15. What is the value of $\sqrt{2}$, carried to three places of decimals? Prove.

16. Find the square root of 10 carried to two places of decimals, and prove.

EXTRACTION OF THE CUBE ROOT, OR DIVISION INTO THREE EQUAL FACTORS.

a. Pointing off Cubes into periods of three figures.

Exemplifications for the Black-board.

1. Write in columns, as follows, on the slate or black-board, the cubes of ·01, ·1, 1, 10, 100, &c., being the smallest significant figure, and the cubes of ·09, ·9, 9, 90, 900, &c., being the greatest significant figure. Write, also, a line of ciphers, and point them off into periods of three figures each, as under:

TABLE OF CUBES.

Cube of ·01 =	Cube of ·09 =
of ·1 =	of ·9 =
of 1 =	of 9 =
of 10 =	of 90 =
of 100 =	of 900 =

PERIODS.

5th 4th 3d 2d 1st.

$0\dot{0}00\dot{0}00\dot{0}00\dot{0}00\dot{0}00\cdot00\dot{0}00\dot{0}$

Suggestive Questions, to be repeated till all can be answered without hesitation.—Of how many figures does the cube of any number of units consist? See 1 and 9 of table of cubes above. *Ans.* Not less than ——, nor more than ——. Which period will the cube of units occupy, then? How many ciphers are there in the cube of any number of tens? See the cube of 10 and of 90 above. Which period, then, will be occupied by the *significant figures* of the cube of tens? Which period will be occupied by those of the hundreds? Of the thousands? &c. Which period will be occupied by those of the tenths? *Ans.* The period to the right of that of the ——. Which period by those of the hundredths? Where, then, should you look for roots of the units? For roots of the thousandths? Of the tenths? Of the tens? Of the hundreds? &c. Why do we divide numbers whose roots are sought, into periods of three figures?

b. To find the Cube Root when it contains an integer of two figures.

2. Find the cube root of 13824, and prove.

```
                          13̇824̇(2
             20³=          8000  4                [Proof
                          ———— 24·24·24=24³=13824
  Divisor, 3·20²=1200)5824 Dividend.
                   ( 3·20²·4  =4800
By 10th prin., 2. <  3·4²·20=   960
                   ( 4³      =   64
                              ————
                              ....
```

Suggestive Questions.—What is the greatest cube in 13? What is its root? Is this root 2 units or 2 tens? Deducting, then, the cube of 20 [8000] from the given cube, what should be looked for next in order in the remainder [5824]? See the exemplification, in Involution, No. 1, cube of 24; or see the 10th principle, p. 166. Which of these numbers is now known? *Ans.* Three times the square of —. If a product [5824] and one of its factors [$3 \cdot 20^2$] be found, how can the other factor be found? See p. 57, 5. What should be looked for next in the remainder [5824]? Is it known? What, lastly, will be found in that remainder? Is 13,824 an exact cube, then?

3. Find the cube root of 373248, and prove by involution.
4. Find the cube root of 19683, and prove by involution.
5. Find the cube root of 262144, and prove by involution.
6. Find the cube root of 166375, and prove by involution.

c. To find the Cube Root, when it contains an integer of more than two figures, &c.

The directions given in treating of the extraction of the *square* root, when it consists of more than two figures, apply almost literally to that of the *cube* root, namely: Proceed with the two periods at the left, as if these were *the whole*, and then, bringing down another period, consider the *two* figures of the root that are thus found as the *tens of the root*, and find the units of the root as before. Should a remainder occur at the close, annex *three* ciphers as a period for tenths, unless there should be decimal places sufficient, and proceed to find the root for the rank of tenths, in the same manner that the root for units was found, and so on, as far as may be considered necessary. The evolution of decimal fractions, in fact, does not differ from that of integers. The following example will make all this sufficiently plain.

7. Find the cube root of 14513286‘7, and prove.

		14513286‘7	(2
	$20^3=$	8000	4
Divisor,	$3\cdot20^2=1200$	)6513	24
By 10th principle, 2.	$3\cdot20^2\cdot4=$	4800	3
	$3\cdot4^2\cdot20=$	960	243
	$4^3\ =$	64	‘9
Divisor,	$3\cdot240^2=172800$	)689286	243‘9
By 10th principle, 2.	$3\cdot240^2\cdot3=$	518400	
	$3\cdot3^2\cdot240=$	6480	
	$3^3\ =$	27	
Divisor,	$3\cdot2430^2=17714700$	)164379700	
By 10th principle, 2.	$3\cdot2430^2\cdot9=$	159432300	
	$3\cdot9^2\cdot2430=$	590490	
	$9^3\ =$	729	
		4356‘181	

Proof, $243‘9^3+4356‘181=14513286‘7$.

8. Find the cube root of 84604519, and prove.

9. Find the cube root of 21024576, and prove.

10. Find the cube root of 28913245, to two places of decimals, and prove.

11. Find the cube root of 21036589, to two places of decimals, and prove.

12. Find the cube root of ‘000729, and prove.

13. Find the cube root of 2 to two decimal places, and prove.

14. Find the cube root of ‘02 to two decimal places, and prove.

15. Find the cube root of 20 to two decimal places, and prove.

16. Find the cube root of 3932586‘4 to two decimal places, and prove.

Practical Exercises in Involution and Evolution.

DEFINITIONS.

I. A figure of three sides is called a *triangle*, and, if one of its corners, or angles, should be a *square* corner, like the angle at B in the annexed figure, it is called a *right angle;* and the figure is called a *right angled triangle*, and the two sides adjoining the right angle are said to be *perpendicular* to each other. The side A C, opposite the right angle, is called the *hypothenuse*. It is shown by Geometry, that the square of the hypothenuse is equal to the *sum* of the squares of the other two sides. It follows that the *difference* between the square of the hypothenuse and that of either of the other sides is equal to the square of the remaining side, since, if 9=4+5, then 9—5=4, and 9—4=5.

II. The round line which forms the boundary of a circle is called its *circumference*. Any straight line which passes through the centre, or middle point, of a circle, and is terminated in both ends by the circumference, is called its *diameter*. Now, we also learn by Geometry, that the areas, or contents, of circles, are not in proportion to their diameters, but to the *squares* of these diameters. Thus, a circle of 6 inches, or 6 feet, in diameter is 4 times as large as one of 3 inches, or 3 feet, in diameter, because the square of 6 [36] is 4 times as large as the square of 3 [9].

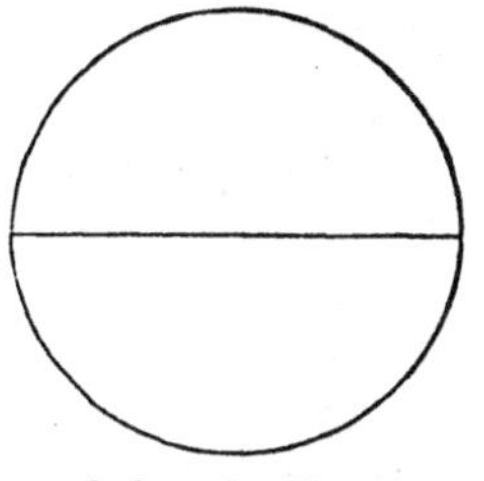

1. If a square field contain 2304 square rods, how many rods does it measure on each side; in other words, what is the square root of 2304? See Definitions 1 and 2, Involution, p. 162.

2. If each side of a square field be 48 rods long, how many square rods does it contain?

3. There are two square fields; the side of one being 20, and of the other 40 rods long. How many square rods in each, and how many times is the one larger than the other?

4. If the sides of one square field be twice as long as that of another square field, how many times is the larger greater than the less?

☞ In order that the pupil may have a clear conception of this fact, let him halve a straight line upon his slate, and form squares of the half and of the whole lines, the one within the other.

5. If each side of a square field measure 25 rods, what will be the length of the side of a square field containing 4 times as many square rods?

6. If the side of a square field measure 50 rods, what will be the length of the side of another square field which contains exactly one-fourth of the number of square rods?

7. A square mile contains 640 acres. How many acres are contained in two miles square? How many square miles in two miles square? *Ans. to 1st question*, 2560 acres.

8. A certain square orchard contains 1600 trees. The trees are in rows, two rods apart each way, and on every side the orchard fence is a rod from the trees. How many acres are there in the orchard, if 160 square rods make an acre?

9. How many trees can be placed in rows in a square field containing 40 acres, the trees to be two rods apart each way, and the outer rows exactly a rod from the fence?

10. A carpenter has a wooden square, one side of which is 4 feet long, and the other 3 feet long. What is the length of a board which will just reach from one end to the other? See Definition 1, above.

11. One of the sides of a carpenter's square is 4 feet long, and a board 5 feet long just reaches from one end of it to the other. What is the length of the other side of the square?

12. A wall is 32 feet high, and a ditch before it is 24 feet wide? What is the length of a ladder that will reach from the top of the wall to the opposite side of the ditch?

13. If a ditch be 24 feet wide, what is the height of a wall that can just be reached by a ladder 40 feet long?

14. If the ladder be 40 feet, and the wall 32 feet, what is the width of the ditch?

15. If a ladder 60 feet long be so placed as to reach a window on one side of a street 36 feet from the ground, and, by turning it over to the other side of the street, without moving its foot, it just reaches a window 48 feet from the ground, what is the width of the street?

☞ If the pupil be at a loss here, let him draw on his slate a horizontal line, to represent the width of the street, and two lines perpendicular to its ends, to represent the walls of the opposite houses, and then complete the figure by another line for the ladder, placing its foot not far from the middle of the street.

16. A certain street is 84 feet wide. How far from the middle of the street must a ladder 60 feet long be placed so as to reach to the top of a wall of the height of 48 feet on one side of the street?

17. The distance across a building between the outer edges of the plates on which the rafters rest is 32 feet, and the height of the ridge above the beam on which they stand is 12 feet. Required the length of the rafters if they project one foot beyond the walls. *Ans.* 21 feet.

18. There is a building 30 feet in length, and 22 in width, and the eaves project beyond the walls one foot on every side. The roof terminates in a point at the centre of the building, and is there supported by a post, the top of which is ten feet above the beams on which the rafters rest. What is the distance from the foot of the post to the corners of the eaves? the length of a rafter reaching to the middle of one *side?* of a rafter reaching to the middle of one *end?* of a rafter reaching to the *corners* of the eaves?

Ans. in order, 20 ft.; 15‘62+ft.; 18‘86+ft; and 23‘36+ft.

19. What is the distance from the centre to each corner of a square field containing 1600 square rods?

Ans. 28‘28+rods.

20. A society of men raised $576 for a certain purpose, each man contributing as many dollars as there were men. What was the number of the society? *Ans.* 24.

21. At another time, their treasurer informed the same society that their funds had been reduced by payments to $39. Whereupon all present made a new contribution, each paying as many dollars as there were members present, when, on a fresh count, the funds amounted to $400. How many members were present? *Ans.* 19.

22. If the diameter of a circle be 2 feet, what will be the diameter of one 4 times as large? *Ans.* 4 feet.

23. What is the distance measured through the centre of a cube from one corner to its opposite corner, the side of the cube being 3 feet? *Ans.* 5‘196 feet.

24. There are two circular pieces of land, the one 100 feet, the other 20 feet in diameter. How many times is the one greater than the other? *Ans.* 25 times.

25. What is the superficies of one side of a cubical block containing 6859 solid inches, and what is the superficies of the whole block? *Ans. to the last question*, 2166 square in.

26. How many times is a globe 2 feet in diameter greater than one that is 1 foot in diameter? *Ans.* 8 times.

27. If a globe of silver, 1 inch in diameter, is worth $6, what is the value of a globe 12 inches in diameter?
Ans. $10368.

28. Find the sum of the roots of all the perfect squares contained between 1 and 100 inclusive. *Ans.* 55.

29. Find the sum of the numbers whose square roots are surds between 1 and 20 inclusive. *Ans.* 180.

30. Find the sum of the numbers whose cube roots are surds between 1 and 20 inclusive. *Ans.* 201.

31. Find the sum of $\sqrt[3]{64}$, $343^{\frac{1}{3}}$, $\sqrt{64}$, $256^{\frac{1}{2}}$, and $8^{\frac{2}{3}}$.
Ans. 39.

MULTIPLICATION AND DIVISION BY EASY NUMBERS.

1. *Multiplication by Division.*

1. Multiply 2647938 by 5 by division (See Oral Arithmetic, Chap. I., Sect. XVIII., 3, 4, 5, p. 60), and prove by multiplication.
2. Multiply 7946287 by 5 by division, and prove as above.
3. Multiply 1678432 by 25 by division (See Oral Arithmetic, pp. 61, 10), and prove as above.
4. Multiply 6238937 by 25 by division, and prove.
5. Multiply 421695 by 25 by division, and prove.
6. Multiply 2834926 by 25 by division, and prove.
7. Multiply 7394845 by 25 by division, and prove.
8. Multiply 84739284 by 125 by division ($125=\frac{1000}{8}$), and prove.
9. Multiply 3462845 by 125 by division, and prove.
10. Multiply 64834921 by 125 by division, and prove.
11. Multiply 1346824 by 125 by division, and prove.

2. *Multiplication by Subtraction.*

12. Multiply 87649324 by 9 by subtraction.

Suggestive Questions.—If a number has been taken 10 times, when it ought to have been taken only 9 times, how many times has it been taken too many? How can the error be rectified?

Then 87649324 =10 times by position,
less 87649324= 1 time,

leaves 788843816= 9 times.

But it is unnecessary to use so many figures, as it may plainly be seen that the process consists simply in *subtracting the right hand figure from* 0, *every other figure from the figure at its right, and lastly* 0 *from the left hand figure.**

On the same principle, a number may be multiplied by 99, 999, or any number of nines, by subtraction, by *supposing* as many ciphers to be annexed to the multiplicand as there are nines in the multiplier, and then *subtracting the original multiplicand* from this product. For instance, if 999 be the multiplier, take each of the three figures at the right from zero, every other figure from the third figure on its right, and zero from the three figures at the left, the reasons for which will distinctly appear from an inspection of the following example.

13. Multiply 47368259 by 999 by subtraction, and prove by adding the complement.

Hence 47368259 =1000 times by position,
less 47368259 =1 time,

leaves 47320890741 =999 times.

14. Multiply 67245896 by 99 by subtraction, and prove by adding the complement.

15. Multiply 34 by 9999 by subtraction, and prove by adding complement.

16. Multiply ‘246 by 9999 by subtraction, and prove by adding complement.

* In the above and in the succeeding exemplification the pupil should omit the 2d line, which is only inserted here to show the reason for the process.

17. Multiply 893254‘25 by 999 by subtraction, and prove by adding complement.

In a similar manner the process of multiplying by 49, 29, or any other two figures ending in 9, may be abridged by using 50, 30, &c., as multiplier, and *at the same time* subtracting the multiplicand once, or adding its complement. Thus,

18. Multiply 7236408 by 69; that is, by 70 with the addition at the same time of the complement of the multiplicand, and prove by multiplying by 69.

```
 7236408
      70
---------
499312152=70—1 time by adding the complement of the
        multiplicand while performing the multiplication.
```

19. Multiply 1423746 by 49 [50 and complement], and prove by 49.

20. Multiply 653492 by 399, as above [by 400 and complement], and prove.

21. Multiply 28‘56 by 29, as above, and prove.

22. Multiply 6734‘5 by 79, as above, and prove.

23. Multiply 6248 by 2499 by division and addition of complement separately, and prove.

24. Multiply 4588‘24 by 499 by division and addition of complement separately, and prove.

25. Multiply 2346 by 8; that is, by *supposing* the multiplicand to be multiplied by 10, and adding double its complement. For example:

```
8 • 2346
--------
  18768
```

Suggestive Questions.—If the multiplicand were multiplied by 10, what would be the right hand figure? If to that we add twice the complement of 6, what would it then be? What would the next figure of the product be if twice the complement of 4 were added to the 6? The next figure of the product + twice the complement of 3? and so on, dropping 2 for *twice* the complement of the rank next above the highest figure of the complement.

26. Multiply 45827 by 38 [by 40 and twice the complement], and prove by dividing by 38.

27. Multiply 372‘65 by 98 [100 and twice the complement], and prove by dividing by 98.

28. Multiply 2536 by 248 by division, adding two complements afterwards, and prove.

29. Multiply 3847 by 78 [80—2], and prove by division.

3. *Multiplication partially or wholly by Addition.*

30. Multiply 62305496 by 11 by addition, and prove.

Suggestive Questions.—If a number has only been taken 10 times, when it ought to have been taken 11 times, how many times too few has it been taken? How shall the error be rectified?

Then 62305496 =10 times by position,
more 62305496= 1 time,

makes 685360456=11 times.

The intelligent pupil will readily perceive that the above depends on a similar principle to that shown in example 12, and that the second line is wholly superfluous, being merely inserted here to show the principle. Of course, in the following exercises, he will omit the superfluity. In fact, it would be preferable that he should even omit the first line also, and write the product by 11 simply by inspection of the book.

31. Multiply 52643889 by 11 by addition, and prove by division.

32. Multiply 846‘25 by 11, and prove by division.

33. Multiply 2345421 by 22, and prove by division.

This, and the 17 succeeding exercises, can be performed without writing any figures on the slate except the product.

34. Multiply 2862‘75 by 33 by addition and multiplication, and prove by division.

35. Multiply 243692801 by 44 by addition, &c., and prove.

36. Multiply 8210432 by 55 as above.

37. Multiply 560438 by 66 as above.

38. Multiply 2439‘004 by 77 as above.

39. Multiply ‘008 by 8‘8 as above.

40. Multiply 3926 by 111 as above.

41. Multiply 62‘25 by 111 as above.

42. Multiply 426832 by 222 as above.

43. Multiply 364852 by 444 as above.

44. Multiply 125896 by 555 as above.

45. Multiply 28·96 by 333 as above.
46. Multiply ·006 by 6·66 as above.
47. Multiply 1236 by 442 as above, and the use of the double complement.
48. Multiply 7216 by 886 as in last exercise.
49. Multiply 18425 by 774 by addition and use of treble complement.
50. Multiply 3215 by 996 as in last exercise.
51. Multiply 7326489 by 4829 by addition.

1.	7326489	
	4829	
1+1=2. .	14652978	multiplied by 10 by position.
2+2=4. .	29305956	" " 1000 by position.
4+4=8. .	58611912	" " 100 by position.
8+1=9	65938401	
	35379615381	Total product.

52. Multiply 6248351 by 3624 by addition, and prove by multiplication.
53. Multiply 1324658 by 18532 as above.
54. Multiply 9024368 by 2936 as above.

4. *Multiplication by Resolution.*

55. Multiply 73250147 by 64328.

	73250147		
	64328		
No. 1.	586001176	multiplied by	8
No. 2.	2344004704	No. 1 by 4 and by 10 by pos. .	320
No. 3.	4688009408	No. 2 by 2 and by 1000 by pos.	64000
	4712035456216	Product by 64328.	64328

56. Multiply 16348792 by 8567. This may be done in two ways: 1st, by 7, by 8=56÷7=8; or 2d, by 7=56÷8=7. Prove by division.

57. Multiply 27364895 by 64816 two ways: 1st, by 8 and by 8=64÷4=16; 2d, by 6 and by 8=48÷3. Prove by division.

58. Multiply 39460582 by 72836 [8·9=72÷2]. Prove by division.

59. Multiply 82964523 by 27315 [3, 9, 5]. Prove by division.

60. Multiply 21879030 by 28442 by resolution. Prove by division.

5. *Division by Multiplication.*

61. Divide 3265 by 5 by multiplication, and prove by division.

```
 3265          5)3265
    2            653 Proof.
 653·0
```

62. Divide 8244 by 5 as above.

63. Divide 91·23 by 5 as above.

64. Divide 26345 by 5 as above.

65. Divide 3276885 by 25 as above.

```
 3276885       25)3276885
       4          131075·4 Proof.
131075·40
```

66. Divide the following numbers severally by 25 by multiplication, and prove by division: 9258, 326725, 8396289.

67. Divide 62845936 by 125 by multiplication, and prove by division.

```
 62845936      125)62845936
        8           502767·488 Proof.
502767·488
```

68. Divide the following numbers severally by 125 by multiplication, and prove by division: 74263485, 29632652, 81297124.

69. Why is it that in dividing integers by 5, the number of decimal places cannot exceed one; in dividing by 25, two; by 125, three?

70. What is the greatest number of decimal places possible in dividing integers by 2? by 4? 8? 16? 32?

CHAPTER III.

THE SHORTENED PROCESSES OF INCREASE AND DECREASE APPLIED TO COMMON FRACTIONS AND DENOMINATE FRACTIONS.

DEFINITIONS.

THERE are three kinds of fractions: Decimal, Common, and Denominate Fractions. All these, as well as integers, have two names or values, namely, the primary, or simple, or absolute value; and the secondary value. The primary name originates alike in all. It is the same with that of the character or characters by which it is represented. The secondary name is derived as follows:

1. In *integers* and *decimal* fractions from their *horizontal position;* that is, from their distance to the right or left of the place of units. (See Chap. I., p. 112, l. 33). Thus, in the number

44·44,

the primary name of each of the figures is the same, namely, *four;* but their secondary names are different; that of the first being *ty* (or *tens*); of the second, *units;* of the third, *tenths;* of the fourth, *hundredths.*

2. In *common* fractions, the secondary name is *written under* the number of the fraction *in figures.* Thus, in $\frac{4}{5}$, the 4 represents the primary, or simple, and the 5 the secondary name and value of the fraction, which, therefore, is called *four fifths.* The chief difference between decimal and common fractions is, that in the former, the integer can be divided only by some power of ten, and hence we can only have tenths, hundredths, &c.; whereas, in common fractions, any number whatever may be used as the divisor. Thus, we have not only $\frac{4}{10}$ (four tenths), $\frac{4}{100}$ (four hundredths), but may use as divisor 5, 6, 18, 356, and so on without end, making $\frac{4}{5}$, $\frac{4}{6}$, $\frac{4}{18}$, $\frac{4}{356}$, &c. From this definition it results, that a common fraction may be changed into an equivalent decimal fraction by *performing the division*

indicated by the horizontal line, while a decimal fraction may be changed into a common one by *writing under it* its secondary name in *figures*, instead of denoting it by a separatrix. Thus, $\frac{3}{5}$='6; and '6=$\frac{6}{10}$=$\frac{3}{5}$. This definition gives us the eleventh principle of arithmetic, as follows:

XI. *An integer or a decimal fraction is changed into a common fraction by expressing its secondary name under it in figures; a common fraction is changed to a decimal one, or to an integer, by performing the division indicated.*

3. *Denominate* fractions are subdivisions of *measures* of any kind, whether of length, surface, or solidity; of weights, money, time, &c. In calculations, their secondary name is written *over* or *beside* them, *in words*, contracted or in full, as gr., or grains, p., or pecks, or in conventional characters, as ʒ, ℥, for drams and ounces. The unit in this kind of fractions is a certain conventional measure, such as pound, bushel, yard, dollar, year, &c., of which the fractions are subdivisions.

Questions by the Teacher.—How many kinds of fractions are there? Name them. What is the primary name of all numbers? The same as that of ——. Whence is the secondary name of integers and decimal fractions derived? Why is this sometimes called their *local* name? How is the secondary name of common fractions ascertained? What is the meaning of the word *vulgar*, frequently used in place of *common* fractions? Why is the word common preferable? *Ans.* Because the word vulgar is now chiefly used in the sense of *mean*, *rude*, *low*. How is the secondary name of denominate fractions ascertained? What is the unit in denominate fractions?

SECTION I.—*Common Fractions.*

[For a full development of the first principles of Common Fractions, see Oral Arithmetic, Chap. II. and III. throughout.]

I. *Change of Form.*

Remarks and Definitions.— Common fractions are capable of assuming, as already noticed, an *infinite* variety of forms. For, as no change of value takes place when *both* terms are multiplied by the same number, it is evident that by multiplication alone they may be infinitely varied. Thus, $\frac{1}{2}$ assumes the forms of $\frac{2}{4}$, $\frac{4}{8}$, $\frac{8}{16}$, and so on without end, by continual multiplication of both terms by 2; and the same fraction, $\frac{1}{2}$, or any other, will assume other endless series of forms by multiplying both terms by 3, 4, 5, 6, or any other number, and all this without the slightest change of value. The same remark applies also to integers and to decimal and denominate fractions, since each of these may assume the form of a common fraction, simply by writing their secondary value under them *in figures.* Thus 6, 18, ·25, 4s. (shillings), may take the fractional forms of $\frac{6}{1}$, $\frac{18}{1}$, $\frac{25}{100}$, $\frac{4}{1}$s., a shape in which they are susceptible of all the variety of form shown above. This capacity of change of form is exceedingly useful in simplifying and abridging the operations of arithmetic.

There are *six* kinds of common fractions, which may be arranged under *two* heads: 1, proper, improper, and mixed; 2, simple, compound, and complex.

1. A *proper* fraction is less than unity, and consequently its numerator is *less* than its denominator, as $\frac{3}{4}$. An *improper* fraction is greater than unity, consequently its numerator is *greater* than its denominator, as $\frac{5}{4}$ or $\frac{17}{5}$. When an integer is separated from the fractional part by division, the *improper* fraction becomes a *mixed* number. Thus, the improper fractions $\frac{5}{4}$, $\frac{17}{5}$, are the same as the mixed numbers $1\frac{1}{4}$, $3\frac{2}{5}$, the integers being separated in the latter by the *partial* performance of the division indicated. Integers in a fractional form, without accompanying fractions, such as $\frac{6}{1}$, or $\frac{12}{4}$, are also considered improper fractions.

2. A *simple* fraction is a fraction in a single expression. It may either be proper, as $\frac{3}{4}$, or improper, as $\frac{5}{4}$. A *compound* fraction is composed of two or more expressions. It is a fraction of a fraction, or a fraction of an integer, and is known by the word *of* between the expressions, a word which was found in Oral Arithmetic, p. 83, 13, always to indicate multiplication in this connection. Thus, $\frac{3}{5}$ of $\frac{5}{8}$, or $\frac{2}{4}$ of $\frac{3}{7}$ of $\frac{9}{10}$, or $\frac{2}{7}$ of 3, are compound fractions. By *performing* the multiplication indicated, they become simple; the first being $\frac{15}{40}$, the second $\frac{54}{280}$, and the third $\frac{6}{7}$. A *complex* fraction is a fraction in which one or both terms are themselves fractional, as $\frac{4\frac{1}{4}}{12}$, $\frac{3\frac{1}{5}}{5\frac{1}{4}}$, $\frac{6}{15\frac{3}{8}}$, and $\frac{\frac{3}{5}}{\frac{7}{8}}$. These, also, become simple by *performing* the operations indicated. Any pupil that is familiar with Chap. III., Oral

Arithmetic, can change their forms to those of simple fractions by the aid of the following questions:

First, $\frac{4\frac{1}{4}}{12}$. How many fourths in 1? How many in 4? In $4\frac{1}{4}$? How much is $\frac{17}{4}$ divided by 12?

Second, $\frac{3\frac{1}{5}}{5\frac{1}{4}}$. How many fifths in $3\frac{1}{5}$? How many fourths in $5\frac{1}{4}$? Divide $\frac{16}{5}$ by $\frac{21}{4}$.

Third, $\frac{6}{15\frac{2}{8}}$. Put 6 in a fractional form. How many eighths in $15\frac{2}{8}$? Divide $\frac{6}{1}$ by $\frac{122}{8}$.

Fourth, $\frac{\frac{3}{5}}{\frac{7}{8}}$. Divide $\frac{3}{5}$ by $\frac{7}{8}$, as indicated by the horizontal line.

a. To change a fraction to its equivalent lowest expression.

1. Change $\frac{28}{35}$ to its lowest expression, by striking out the prime factors common to both terms.

$$\frac{28}{35}=\frac{2\cdot 2\cdot 7}{5\cdot 7}$$

Suggestive Questions.—What factors are the same in both terms? What will the fraction be if the factor 7 be omitted in both terms? Is $\frac{4}{5}$ the lowest term of the fraction?

2. Change $\frac{275}{440}$ to its lowest expression.

$$\frac{275}{440}=\frac{5\cdot 5\cdot 11}{2\cdot 2\cdot 2\cdot 5\cdot 11}$$

Suggestive Question.—When the factors common to both terms are stricken out, what will be the lowest expression of the fraction?

3. Change $\frac{369}{1296}$ to its lowest expression by inspection merely.
4. Change $\frac{250}{1725}$ to its lowest expression by inspection.
5. Change $\frac{29}{58}$ to its lowest expression by inspection.
6. Change $\frac{27}{108}$ to its lowest expression by inspection.
7. Change $\frac{9}{84}$ to its lowest terms by inspection.
8. Change $\frac{1644}{2192}$ to its lowest terms by analysis.
9. Change $\frac{315}{405}$ to its lowest terms by inspection.

b. To change an integer or a decimal fraction to the form of an equivalent common fraction.

1. Change 42 to the form of an equivalent common fraction. See p. 193, XI.
2. Change 3·75 to an equivalent common fraction.
3. Change 7·07, 2·5, ·003, 4·25, 265, severally to the form of equivalent common fractions.

c. To change a common fraction, whether proper or improper, to a whole or mixed number, or to a decimal fraction.

1. Change $\frac{4}{100}$, $\frac{5}{10}$, $\frac{16}{10}$, $\frac{125}{10}$, severally, to decimal fractions, or mixed numbers; that is, *perform the division indicated*, and prove by rechange to common fractions. See 11th principle, p. 193.
2. Change $\frac{210}{10}$, $\frac{4500}{100}$, and $\frac{33}{1}$, severally, to whole numbers, and prove by rechange to common fractions.
3. Change $\frac{3}{4}$, $\frac{5}{8}$, $\frac{1}{2}$, $\frac{6}{24}$, $\frac{9}{15}$, and $\frac{4}{2}$, severally, to decimal fractions, and prove by rechange to their original form.
4. Change $\frac{9}{4}$, $\frac{19}{5}$, $\frac{37}{4}$, and $\frac{45}{8}$, severally, to mixed numbers, and prove by rechange to their original form.

DEFINITIONS.

I. Every decimal fraction consists of a specified number of tenths, or of some power of tenths, such as hundredths, thousandths, &c.; consequently every fraction that cannot be expressed in tenths, or one of the powers of tenths, cannot be accurately expressed in a decimal form. It follows, then, that in changing common fractions to a decimal form, the division will never terminate, but go on to infinity in every case where any prime factor other than 2 or 5 (prime factors of 10) remains in the denominator after the fraction has been brought to its lowest terms. For instance, if 3 or 7 (or any other prime factor but 2 and 5) should be a factor in the denominator and not in the numerator, an undivided remainder would always occur in the division, and, by adding a cipher, the quotient would form an endless series of figures. Thus, if we perform the division indicated in $\frac{1}{3}$, it gives ·333, &c., without end; and in $\frac{12}{37}$ it gives ·324324324, &c., a period of three figures endlessly recurring. Such decimals as these, which continually recur in periods of one or more figures in the same order, are called *circulating decimals*. They are distinguished from ordinary decimals by a dot placed over the first and last figure of the circulating period. Thus, $\frac{1}{3}$ is expressed by $\dot{3}$, and $\frac{12}{37}$ by $\dot{3}2\dot{4}$. The set of figures which repeats is called a *repetend.*

II. When a period begins with the first decimal figure, it is called a *simple* repetend. But when other decimal figures occur before the period commences, it is called a *compound* repetend. Thus, $\frac{1}{3}$='333, &c., forms a simple, and $\frac{1}{6}$='1666, &c., forms a compound repetend.

1. *Decimal Fractions with Simple Repetends.*

1. Change the following common fractions to equivalent decimal fractions carried to twelve or fifteen places, allowing the operation to remain on the slate till examined by the following questions:

1. $\frac{1}{9}$= , &c.
2. $\frac{1}{99}$= , &c.
3. $\frac{1}{999}$= , &c.
4. &c. , &c.

Suggestive Questions.—What decimal fraction is equivalent to $\frac{1}{9}$? Then what decimal fraction is equivalent to $\frac{2}{9}$, or 2 times $\frac{1}{9}$? What to $\frac{3}{9}$? To $\frac{4}{9}$? &c., up to $\frac{8}{9}$? If, then, it were required to change a decimal fraction of one figure continually repeated, that is, a *circulating decimal* with one figure for *repetend,* into a common fraction, what would be the numerator? What the denominator?

2. What decimal fraction is equivalent to $\frac{1}{99}$? What to $\frac{2}{99}$, or 2 times $\frac{1}{99}$? To $\frac{4}{99}$? To $\frac{5}{99}$? To $\frac{7}{99}$? To $\frac{15}{99}$? To $\frac{27}{99}$? To $\frac{45}{99}$? &c. If, then, it be required to change a circulating decimal with two figures for a repetend into a common fraction, what will be the numerator? What the denominator?

3. What decimal fraction is equivalent to $\frac{1}{999}$? What to $\frac{2}{999}$? To $\frac{3}{999}$? To $\frac{4}{999}$? &c. To $\frac{25}{999}$? To $\frac{125}{999}$? To $\frac{605}{999}$? To $\frac{872}{999}$? &c.? If, then, it be required to change a circulating decimal with three figures for a repetend into a common fraction, what would be the numerator? What the denominator?

4. In general, then, if it be required to change a circulating decimal with any number of figures for a repetend, what would be the numerator? What the denominator? May it not, then, be considered as the 12th principle of arithmetic, that,

*

XII. *Circulating decimals, with a simple repetend, are changed to common fractions by using the repetend as numerator, and as many 9s as there are figures in the repetend as denominator.*

5. Change $\frac{3}{9}$, $\frac{6}{9}$, and $\frac{8}{9}$, severally, to circulating decimals, without any formal division, and prove by rechange to common fractions by the 12th principle above.

6. Change $\frac{7}{99}$, $\frac{24}{99}$, and $\frac{37}{99}$, severally to circulating decimals by inspection, and prove by rechange to common fractions.

7. Change $\frac{2}{999}$, $\frac{7}{999}$, $\frac{25}{999}$, $\frac{164}{999}$, $\frac{768}{999}$, severally to circulating decimals by inspection, and prove by rechange to common fractions.

2. *Decimal Fractions with Compound Repetends.*

1. Change $\frac{1}{6}$ to the form of a decimal fraction, and prove by rechange to its original form.

$$\tfrac{1}{6} = .1\dot{6} = \tfrac{1}{10} + \tfrac{6}{90} = \tfrac{6}{90} + \tfrac{9}{90} = \tfrac{15}{90} = \tfrac{1}{6}.$$

Suggestive Questions.—Of what denomination is the 1 in the decimal fraction? Of what denomination would the 6 with a dot over it have been, had it stood in the place of the 1? What, then, is its denomination standing one rank more to the right? What prime factor in the denominator of $\frac{1}{6}$ causes it to repeat? Why are the two parts of the decimal changed separately to common fractions?

2. Change $\frac{1}{24}$ to a decimal fraction, first determining what figures, if any, in the denominator will cause it to repeat, and prove by rechange to its original form.

$$\tfrac{1}{24} = .041\dot{6} = \tfrac{41}{1000} + \tfrac{6}{9000} = \tfrac{369}{9000} + \tfrac{6}{9000} = \tfrac{375}{9000} = \tfrac{1}{24}.$$

Suggestive Questions.—Why is the $\dot{6}$ changed to $\frac{6}{9000}$ in place of $\frac{6}{9}$? What prime factor in $\frac{1}{24}$ causes the decimal to repeat? From the last two demonstrations, then, may it not be legitimately inferred as a principle in arithmetic, that,

XIII. *Circulating decimals, with compound repetends, may be changed to common fractions, by changing separately the repetends and the figures that precede them to common fractions, and then adding them together.*

3. Change the following common fractions to a decimal form, first determining what prime factors in the denominators, if any, will cause them to repeat. Prove by a rechange to their original form: $\frac{1}{12}$, $\frac{1}{14}$, $\frac{1}{15}$, $\frac{1}{18}$, $\frac{1}{22}$, $\frac{1}{26}$.

d. To change a mixed number to an equivalent improper fraction.

1. Change $4\frac{5}{8}$ to an improper fraction, and prove by rechange to its original form. See definition 1 to this section, p. 194.

2. Change $9\frac{3}{4}$, $62\frac{4}{5}$, $81\frac{5}{8}$, $6\frac{7}{8}$, severally to improper fractions, and prove by rechange to their original form.

3. Change $14\frac{2}{5}$, $27\frac{9}{15}$, $38\frac{4}{17}$, and $8\frac{4}{9}$, severally to improper fractions, and prove by rechange to their original form.

e. To change a compound fraction to an equivalent simple one, or to a whole or a mixed number.

1. Change $\frac{3}{4}$ of $\frac{4}{5}$ of $\frac{10}{15}$ to an equivalent simple fraction.

$$\frac{3\cdot4\cdot10}{4\cdot5\cdot15}=\frac{120}{300}=\frac{2}{5}$$

Suggestive Questions.—What is the meaning of the word *of* in common fractions? What is indicated by the period placed between numbers? What, then, is the process for simplifying compound fractions?

Remark.—By cancelling, that is, by striking out the prime factors common to both terms in the compound fraction, the intermediate multiplication is always shortened, and sometimes, as in the present instance, rendered wholly unnecessary. For, if 10 in the numerator is mentally resolved into its two factors, 2 and 5, it will be perceived that the 3 and 5 in the numerator will cancel the 15 in the denominator; and, as the two 4s destroy each other, nothing remains but 2 and 5, making $\frac{2}{5}$.

2. Change $\frac{14}{5}$ of $\frac{3}{7}$ of $\frac{2}{6}$ of $\frac{5}{2}$ to an equivalent whole number, cancelling by inspection. *Ans.* 1.

3. Change $\frac{5}{8}$ of $\frac{2}{4}$ of $\frac{16}{5}$ of $\frac{9}{10}$ to an equivalent simple fraction, cancelling by inspection.

4. Change $\frac{3}{4}$ of $\frac{9}{2}$ of $\frac{4}{5}$ of $\frac{5}{3}$ to an equivalent mixed number, cancelling by inspection. *Ans.* $4\frac{1}{2}$.

5. Change $\frac{2}{7}$ of $\frac{4}{5}$ of $\frac{3}{6}$ of $\frac{8}{15}$ to an equivalent simple frac-

tion, by cancelling two figures in the numerator and one in the denominator. *Ans.* $\frac{32}{525}$.

6. Change $\frac{5}{8}$ of $\frac{7}{9}$ of $\frac{3}{5}$ to an equivalent simple fraction. *Ans.* $\frac{7}{24}$.

f. To change fractions of different denominators to equivalent fractions with a common denominator.

1. Change $\frac{3}{7}$ and $\frac{4}{5}$ to equivalent fractions with a common denominator. See Oral Arithmetic, Chap. III., Sect. III., p. 89.

2. Change $\frac{1}{4}$ and $\frac{7}{9}$ to equivalent fractions with a common denominator.

3. Change $\frac{3}{5}$ and $\frac{4}{20}$ to equivalent fractions with a common denominator. 1st. By multiplication; 2d, by division; in both cases by inspection.

☞ Whenever an operation can be performed by division as well as by multiplication, the former should always be chosen, since it leaves the result in a more simple form.

4. Change $\frac{3}{4}$ and $\frac{3}{20}$ to equivalent fractions with a common denominator by inspection. Can this be done by division? Why? Must both fractions or only one be changed, in order to bring them to the same denomination?

5. Change $\frac{4}{10}$ and $\frac{15}{5}$ by inspection to equivalent fractions with a common denominator. 1st, By multiplication; 2d, by division.

6. Change $\frac{1}{4}$ and $\frac{3}{36}$ to equivalent fractions with a common denominator by inspection. By division and multiplication.

g. To change fractions of different denominators to equivalent fractions with the least common denominator.

1. Change $\frac{3}{4}$ and $\frac{5}{6}$ to equivalent fractions with the least common denominator, by inspection. (The factor 2 in the second denominator may be omitted, as it is to be found in the first. See Oral Arithmetic, Chap. III., Sect. III., p. 89.) Prove by changing each fraction to its lowest denomination.

2. Change $\frac{7}{8}$ and $\frac{5}{12}$ to equivalent fractions with least common denominator. What factors may be omitted in 12? Why? Prove as in last example.

3. Change $\frac{3}{14}$, $\frac{1}{2}$, $\frac{5}{7}$, $\frac{2}{3}$, $\frac{5}{6}$, $\frac{8}{21}$, to equivalent fractions with least common denominator by inspection. The 2d, 3d, 5th,

and 6th denominators may be omitted. Why? Point them out in the 1st and 4th. Prove, as in last example.

Remark. — Hitherto fractions have been changed to equivalent fractions with least common denominator by *inspection* merely. When the denominators are large and numerous, however, this is somewhat difficult for the unpractised student. It may be proper, therefore, to show how such changes may be effected by *calculation* on the slate or black-board, as follows:

4. Change $\frac{5}{80}$, $\frac{7}{24}$, $\frac{5}{12}$, $\frac{3}{16}$, and $\frac{4}{25}$, to equivalent fractions with the least common denominator. Prove as in last example.

By analyzing the denominators into their prime factors, we have

a	*b*	*c*	*d*	*e*
$\frac{5}{80}$	$\frac{7}{24}$	$\frac{5}{12}$	$\frac{3}{16}$	$\frac{4}{25}$
2•2•2•2•5	$\underline{2•2•2•3}$	$\underline{2•2•3}$	$\underline{2•2•2•2}$	$\underline{5}$•5

Suggestive Questions.—Are all the underlined factors to be found in the denominators of the fractions marked *a* and *b*? Should they be omitted, then, in finding the *lowest* common denominator? What is the product of the factors that are not underlined? (80•3•5.) Has this product every factor contained in all the given denominators? Will it form their *common* denominator, then? Does it contain no more factors than they do? Will it form, then, their *lowest* common denominator?

If, then, the fraction *a* is to be changed to the denomination of 1200, and one of its factors (80) is given, how shall the other factor be found? Should the second factor for each of the other denominators be found by the same process (division)? Probably, then, the numbers cannot be arranged more conveniently than as follows:

Divisor factor, or given denominator of	General product, or dividend, or common denominator.	Quotient factors.	Gives
a 80		15 for *a*	*a* $\frac{75}{1200}$
b 24		50 for *b*	*b* $\frac{350}{1200}$
c 12	1200	100 for *c*	*c* $\frac{500}{1200}$
d 16		75 for *d*	*d* $\frac{225}{1200}$
e 25		48 for *e*	*e* $\frac{192}{1200}$

5. Change $\frac{11}{15}$, $\frac{3}{4}$, $\frac{7}{12}$, and $\frac{5}{8}$, to least common denominator. Prove as by last example.

$\frac{11}{15}$	$\frac{3}{4}$	$\frac{7}{12}$	$\frac{5}{8}$
3•5	2•2	2•2•3	2•2•2=120.

$$\left.\begin{matrix}15\\4\\12\\8\end{matrix}\right\} 120 \left\{\begin{matrix}8\\30\\10\\15\end{matrix}\right. \left\{\begin{matrix}\frac{88}{120}\\ \frac{90}{120}\\ \frac{70}{120}\\ \frac{75}{120}\end{matrix}\right.$$

6. Change $\frac{3}{4}$, $\frac{4}{35}$, $\frac{1}{20}$, $\frac{3}{14}$, by calculation, to equivalent fractions with least common denominator. Prove by changing each fraction severally to its lowest denomination.

7. Change $\frac{5}{7}$, $\frac{3}{13}$, $\frac{2}{6}$, and $\frac{1}{4}$, by calculation, to equivalent fractions with least common denominator. Prove as in last example.

8. Change $\frac{2}{5}$, $\frac{3}{5}$, $\frac{4}{20}$, and $\frac{5}{8}$, by calculation, to equivalent fractions with least common denominator. Prove as in last example.

9. Change $\frac{2}{5}$, $\frac{5}{6}$, $\frac{1}{4}$, $\frac{2}{9}$, $\frac{4}{15}$, and $\frac{5}{36}$, by calculation, to equivalent fractions with least common denominator. Prove as in last example.

10. Change $\frac{2}{105}$, $\frac{6}{5}$, $\frac{2}{3}$, $\frac{5}{7}$, $\frac{1}{2}$, by inspection, to equivalent fractions with least common denominator. One only of these denominators is a factor in the least common denominator. Which is it? Prove as in last example.

11. Change $\frac{3}{5}$, $\frac{5}{6}$, $\frac{4}{9}$, $\frac{7}{8}$, $\frac{8}{27}$, $\frac{3}{4}$, by inspection, to equivalent fractions with least common denominator. Prove as in last example.

12. Change $\frac{4}{7}$, $\frac{3}{5}$, $\frac{5}{14}$, $\frac{9}{11}$, $\frac{7}{91}$, $\frac{8}{35}$, by calculation, to equivalent fractions with least common denominator. Prove as in last example.

13. Change $\frac{4}{5}$, $\frac{3}{4}$, $\frac{5}{6}$, $\frac{3}{18}$, $\frac{11}{24}$, by calculation, to equivalent fractions with least common denominator. Prove as by last example.

14. Repeat examples 4, 5, 6, 7, 8, 9, 12, and 13, by inspection, and prove as above.

h. To change a complex fraction to an equivalent simple one.

1. Change $\frac{\frac{3}{8}}{\frac{2}{5}}$ to an equivalent simple fraction; in other words, *perform partially the division indicated.*

2. Change $\frac{\frac{2}{7}}{\frac{4}{9}}$ to an equivalent simple fraction in its lowest expression, performing both changes at one operation by inspection. *Ans.* $\frac{9}{14}$.

3. Change $\frac{4}{\frac{3}{4}}$ and $\frac{\frac{2}{5}}{7}$ severally to equivalent simple fractions. The manner of doing this will be readily perceived, if the integers 4 and 7 be each placed in a fractional form.

Ans. $\frac{16}{3}$ and $\frac{2}{35}$.

Suggestive Questions.—What terms must be multiplied together to change $\frac{\frac{4}{1}}{\frac{3}{4}}$ to a simple fraction? What terms, then, in the same fraction in the other form $\left(\frac{4}{\frac{3}{4}}\right)$? In changing complex fractions of three terms, then, to simple ones, which terms are factors of the new numerator, when the upper term is an integer; the first and second, the second and third, or the first and third?

What terms are factors of the new denominator in changing $\frac{\frac{2}{5}}{\frac{7}{1}}$ to a simple fraction? What terms, then, in the same fraction in another form $\frac{\frac{2}{5}}{7}$, when the lower term is a whole number?

Repeat the rule developed by these illustrations. *Ans.* In changing complex fractions of three terms to simple fractions, when the upper term is an integer, the — and — terms are factors of the —, and the other term is the —; but, when the lower term is an integer, the — and — terms are factors of the —, and the other term is the —.

Remark.—The pupil will remove all difficulty in simplifying complex fractions of three terms by inspection, by merely placing the integer, mentally, in a fractional form.

4. Change $\frac{\frac{3}{8}}{7}$ to an equivalent simple fraction by inspection.

Ans. $\frac{3}{56}$.

5. Change $\frac{2}{\frac{4}{9}}$ to an equivalent mixed number by inspection.

Ans. $4\frac{1}{2}$.

6. Change $\frac{3}{\frac{2}{7}}$ to an equivalent mixed number by inspection.

Ans. $10\frac{1}{2}$.

7. Change $\frac{\frac{4}{5}}{8}$ to an equivalent simple fraction by inspection.

Ans. $\frac{1}{10}$.

8. Change $\frac{\frac{2}{3}}{\frac{9}{27}}$ to an equivalent whole number by inspection.

Ans. 2.

9. Change $\frac{\frac{4}{5}}{52}$ to an equivalent simple fraction, in its lowest terms, at one operation, by inspection. *Ans.* $\frac{1}{65}$.

10. Change $\frac{4\frac{3}{7}}{5\frac{1}{4}}$ and $\frac{2\frac{1}{2}}{7\frac{3}{9}}$ to equivalent simple fractions in their lowest terms. *Ans.* $\frac{124}{147}$ and $\frac{15}{44}$.

i. To change an integer, a decimal fraction, or a common fraction, to an equivalent common fraction of a specified numerator or denominator.

It has already been shown that an integer, or a decimal fraction, assumes the form of a common fraction by *writing its denomination* under it *in figures*. As *any* number, then, readily assumes this form, the proposition becomes, simply, to *supply* a deficient numerator or denominator in a common fraction, where an equivalent fraction is given with a *different* denominator or numerator.

CASE I. *Where the given Numerator or Denominator of the imperfect fraction is a* FACTOR *of the Numerator or Denominator of the perfect one.*

1. Change $\frac{6}{9}$ to an equivalent fraction with 2 for numerator; or, in other words, supply the deficient denominator in the following equivalent fractions:

$$\frac{2}{\ }=\frac{6}{9}$$

Suggestive Questions.—Is 2 a factor of 6? What is the other factor of 6? How, then, shall $\frac{6}{9}$ be changed to an equivalent fraction with 2 for denominator? *Ans.* By —— both terms by ——.

2. Supply the deficient denominators in each of the following pairs of fractions by inspection:

$\frac{2}{\ }=\frac{10}{25}$; $\frac{4}{\ }=\frac{8}{24}$; $\frac{1}{\ }=\frac{3}{45}$; $\frac{7}{\ }=\frac{14}{84}$; $\frac{7}{\ }=\frac{14}{26}$; $\frac{6}{\ }=\frac{5\cdot12}{8\cdot15}$;

3. Change $\frac{9}{6}$ to an equivalent fraction with 2 for denominator; in other words, supply the deficient numerator in the following equivalent fractions:

$$\frac{\ \ }{2}=\frac{9}{6}$$

Suggestive Questions.—Is 2 a factor of 6? What is the other factor of 6? How, then, shall $\frac{9}{6}$ be changed to an equivalent fraction with 2 for numerator?

4. Supply the deficient numerators in the following pairs of equivalent fractions by inspection:

$$\frac{\ \ }{2}=\frac{25}{10};\ \frac{\ \ }{4}=\frac{24}{8};\ \frac{\ \ }{1}=\frac{45}{3};\ \frac{\ \ }{7}=\frac{84}{14};\ \frac{\ \ }{7}=\frac{26}{14};\ \frac{\ \ }{6}=\frac{8\cdot15}{5\cdot12};\ \frac{\ \ }{9}=\frac{4\cdot28}{3\cdot42}$$

CASE II. *Where the Numerator or Denominator of the imperfect fraction is a* MULTIPLIER *of the Numerator or Denominator of the perfect one.*

1. Change $\frac{5}{9}$ to an equivalent fraction with 45 for numerator; that is, supply the deficient denominator.

$$\frac{45}{\ \ }=\frac{5}{9}$$

Suggestive Questions.—Is 45 a multiple of 5? How, then, may $\frac{5}{9}$ be changed to an equivalent fraction with 45 for numerator?

2. Supply the deficient denominators in the following pairs of equivalent fractions by inspection:

$$\frac{18}{\ \ }=\frac{3}{4};\ \frac{24}{\ \ }=\frac{6}{15};\ \frac{35}{\ \ }=\frac{5}{14};\ \frac{42}{\ \ }=\frac{7}{9};\ \frac{65}{\ \ }=\frac{13}{16}$$

3. Change $\frac{9}{5}$ to an equivalent fraction with 45 for denominator; that is, supply the deficient numerator.

$$\frac{\ \ }{45}=\frac{9}{5}$$

Suggestive Questions.—Is 45 a multiple of 5? How, then, may $\frac{9}{5}$ be changed to an equivalent fraction with 45 for denominator?

4. Supply the deficient numerators in the following pairs of equivalent fractions by inspection:

$$\frac{\quad}{18}=\frac{4}{3};\ \frac{\quad}{24}=\frac{15}{6};\ \frac{\quad}{35}=\frac{2}{5};\ \frac{\quad}{42}=\frac{5}{7};\ \frac{\quad}{65}=\frac{9}{13}$$

CASE III. *Where the Numerators or the Denominators are prime to each other; that is, where neither of the respective Numerators or Denominators are Factors the one of the other.*

1. Change $\frac{6}{18}$ to an equivalent fraction with 7 for numerator, so as to supply the deficient denominator.

$$\frac{7}{\quad}=\overset{a}{\frac{6}{18}}=\overset{b}{\frac{7\cdot6}{7\cdot18}}=\overset{c}{\frac{7\cdot6}{126}}=\overset{d}{\frac{7}{21}}$$

Suggestive Questions.—Are the fractions marked *a*, *b*, *c*, *d*, equivalent? How is fraction *a* changed to that of *b*? *b* to *c*? *c* to *d*?

2. Repéat the above operation, omitting fraction *c*. Repeat it once more, omitting *b* and *c*.

3. Change $\frac{18}{6}$ to an equivalent fraction with 7 for denominator, so as to supply the deficient numerator.

$$\frac{\quad}{7}=\overset{a}{\frac{18}{6}}=\overset{b}{\frac{7\cdot18}{7\cdot6}}=\overset{c}{\frac{126}{7\cdot6}}=\overset{d}{\frac{21}{7}}$$

Suggestive Questions.—Are the fractions, marked *a*, *b*, *c*, *d* equivalent? How is fraction *a* changed to that of *b*? *b* to *c*? *c* to *d*?

4. Repeat the 3d example, omitting fraction *c*. Repeat it once more, omitting *b* and *c*.

Remark.—Every fraction may be changed into an equivalent one, with a specified denominator or numerator, by the method elucidated above. But frequently it can be done in a more simple manner, as follows: by bringing it in under Case I. or II., where the given numerator or denominator, though neither factors nor multipliers the one of the other, still contain one or more common factors. Thus, 16 and 6 are neither factors nor multipliers one of the other, yet they contain a common factor, 2. An instance like this happens very commonly where the complete fraction is compound. Often, indeed, the whole process consists in *cancelling* such *factors in the complete frac-*

tion as are not to be found in the given term of the imperfect fraction, as shown in Examples 5 and 6, below.

5. Supply denominators in the following pairs of fractions: $\frac{16}{\ }=\frac{6}{9}$; $\frac{14}{\ }=\frac{21}{27}$; and prove by bringing the newly-completed fraction to the denomination of the given complete one.

Suggestive Questions.—How can $\frac{6}{9}$ be changed to $\frac{2}{3}$? Under what Case will such a change bring the first pair of fractions? Again: how may $\frac{6}{9}$ be changed to $\frac{48}{72}$? Under what Case will such a change bring them?

How can $\frac{21}{27}$ be changed to $\frac{7}{9}$? Under what Case will such change bring the second pair of fractions? Again: how may $\frac{21}{27}$ be changed to $\frac{42}{54}$? Under what Case will such a change bring them?

6. Supply numerators in the following pairs of fractions: $\frac{\ }{16}=\frac{9}{6}$; $\frac{\ }{14}=\frac{27}{21}$; and prove by bringing the newly completed fractions to the denomination of the given complete ones.

Suggestive Questions.—How can $\frac{9}{6}$ be changed to $\frac{3}{2}$? Under what Case will the change bring the first pair of fractions? How may $\frac{9}{6}$ be changed to $\frac{72}{48}$? Under what Case will such change bring them?

How can $\frac{27}{21}$ be changed to $\frac{9}{7}$? Under what Case will such change bring the second pair of fractions? How may $\frac{27}{21}$ be changed to $\frac{54}{42}$? Under what Case will such change bring them?

7. Supply the denominator in $\frac{16}{\ }=\frac{24}{6}$ of $\frac{20}{200}$ of $\frac{6}{8}$ of $\frac{4}{6}$; and prove by restoring the completed fraction to the denomination of the given complete fraction; that is, change the answer, $\frac{16}{80}$, into an equivalent fraction whose denomination shall be 57600 [=6•200•8•6]. Should the new numerator be 11520 [=24•20•6•4], it accords with the given numerator, and thus proves the process to be correct.

8. Supply the numerator in $\frac{\ }{16}=\frac{6}{24}$ of $\frac{200}{20}$ of $\frac{8}{6}$ of $\frac{6}{4}$, and prove by restoring the completed fraction to the denomination of the given complete fraction; that is, change the answer, $\frac{80}{16}$, into an equivalent fraction whose numerator shall be 57600 [=6•200•8•6].—Should the new denominator be 11520 [=24•20•6•4], it accords with the given denominator, and thus proves the process to be correct.

No. 7. $$\frac{2\cdot2\cdot2\cdot2}{\ }=\frac{2\cdot2\cdot2\cdot\cancel{3}}{2\cdot\cancel{3}}\times\frac{2\cdot\cancel{2}\cdot\cancel{5}}{\cancel{2}\cdot\cancel{2}\cdot\cancel{2}\cdot\cancel{5}\cdot5}\times\frac{\cancel{2}\cdot\cancel{3}}{2\cdot2\cdot2}\times\frac{\cancel{2}\cdot\cancel{2}}{\cancel{2}\cdot\cancel{3}}$$

By resolving both fractions as above into their prime factors, and striking out [or marking] all those that occur in both terms of the complete fraction, and are not found in the imperfect one, the required denominator appears to be $2\cdot 2\cdot 2\cdot 5=80$.

By following a similar process for No. 8, the required numerator would also be found to be 80.

But processes like these may be rendered much more simple by cancelling *while transferring* the compound fraction to the slate, reserving, of course, such factors as are found in the imperfect fraction. Thus,

No. 7. $\frac{16}{\quad}=\frac{4\cdot 1\cdot \not{6}\cdot 4}{1\cdot 10\cdot 8\cdot \not{6}}$ No. 8. $\frac{\quad}{16}=\frac{1\cdot 10\cdot 8\cdot \not{6}}{4\cdot 1\cdot \not{6}\cdot 4}$

giving 80 as before for the required deficient term, when the two superfluous 6s are struck out.

9. Supply the denominator in $\frac{18}{\quad}=\frac{14\cdot 27}{16\cdot 6}$, and prove.

Prime factors $\frac{2\cdot 3\cdot 3}{\quad}=\frac{2\cdot 7 \quad \times 3\cdot 3\cdot \bar{\not{3}}}{2\cdot 2\cdot 2\cdot 2\times 2\cdot \underline{\not{3}}}$

10. Supply the numerator in $\frac{\quad}{18}=\frac{16\cdot 6}{14\cdot 27}$, and prove.

Prime factors $\frac{\quad}{2\cdot 3\cdot 3}=\frac{2\cdot 2\cdot 2\cdot 2\times 2\cdot \bar{3}}{2\cdot 7 \quad \times 3\cdot 3\cdot \underline{3}}$

Here, cancelling the marked 3s, and dividing by the superfluous 7, gives $4\frac{4}{7}$ as the required deficient term.

11. Supply the denominator in $\frac{24}{\quad}=\frac{16\cdot 5\cdot 14}{18\cdot 35\cdot 16}$, and prove.

12. Supply the numerator in $\frac{\quad}{24}=\frac{18\cdot 35\cdot 16}{16\cdot 5\cdot 14}$, and prove.

Prime factors for No. 11, $\frac{2\cdot 2\cdot 2\cdot 3}{\quad}=\frac{2\cdot 2\cdot 2\cdot \not{2}\times \not{5}\times \not{2}\cdot \not{7}}{2\cdot 3\cdot 3\times \not{5}\cdot \not{7}\times 2\cdot 2\cdot \not{2}\cdot \not{2}}$

Here, as there is a 3 in the given term of the imperfect fraction, and none in the corresponding term of the perfect one, it must be supplied in both terms of the latter; that is, the fraction must be multiplied by 3, giving, after proper cancellation, as marked above, 216 for the required deficient terms. Or the operation may be more simply performed by cancelling whilst transferring to the slate, as under:

$$\frac{24}{\quad}=\frac{8\cdot 1\cdot \not{7}}{9\cdot \not{7}\cdot 8}$$

By striking out the two 7s, and inserting 3, as before, we have again 216 as the deficient term.

No. 12 does not differ from the above, except in the reversal of the terms.

13. Supply the denominator in $\frac{15}{\quad}=\frac{16\cdot14\cdot17}{25\cdot34\cdot28}$, and prove.

14. Supply the numerator in $\frac{\quad}{15}=\frac{25\cdot34\cdot28}{16\cdot14\cdot17}$, and prove.

Here, as the corresponding terms of the equivalent fractions have no common factor, no advantage would result from resolving them into their prime factors. But, as 17 and 34, and also 14 and 28, admit of cancellation, and as the two 2s thence arising may be cancelled with 16, the complete fraction thus becomes $\frac{4}{25}$, giving $93\frac{3}{4}$ for the deficient term.

15. Supply the denominator in $\frac{7}{\quad}=\frac{4\cdot2\cdot13}{5\cdot3\cdot17}$, and prove.

16. Supply the numerator in $\frac{\quad}{7}=\frac{5\cdot3\cdot17}{4\cdot2\cdot13}$, and prove.

Here, as no factor is common to the equivalent fractions, and the perfect fraction admits of no cancellation, the problem can only be solved as in Example 1 of this Case, namely, by placing 7 in both terms of the perfect fraction; that is, by multiplying by 7; and then dividing by the factors 4•2•13, or their product, 104.

Suggestive Questions.—How can both terms of the perfect fraction be multiplied by 7? How can the other factors of the numerator on the left, and of the denominator on the right (4•2•13), be removed, thus leaving 7 as sole numerator [or denominator]? Will this solve the question?

17. Supply denominators in each of the following pairs of equivalent fractions: $\frac{56}{\quad}=\frac{16\cdot7}{48\cdot3}$ (solely by division of perfect fractions); $\frac{18}{\quad}=\frac{27}{15}$ (division and multiplication); $\frac{20}{\quad}=\frac{12}{15}$ (division and multiplication); $\frac{90}{\quad}=\frac{1\frac{1}{8}}{\frac{3}{4}}$ (change mixed number to improper fraction, and $\frac{3}{4}$ to eighths.) Prove as before.

18. Supply the numerators in each of the following pairs of equivalent fractions: $\frac{\quad}{56}=\frac{48\cdot3}{16\cdot7}$ (solely by division of perfect fraction); $\frac{\quad}{18}=\frac{15}{27}$ (division and multiplication); $\frac{\quad}{20}=\frac{15}{12}$ (division and multiplication); $\frac{\quad}{90}=\frac{\frac{3}{4}}{1\frac{1}{8}}$ (change $\frac{3}{4}$ to eighths, and mixed number to improper fraction.) Prove as before.

19. Supply the denominator $\frac{5}{\quad}=\frac{1\cdot3\cdot4\cdot3\cdot2}{11\cdot4\cdot7\cdot5\cdot3}$, and prove.

20. Supply the numerator, $\frac{\quad}{5}=\frac{11\cdot4\cdot7\cdot5\cdot3}{1\cdot3\cdot4\cdot3\cdot2}$, and prove.

Multiply by 5, and divide by 3·4·6, or their product. Why?

21. Supply the denominators: $\frac{4}{\quad}=\frac{69\cdot16}{644\cdot3}$; $\frac{15}{\quad}=\frac{19}{16}$; $\frac{504}{\quad}=\frac{5}{9}$; $\frac{12}{\quad}=\frac{5\cdot18\cdot2}{6\cdot30\cdot5}$; $\frac{7}{\quad}=\frac{3\cdot6}{4\cdot18}$; $\frac{350}{\quad}=\frac{9\cdot15}{12\cdot6}$; $\frac{16}{\quad}=\frac{5}{6}$; $\frac{5}{\quad}=\frac{7}{686}$; $\frac{5}{\quad}=\frac{4}{320}$. Prove each as before.

22. Supply the numerators: $\frac{\quad}{4}=\frac{644\cdot3}{69\cdot16}$; $\frac{\quad}{15}=\frac{16}{19}$; $\frac{\quad}{504}=\frac{9}{5}$; $\frac{\quad}{12}=\frac{6\cdot30\cdot5}{5\cdot18\cdot2}$; $\frac{\quad}{7}=\frac{4\cdot18}{3\cdot6}$; $\frac{\quad}{350}=\frac{12\cdot6}{9\cdot15}$; $\frac{\quad}{16}=\frac{6}{5}$; $\frac{\quad}{5}=\frac{686}{7}$; $\frac{\quad}{5}=\frac{320}{4}$. Prove each as before.

23. Change $\frac{6}{24}$ of 10 of $\frac{4}{3}$ of $\frac{3}{4}$ to an equivalent fraction, with 16 as denominator, by inspection, and prove.

24. Change $\frac{24}{6}$ of $\frac{1}{10}$ of $\frac{3}{4}$ of $\frac{4}{3}$ to an equivalent fraction, with 16 as numerator, by inspection, and prove.

25. Change $\frac{21}{7}$ of $\frac{3}{14}$ to an equivalent fraction, with 56 as denominator, by inspection, and prove.

26. Change $\frac{7}{21}$ of $\frac{14}{3}$ to an equivalent fraction, with 56 as numerator, by inspection, and prove.

27. Change $\frac{3}{16}$ of $\frac{10304}{1104}$ to an equivalent fraction, with 4 as denominator, ascertaining the sole divisor by inspection, and prove.

28. Change $\frac{16}{3}$ of $\frac{1104}{10304}$ to an equivalent fraction, with 4 as numerator, ascertaining the sole divisor by inspection, and prove.

29. Change $\frac{100}{75}$ of $\frac{12}{8}$ to an equivalent fraction, with 4 as denominator, ascertaining the sole divisor by inspection, and prove.

30. Change $\frac{75}{100}$ of $\frac{8}{12}$ to an equivalent fraction, with 4 as numerator, ascertaining the sole divisor by inspection, and prove.

31. Change $\frac{3}{4}$ into an equivalent fraction, with $\frac{5}{11}$ as denominator, and prove by again resolving it into a simple fraction of lowest denomination.

32. Change $\frac{4}{3}$ into an equivalent fraction, with $\frac{11}{5}$ as numerator, and prove by again resolving it into a simple fraction of lowest denomination.

33. Change $\frac{3}{\frac{4}{5}}$ and $\frac{\frac{3}{4}}{5}$ into equivalent simple fractions, and

34. Change $\frac{\frac{4}{5}}{3}$ and $\frac{5}{\frac{3}{4}}$ into equivalent simple fractions, and prove by again resolving the

prove by again resolving the first into a fraction whose denominator shall be $\frac{4}{5}$, and the second into one whose denominator shall be 5.

35. Change $\dfrac{2}{\frac{5}{8}}$ and $\dfrac{\frac{1}{7}}{9}$ into equivalent simple fractions, and prove by again resolving the first into the denomination of $\frac{5}{8}$, and the second into the denomination of 9ths.

first into a fraction whose numerator shall be $\frac{4}{5}$, and the second into one whose numerator shall be 5.

36. Change $\dfrac{\frac{5}{8}}{2}$ and $\dfrac{9}{\frac{1}{7}}$ into equivalent simple fractions, and prove by again resolving the first into a fraction whose numerator shall be $\frac{5}{8}$, and the second into one whose numerator shall be 9.

☞ A remarkable property of numbers is developed by the above examples, namely, that any number whatever may be expressed by a common fraction, whose numerator (or whose denominator) shall consist of any specified number, whether whole or fractional.

II.—*Addition and Subtraction of Common Fractions.*

Suggestive Questions.—Can numbers of different denominations be added together or subtracted? See Oral Arithmetic, Chap. III., Sect. IV., p. 91. What previous operation is necessary?

Exercises for the Slate or Black-board.

1. Find the sum and the difference of $\frac{3}{5}$ and $\frac{4}{20}$.

Ans. Sum $\frac{4}{5}$; Diff. $\frac{2}{5}$.

2. Find the sum of the five following fractions, and prove the operation by subtracting from it the sum of the last four: $\frac{2}{7}$, $\frac{3}{5}$, $\frac{3}{4}$, $\frac{6}{35}$, $\frac{4}{70}$. What fraction will be left?

3. Add $\frac{5}{7}$, $\frac{3}{15}$, $\frac{5}{14}$, $\frac{6}{20}$, $\frac{9}{25}$, and prove by subtracting the sum of the first four.

4. Add $7\frac{2}{5}$, $3\frac{5}{8}$, $9\frac{3}{12}$, $5\frac{3}{7}$, and prove by subtracting the sum of the last three. [Change the fractional parts of these numbers to the same denomination, add them, carrying what integers they may contain to the given integers.]

5. What is the sum and difference of $\dfrac{3\frac{3}{8}}{6\frac{2}{5}}$ and $\dfrac{2\frac{1}{4}}{7\frac{1}{7}}$.

Ans. Sum, $\frac{5391}{6400}$; Diff., $\frac{1359}{6400}$.

6. Add $\frac{4}{5}$, $\frac{3}{8}$, $\frac{9}{15}$, and $\frac{7}{8}$, and prove the operation by giving

these 4 fractions a decimal form, and changing their sum into a common fraction, which, of course, will be equivalent to the sum of the four common fractions.

7. Add $\frac{3}{8}$, $\frac{3}{6}$, and $\frac{7}{28}$, and prove by the same process as in the last example.

III.—*Multiplication and Division of Common Fractions.*

[See Oral Arithmetic, Chap. III., Sect. V., p. 93.]

Exercises for the Slate or Black-board.

1. Multiply $\frac{5}{12}$ by 4, $\frac{8}{35}$ by 7, $\frac{2}{24}$ by 3, and $\frac{4}{49}$ by 7, all by division; or, which is the same thing, by cancelling what would otherwise be equal factors in both terms of the product. Prove the operations by performing them by multiplication, and bringing each fraction to its lowest denomination.

2. Divide $\frac{5}{6}$ by $\frac{5}{12}$, or, which is the same thing, change $\frac{\frac{5}{6}}{\frac{5}{12}}$ into a simple fraction, or integer. *Ans.* 2.

3. Divide $\frac{1}{8}$ by $\frac{1}{7}$; $\frac{2}{8}$ by $\frac{5}{8}$; $\frac{7}{8}$ by $\frac{7}{9}$.

☞ Remark, that in the above example, and in all other cases of division where the numerators (or where the denominators) are alike in the divisor and dividend, the quotient is formed by removing the like term in the dividend, and putting the unlike term of the divisor in its place. Thus, $\frac{1}{8}\div\frac{1}{7}=\frac{7}{8}$; $\frac{2}{8}\div\frac{5}{8}=\frac{2}{5}$. Why is this so? Perform the above example in the usual method, and see.

4. Divide $\frac{14}{99}$ by $\frac{17}{99}$; $\frac{17}{45}$ by $\frac{29}{45}$; $\frac{33}{47}$ by $\frac{33}{25}$; $\frac{35}{79}$ by $\frac{24}{79}$; each at a glance.

5. Divide $\frac{5}{8}$ by $\frac{15}{32}$, by cancelling what would otherwise be equal factors in the quotient. *Solution.* $\frac{5}{8}\div\frac{15}{32}=\frac{5}{8}\times\frac{32}{15}=\frac{5\cdot32}{8\cdot15}=\frac{4}{3}$; or, omitting the superfluous steps, $\frac{5\cdot32}{8\cdot15}=\frac{4}{3}$.

6. Divide $\frac{14}{36}$ by $\frac{21}{27}$ by inspection, first cancelling equal factors in the numerators, and equal factors in the denominators. Why? *Ans.* $\frac{1}{2}$.

7. Divide $\frac{35}{81}$ by $\frac{7}{9}$; $\frac{5}{24}$ by $\frac{15}{16}$; $\frac{6}{57}$ by $\frac{4}{19}$; $\frac{12}{125}$ by $\frac{4}{25}$; by cancelling as in the preceding example. Prove by reproducing the dividend as in division of integers; that is, by considering the divisor and quotient as factors of the dividend.

8. Change $\frac{3}{5}$ of $\frac{5}{8}$ of $\frac{7}{6}$ of $\frac{2}{3}$ of $\frac{8}{7}$ of $\frac{1}{2}$, by cancellation, into a simple fraction, by inspection, and prove by dividing the

result by each of these several factors except $\frac{5}{8}$, employing cancellation in the division wherever practicable.

9. Multiply $6\frac{3}{4}$ by $4\frac{2}{3}$; $34\frac{2}{3}$ by 6; $27\frac{3}{4}$ by 8; $88\frac{4}{5}$ by $\frac{5}{12}$.

It is usual to change mixed numbers into improper fractions before multiplying them. But, in some of these, as well as in many other cases, it will be found quite as easy, and much shorter, to multiply without any such change. For example,

$$\begin{array}{r} 6\frac{3}{4} \\ 4\frac{2}{3} \\ \hline 27\ =4\times6\frac{3}{4} \\ 4\frac{1}{2}=\frac{2}{3}\times6\frac{3}{4} \\ \hline 31\frac{1}{2}=4\frac{2}{3}\times6\frac{3}{4} \\ \hline \end{array}$$

But in division, where the divisor is a mixed number, and the dividend is either a mixed number or an integer, it will be found most convenient to change both to the form of improper fractions.

10. Divide $3\frac{3}{4}$ by $4\frac{5}{8}$; also 16 by $5\frac{2}{7}$; and prove by multiplication.

RAPID AND CONCISE METHODS OF COMPUTING WITH COMMON FRACTIONS.

In all treatises on arithmetic, the pupil is directed to bring fractions that are to be added or subtracted to a common *denominator.* But, when the fractions do not exceed two in number, these operations can frequently be performed much more rapidly and quite as correctly by bringing them to a common *numerator*, as will appear from the following exemplifications and exercises:

I.—*Addition by a Common Numerator.*

Exemplification for the Black-board.

1. Find the sum of $\frac{1}{7}$ and $\frac{1}{5}$.

$$(\tfrac{1}{7}+\tfrac{1}{5})=(\tfrac{5}{35}+\tfrac{7}{35})=\tfrac{12}{35}.$$

Suggestive Questions.—1. Compare the new numerator with the given denominators, and say what relation it bears to them.

Is it their sum, difference, product, or quotient? The given numerators remaining the same (1), would the new numerator be the *sum* of the given denominators, whatever might be their numbers? 2. What relation does the new denominator bear to the given denominators; is it their sum, difference, product, or quotient? Would it be so whatever were the numbers of the given denominators? How, then, can any two fractions of different denominations be added by simple inspection if their numerators be 1?

2. Add $\frac{1}{3}$ and $\frac{1}{4}$ by inspection; that is, by using the sum and product of their denominators for the sum of the two fractions, and prove by addition in the old method.

3. Add, in the same manner, $\frac{1}{4}$ and $\frac{1}{7}$; $\frac{1}{5}+\frac{1}{8}$; $\frac{1}{3}+\frac{1}{8}$; $\frac{1}{7}+\frac{1}{2}$; $\frac{1}{14}+\frac{1}{9}$; and prove each by addition in the old method.

Suggestive Questions.—If the sum of 1 seventh and 1 fifth be 12 thirty-fifths, what will be the sum of 4 sevenths and 4 fifths; that is, how many times will the amount be greater than the other? How many times will the sum of 3 sevenths and 3 fifths be greater than the sum of 1 seventh and 1 fifth? How, then, can you add two fractions by inspection, when their equal numerators are greater than 1?

4. Add, by inspection, $\frac{4}{5}$ and $\frac{4}{9}$ [$\frac{14}{45}\times4=\frac{56}{45}$], or, omitting superfluous figures [$\frac{4}{5}+\frac{4}{9}=\frac{56}{45}$].

5. Add $\frac{3}{7}$ and $\frac{3}{4}$ by inspection, and prove by addition in the old method.

6. Add and prove in the same manner, $\frac{4}{5}$ and $\frac{4}{25}$; $\frac{4}{7}$ and $\frac{4}{13}$; $\frac{4}{9}$ and $\frac{4}{5}$; $\frac{3}{8}$ and $\frac{3}{4}$; $\frac{7}{16}$ and $\frac{7}{10}$; $\frac{6}{7}$ and $\frac{6}{15}$.

Remark.—When both terms are unlike in the fractions to be added, a single glance will generally show whether it be easier to make the numerators or denominators common; and, by having a choice, it will very rarely be necessary to change both fractions before they are added. The numerators may be altered as follows: $(\frac{3}{7}+\frac{9}{5})=(\frac{9}{21}+\frac{9}{5})$, by multiplication; $(\frac{4}{5}+\frac{2}{7})=(\frac{4}{5}+\frac{4}{14})$; $(\frac{5}{8}+\frac{45}{69})=(\frac{15}{24}+\frac{15}{23})$ by division. In all these cases, it will be perceived that it is a more simple operation to change the numerators than the denominators.

7. Add, by inspection, $\frac{2}{5}$ and $\frac{8}{9}$, and prove by the old method.

8. Add, by inspection, $\frac{1}{2}$ and $\frac{6}{7}$; $\frac{2}{5}$ and $\frac{4}{18}$; $\frac{2}{3}$ and $\frac{3}{4}$ (changing numerators to 6); $\frac{3}{5}$ and $\frac{21}{19}$; and $\frac{5}{6}$ and $\frac{10}{16}$.

II.—*Subtraction by a Common Numerator.*

[Subtraction and Addition by this method differ in no respect, save that the *difference*, instead of the *sum* of the given denominators, constitutes the *new numerator*.]

1. Find the difference of $\frac{1}{3}$ and $\frac{1}{4}$ by inspection, and prove by the old method.

2. Find the difference of $\frac{1}{7}$ and $\frac{1}{8}$; $\frac{1}{9}$ and $\frac{1}{2}$; $\frac{1}{16}$ and $\frac{1}{14}$; $\frac{1}{3}$ and $\frac{1}{11}$, as above, and prove.

3. Find the difference of $\frac{2}{5}$ and $\frac{2}{3}$, $\frac{5}{7}$ and $\frac{5}{11}$, $\frac{7}{13}$ and $\frac{7}{9}$, $\frac{4}{5}$ and $\frac{4}{7}$, as above, and prove.

4. Find the difference of $\frac{2}{5}$ and $\frac{3}{4}$, $\frac{1}{6}$ and $\frac{5}{4}$, $\frac{2}{8}$ and $\frac{12}{19}$, $\frac{3}{7}$ and $\frac{4}{9}$, as above, and prove.

III.—*Multiplication.*

Where the numerator of one factor is equal to the denominator of the other.

Exemplification for the Black-board.

1. What are the several products of $\frac{4}{5}$ by $\frac{5}{17}$, and $\frac{6}{13}$ by $\frac{5}{6}$, in their lowest denominations?

$$\frac{4}{5}\times\frac{5}{17}=\frac{4}{17}. \qquad\qquad \frac{6}{13}\times\frac{5}{6}=\frac{5}{13}.$$

Suggestive Questions.—Which of the numbers in the two given factors are retained in the product; the equal or the unequal? Do the unequal numbers retain their respective places, or are they reversed in the product?

2. Multiply $\frac{5}{9}$ by $\frac{3}{5}$ by striking out the equal numbers, and prove by division.

3. Multiply $\frac{2}{5}$ by $\frac{5}{7}$, $\frac{3}{8}$ by $\frac{8}{15}$, $\frac{4}{7}$ by $\frac{7}{12}$, $\frac{9}{17}$ by $\frac{17}{26}$, by inspection, and prove by division.

When all the numbers in the two factors are unequal.

Exemplification for the Black-board.

1. Multiply $\frac{5}{6}$ by $\frac{3}{17}$, $\frac{2}{5}$ by $\frac{15}{24}$, $\frac{3}{4}$ by $\frac{7}{5}$, by inspection, and prove by division.

Suggestive Questions.—How can $\frac{3}{17}$ be changed to a fraction with 6 for a numerator? How can $\frac{15}{24}$ be changed to a fraction with 5 for a numerator? How can $\frac{7}{5}$ be changed to a fraction with 3 for a denominator?

2. Multiply $\frac{9}{10}$ by $\frac{12}{27}$, $\frac{6}{7}$ by $\frac{1}{8}$, $\frac{2}{5}$ by $\frac{3}{7}$, $\frac{6}{13}$ by $\frac{4}{9}$, by inspection, and prove by division.

IV.—*Division.*

Where the two numerators or the two denominators are equal.

1. Divide, in the usual manner, $\frac{2}{5}$ by $\frac{2}{8}$, and $\frac{3}{7}$ by $\frac{8}{7}$.

Suggestive Questions —Comparing the numbers in the given divisor and dividend of both problems with those of the quotients: which are removed, those that are equal, or those that are unequal? Where is the remaining number of the divisor found in the quotient? *Ans.* In the vacant place in the —.

2. Divide $\frac{5}{6}$ by $\frac{7}{6}$ by inspection, and prove by multiplication by inspection.

3. Divide $\frac{3}{5}$ by $\frac{3}{7}$, $\frac{5}{8}$ by $\frac{3}{8}$, $\frac{14}{15}$ by $\frac{14}{17}$, $\frac{9}{17}$ by $\frac{7}{17}$, $\frac{23}{26}$ by $\frac{23}{24}$, $\frac{15}{19}$ by $\frac{4}{19}$, all by inspection, and prove as above.

Where all the given terms are unequal.

1. Divide $\frac{3}{5}$ by $\frac{6}{7}$, $\frac{4}{7}$ by $\frac{12}{21}$, $\frac{3}{8}$ by $\frac{5}{9}$, by inspection, and prove by multiplication.

Suggestive Questions.—How can $\frac{3}{5}$ be changed to an equivalent fraction, with 6 for a numerator? How can $\frac{12}{21}$ be changed to an equivalent fraction, with 7 for a denominator? How can $\frac{3}{8}$ be changed to an equivalent fraction, with 5 for a numerator?

2. Divide $\frac{4}{15}$ by $\frac{8}{17}$, $\frac{2}{5}$ by $\frac{8}{24}$, $\frac{5}{7}$ by $\frac{11}{24}$, by inspection, and prove as above.

3. Divide $\frac{5}{6}$ by $\frac{4}{9}$, $\frac{3}{4}$ by $\frac{2}{3}$, $\frac{5}{8}$ by $\frac{35}{63}$, by inspection, and prove.

V.—*Multiplication and Division by Addition or Subtraction.*

Exemplifications for the Black-board.

1. Multiply 32 by $\frac{15}{16}$.

Suggestive Questions.—If 32 were to be multiplied by $\frac{16}{16}$ (=1) instead of $\frac{15}{16}$, how much too large would the number be? *Ans.* $\frac{1}{16}$ too large. Then what portion of 32 must be subtracted to make the process correct?

$$32\times\tfrac{15}{16}=32\times\tfrac{16}{16}\text{ (or 1)}-\tfrac{1}{16}\text{; therefore,}$$

$32\times\frac{16}{16}=32$
$32\times\frac{1}{16}=\ 2$ or, omitting superfluous figures, $32\times\frac{15}{16}=30$.

$32\times\frac{15}{16}=30$

2. Multiply 54 by $\frac{12}{14}$. $\frac{2}{14}$ of $54=7\frac{5}{7}$; therefore,

$$54\times\frac{14}{14}=54$$
$$54\times\frac{2}{14}=\ 7\frac{5}{7}$$
$$54\times\frac{12}{14}=46\frac{2}{7}$$

3. Multiply $7\frac{3}{8}$ by $\frac{4}{5}$.

$$7\frac{3}{8}\times\frac{5}{5}=7\frac{3}{8}$$
$$7\frac{3}{8}\times\frac{1}{5}=1\frac{19}{40}$$
$$7\frac{3}{8}\times\frac{4}{5}=5\frac{36}{40} \text{ or } 5\frac{9}{10}$$

4. Multiply $24\frac{6}{11}$ by $\frac{7}{16}$.

$$24\frac{6}{11}\times\frac{8}{16}=12\ \frac{3}{11}$$
$$24\frac{6}{11}\times\frac{1}{16}=\ 1\frac{94}{176}$$
$$24\frac{6}{11}\times\frac{7}{16}=10\frac{130}{176}$$

5. Multiply each of the following numbers, namely, 216, 325, 84, 125, 64, 236, 27, $\frac{5}{8}$, $7\frac{3}{4}$, $9\frac{1}{16}$, severally by $\frac{11}{12}$, by $\frac{24}{25}$, by $\frac{23}{25}$, by $\frac{7}{8}$, by $\frac{6}{8}$, by $\frac{4}{5}$, by $\frac{3}{5}$, by $\frac{14}{15}$, by $\frac{13}{15}$, by $\frac{9}{10}$, by $\frac{7}{8}$, by $\frac{6}{5}$, by $\frac{11}{10}$, by $\frac{26}{25}$. As the last three factors are improper fractions, the process consists of addition. Why? Try the problem, and see. Prove each of the above by multiplying in the usual manner.

Exemplifications for the Black-board.

6. Divide 42 by $\frac{7}{8}$. $42\times\frac{8}{7}=42+\frac{1}{7}$ of $42=48$.
7. Divide 37 by $\frac{5}{7}$. $37\times\frac{7}{5}=37+\frac{2}{5}$ of $37=47\frac{4}{5}$.
8. Divide $2\frac{1}{7}$ by $\frac{8}{9}$. $2\frac{1}{7}\times\frac{9}{8}=2\frac{1}{7}+\frac{1}{8}$ of $2\frac{1}{7}=2\frac{23}{56}$.
9. Divide 24 by $\frac{7}{6}$. $24\times\frac{6}{7}=24-\frac{1}{7}$ of $24=20\frac{4}{7}$.

10. Divide each of the following numbers, namely, 75, 254, 36, $5\frac{1}{4}$, 28, 60, $3\frac{5}{8}$, $7\frac{3}{4}$, 92, 46, and 316, severally by $\frac{7}{8}$, by $\frac{15}{16}$, by $\frac{17}{16}$, by $\frac{8}{9}$, by $\frac{2}{9}$ $(\frac{1}{3}-\frac{1}{9})$, by $\frac{3}{5}$, by $\frac{14}{16}$, by $\frac{7}{16}$ $(\frac{1}{2}-\frac{1}{16})$, by $\frac{9}{16}$ $(\frac{8}{16}+\frac{1}{16})$, by $\frac{49}{50}$, by $\frac{24}{50}$ $(\frac{25}{50}-\frac{1}{50})$, by $\frac{26}{50}$ $(\frac{25}{50}+\frac{1}{50})$, by $\frac{25}{27}$, and by $\frac{29}{27}$. Prove each by division by the old method.

11. Perform problems from 1 to 10, immediately above, by inspection; that is, omit all superfluous steps, as follows:

No. 1. $32\times\frac{15}{16}=30$. No. 2. $54\times\frac{12}{14}=46\frac{2}{7}$.
No. 3. $7\frac{3}{8}\times\frac{4}{5}=5\frac{9}{10}$. No. 4. $26\frac{6}{11}\times\frac{7}{16}=10\frac{130}{176}$, &c.

The above rapid and concise methods will furnish excellent exercise for the pupil, giving employment both to his thinking and active faculties. Some few of the computations may be found more operose than by the usual methods; but a little practice will enable the student at a glance to tell which will be the simplest mode, and to choose accordingly.

Involution and Evolution of Common Fractions.

1. What is a square? See Involution, Def. 2, p. 162. How much is $\frac{3}{4}$ of $\frac{3}{4}$? Is $\frac{9}{16}$, then, the square of $\frac{3}{4}$? If a fraction, then, be squared by multiplying each of its terms by itself, how can the square root of a fraction be found? *Ans.* By dividing each of its terms into —— equal factors, or finding the square root of each term. What, then, is the square root of $\frac{9}{16}$?

2. What is a cube? See Involution, Def. 4, p. 163. How much is $\frac{3}{4}$ of $\frac{3}{4}$ of $\frac{3}{4}$? Is $\frac{27}{64}$, then, the cube of $\frac{3}{4}$? How is a fraction involved to the third power, or cubed, then? If a fraction, then, be involved to the third power, or cubed, by multiplying each of its terms twice by itself, how can the cube root of a fraction be found? *Ans.* By dividing each of its terms into —— equal factors, or by finding the cube root of each term. What, then, is the cube root of $\frac{27}{64}$?

Remark.—Fractions should always be placed in their most simple form before attempting to find their roots; that is, compound fractions should be changed to simple ones, mixed numbers to improper fractions, and every fraction should be in its lowest terms, as any other course would unnecessarily multiply figures. If either term has no exact root, an approximation may be found by putting the common fraction in a decimal form.

3. Find the square root of each of the following fractions: $\frac{25}{81}$, $\frac{16}{49}$, $\frac{36}{121}$, and prove by involution.

4. Find the cube roots of $\frac{125}{4096}$, $\frac{2744}{13824}$, $\frac{8}{125}$, and $\frac{216}{729}$, and prove by involution.

5. Find the square roots of $\frac{6}{8}$ of $\frac{24}{72}$; also of $\frac{4}{5}$ of $\frac{9}{20}$, and prove by involution.

6. Find the cube root of $\frac{3375}{2744}$, $\frac{8}{1728}$, and $\frac{27}{64}$, and prove by involution.

Remark.—Sometimes the exact square or cube root of a common fraction, both of whose terms are surds, can be found by changing their form. Thus, $\sqrt[3]{\frac{16}{128}}=\sqrt[3]{\frac{64}{512}}=\frac{4}{8}=\frac{1}{2}$; and $\sqrt{\frac{2}{8}}=\sqrt{\frac{1}{4}}=\frac{1}{2}$.

7. Find the exact square root of $\frac{1}{2\frac{1}{4}}$, and the exact cube root of $\frac{81}{375}$, and prove by involution. *Ans.* $\frac{2}{3}$ and $\frac{3}{5}$.

8. Find the exact square root of $\frac{6\frac{1}{8}}{6\frac{1}{4}}$, and the exact cube root of $\frac{648}{2187}$.

Practical Exercises on Fractional Quantities.

1. A tradesman, taking an account of stock, desired his clerk to ascertain the amount of the following remnants of calico: $3\frac{1}{6}$ yards, at $6\frac{1}{4}$ cents per yard; $2\frac{1}{8}$ yards, at $8\frac{1}{3}$ cents; $5\frac{1}{4}$ yards, at 9 cents; and $2\frac{1}{8}$ yards, at $7\frac{1}{2}$ cents. What was the amount? *Ans.* \$1‘00$\frac{1}{16}$.

2. A lady bought $5\frac{7}{8}$ yards of cotton. How much would remain after using $2\frac{3}{5}$ yards? *Ans.* $3\frac{11}{40}$.

3. A merchant, owning $\frac{3}{5}$ of a ship, sold $\frac{5}{8}$ of his share for \$4500. What portion of the ship did he sell? and what portion remained in his possession? *Ans.* $\frac{3}{8}$; and $\frac{9}{40}$.

4. If $\frac{5}{8}$ of $\frac{3}{5}$ of a ship be worth \$4500, what is $\frac{1}{8}$ of the vessel worth? and what is the value of the whole ship at that rate? *Ans. to the last question*, \$12,000.

5. The purchaser of the part of the vessel mentioned above, wishing to have the whole in his own hands, offered the owners to take the remainder of the ship at the same rate (that is, at \$4500 for $\frac{5}{8}$ of $\frac{3}{5}$). What would be the amount of this second purchase? *Ans.* \$7500.

6. If 4 yards of cloth cost a certain sum, what portion of that sum will 1 yard cost? If 1 yard cost $\frac{1}{4}$ of the sum, what portion of it will 7 yards cost? By what fraction, then, must the price of 4 yards be multiplied to ascertain the price of 7

yards? Then, if 4 yards of cloth cost $12, what will 7 yards of the same cloth cost? Of what cancellation is $\frac{7}{4}$ of 12 susceptible? Cancel, and ascertain the result by inspection.

7. If 7 yards of cloth cost a certain sum, what portion of that sum will 1 yard cost? What portion, then, will 4 yards cost? By what fraction, then, must the price of 7 yards of cloth be multiplied to ascertain the price of 4 yards? If 7 yards of cloth, then, cost $21, what will 4 yards cost? Of what cancellation is $\frac{4}{7}$ of 21 susceptible? Cancel, and ascertain the result by inspection.

☞ In the last two examples, 1 yard requires LESS (that is, costs less), than 4 or 7 yards, and is represented by $\frac{1}{4}$ or $\frac{1}{7}$. But it frequently happens that 1 requires MORE than a larger number, and consequently is represented by an improper fraction, as $\frac{4}{1}$ or $\frac{7}{1}$, as will plainly appear from the two following examples. As this inversion, as it is called by mathematicians, frequently occurs in computations of this nature, it is requisite that pupils, when forming the factor fraction, should ask themselves LESS or MORE? at least, until the subject has become very familiar to them. This question is inserted into a few of the examples that follow, to show how and where it should be introduced. The pupil himself should introduce it into *all* the others.

8. If a piece of work can be finished in a certain number of days by 5 men, in what time can it be done by 1 man? In LESS or MORE time; that is, in $\frac{1}{5}$ or $\frac{5}{1}$ of the time? If 1 man require $\frac{5}{1}$ times longer to finish it than 5 men, how much time will 6 men require? LESS or MORE; that is, $\frac{1}{6}$ or $\frac{6}{1}$ the time? What is $\frac{1}{6}$ of $\frac{5}{1}$? By what fraction, then, must the time required by 5 men be multiplied to give the time required by 6 men? If, then, 5 men can do a piece of work in 10 days, in what time will 6 men perform it? Ascertain the result by inspection, as follows:

$$(\tfrac{5}{6}\times10)=8\tfrac{1}{3} \text{ days.}$$

9. If 6 men can do a piece of work in a certain number of days, in what time can 1 man do it? LESS or MORE? in $\frac{1}{6}$ or $\frac{6}{1}$? If 1 man require $\frac{6}{1}$ longer than 6 men, what time will be necessary for 5 men? LESS or MORE? $\frac{1}{5}$ or $\frac{5}{1}$? What is $\frac{1}{5}$ of $\frac{6}{1}$? By what fraction, then, must the time required by 6 men be multiplied to give the time required by 5 men? If, then,

6 men can do a piece of work in $8\frac{1}{3}$ days, in what time can 5 men perform it? Cancel, and ascertain the result by inspection, as follows:

$$(\tfrac{6}{5}\times\tfrac{25}{3})=2\times5=10 \text{ days.}$$

10. If 8 men, in a certain time, can make 24 rods of wall, how many men will be required for 18 rods in the same time? LESS or MORE? $\frac{18}{24}$ or $\frac{24}{18}$ of 8? Cancel, and ascertain the result by inspection.

11. If 4 lbs. of tea cost $2‘50, what will be the cost of 24 lbs.? LESS or MORE?

12. If 24 lbs. of tea cost $15, what will be the cost of 4 lbs.? LESS or MORE?

13. If 16 lbs. of sugar cost $1‘28, what will 54 lbs. cost?

14. If 54 lbs. of sugar cost $4‘32, what will 16 lbs. cost?

15. If 6 bushels of turnips cost $1‘50, what will 33 bushels cost?

16. If 33 bushels of turnips cost $8‘25, what will 6 bushels cost?

17. How many men must be employed to finish a piece of work in 8 days, if 4 men can do it in 24 days?

18. If 12 men can finish a piece of work in 8 days, how many men will be able to finish it in 24 days?

19. If 4 men can do a piece of work in 24 days, in how many days can 12 men do it?

20. If 12 men can perform a piece of work in 8 days, in how many days can 4 men do it?

21. If 15 cords of wood cost $50, what would 27 cords of the same wood cost?

22. If 27 cords of wood cost $90, what would be the cost of 15 cords of the same wood?

23. If 60 bushels of potatoes can be exchanged for 25 bushels of rye, how much rye can be had for 200 bushels of potatoes?

24. If $83\frac{1}{3}$ bushels of rye can be exchanged for 200 bushels of potatoes, how much rye can be had for 60 bushels of potatoes?

25. If 6 men can cut 24 acres of grain in 5 days, in how many days could 4 men have cut the same field?

26. If 4 men take $7\frac{1}{2}$ days to cut a certain field of grain, in what time could 6 men cut it?

19*

27. If 5 days be required for 6 men to reap a certain field, how many men could reap it in $7\frac{1}{2}$ days?

28. If $\frac{4}{5}$ of a bushel of grain cost $\$\frac{75}{100}$, what will $18\frac{2}{5}$ bushels cost? Simplify the money term by performing the division indicated, and the other two terms by multiplying each by 5, and dividing by 4. Why?

29. If $18\frac{2}{5}$ bushels of grain cost $\$17\frac{1}{4}$, how much will $\frac{4}{5}$ of a bushel cost? Simplify as above.

30. Bought 5000 planks, of 15 feet long, 1 foot wide, and $2\frac{1}{2}$ inches thick. To how many planks of $12\frac{1}{2}$ feet long, 1 foot wide, and $1\frac{3}{4}$ inches thick, are they equivalent? Make the fractional quantities disappear, by quadrupling the length and thickness of the planks. Why?

31. A carpenter exchanged $8571\frac{3}{7}$ planks, each $12\frac{1}{2}$ feet long, 1 foot wide, and $1\frac{3}{4}$ inches thick, for some that were 15 feet long, 1 foot wide, and $2\frac{1}{2}$ inches thick. How many ought he to receive?

32. If 8 men, in a certain time, make 24 rods of wall, how many men will be required to build 18 rods in the same time? LESS or MORE? $\frac{18}{24}$ or $\frac{24}{18}$ of 8? Again; if 8 men can make the 18 rods in 6 days, how many men can make it in 3 days? LESS or MORE? $\frac{3}{6}$ or $\frac{6}{3}$? Now, if a change in the *length* of wall requires $\frac{18}{24}$ the number of men, and the change of *time* $\frac{6}{3}$ the number, what are the factors of both changes, as in the following statement?

If 8 men can build 24 rods of wall in 6 days, how many men can build 18 rods in 3 days? Resolve into primes, and cancel as follows:

$$\left(\overset{\cancel{2}\cdot\cancel{2}\cdot\cancel{2}}{8} \times \frac{\overset{2\cdot\cancel{3}\cdot\cancel{3}}{18}}{\underset{\cancel{2}\cdot\cancel{2}\cdot\cancel{2}\cdot\cancel{3}}{24}} \times \frac{\overset{2\cdot3}{6}}{\underset{\cancel{3}}{\cancel{3}}}\right) = 12 \text{ men.}$$

☞ Observe here that the *number of men* depends upon two circumstances,—the number of rods, and the number of days.

33. If 12 men can build 18 rods of wall in 3 days, *what number of rods* can be built by 8 men in 6 days? Resolve and cancel. The number of rods will be affected by what fraction? To know which term is numerator, ask, for each fraction, MORE or LESS rods?

$$(18\times\frac{8}{12}\times\frac{6}{3})=4\times6=24 \text{ rods.}$$

☞ Look at the statement within parentheses, and say whether the fractions could not as easily be cancelled before writing them, so that the computation could be readily performed by inspection merely. Thus,

$$(18\times\frac{2}{3}\times2)=24 \text{ rods, as before.}$$

34. If 18 rods of wall can be built by 12 men in 3 days, in *what time* can 8 men build 24 rods?

35. If 24 rods of wall can be built by 8 men in 6 days, how many rods can be built by 12 men in 3 days?

36. If 6 men build a wall 20 feet long, 6 feet high, and 4 feet thick, in 16 days, in *how many days* will 24 men build one 200 feet long, 8 feet high, and 6 feet thick? The number of days is modified by the number of men, and by the length, the height, and the thickness of the wall. Cancel the four fractions mentally before writing them, and place them together in the form of a simple fraction, the better to bring them under the eye.

By inspection, $\left(16\times\frac{\overset{\not{2}\cdot5}{1\cdot\not{1}\not{0}\cdot\not{4}\cdot\not{3}}}{\not{4}\cdot1\cdot\not{3}\cdot\not{2}}\right)=80$ days, the compound fraction being equal to 5.

37. If a wall 200 feet long, 8 feet high, and 6 feet thick, can be built by 24 men in 80 days, *how many men* will build one 20 feet long, 6 feet high, and 4 feet thick, in 16 days? By cancelling before writing, we have

$$\left(24\times\frac{1\cdot3\cdot2\cdot5}{10\cdot4\cdot3\cdot1}\right)=6 \text{ men.}$$

Here the factors 3·2·5 balance the divisors 3·10, leaving 4 as divisor to the 24=6.

38. If 24 men can build a wall 200 feet long, 8 feet high, and 6 feet thick, in 80 days, in *what time* will 6 men build one 20 feet long, 6 feet high, and 4 feet thick?

39. If 6 men can build a wall 20 feet long, 6 feet high, and

4 feet thick, in 16 days, how many men will be necessary to build one 200 feet long, 8 feet high, and 6 feet thick, in 80 days?

40. If 60 bushels of oats serve for 15 horses for 16 days, how long will 24 bushels last 8 horses at the same rate? Cancel whilst writing the fraction, and ascertain the result by inspection.

41. If 24 bushels of oats serve for 8 horses for 12 days, how many bushels will be wanted for 15 horses for 16 days at the same rate? By inspection, after cancelling.

42. If 8 horses eat 24 bushels of oats in 12 days, how many horses may be fed on 60 bushels for 16 days at the same rate?

43. If 15 horses require 60 bushels of oats for 16 days, how many horses can be fed on 24 bushels for 12 days at the same rate?

44. If the interest on $100 for 12 months be $6, what will be the interest of $250 for 8 months?

45. If the interest of $250 for 8 months be $10, what will be the interest of $100 for 12 months?

46. What principal (or sum lent at interest) will gain $10 in 8 months, if $100 gain $6 in 12 months?

47. The sum of $250 was put at interest until it had gained $10, at the rate of $6 interest for every $100 for 12 months. How long was the $250 lent?

48. If $100 gain $6 interest in 12 months, how much will $356 gain in 4 months?

49. If $6 be the interest of $100 for 12 months, in what time will $356 gain $7'12.

50. What principal will gain $7'12 in 4 months, if $6 be the interest for 12 months for $100?

51. What will be the interest of $100 for 12 months, if $356 gain $7'12 in 4 months?

52. What will be the interest of $450 for 24 days, if $100 gain $6 in one year?

☞ In calculating interest for days, it is customary to consider the year as 360 days, and the months as 30 days each, unless the months are designated.

53. If the interest of $450 for 24 days be $1'80, what will be the interest of $100 for one year?

54. Nine merchants associated to build a steamboat, for which they advanced equal sums of money. After a while, one of the partners purchased the shares of 6 of the others;

but afterwards, being pressed for money, sold to a friend one-fifth of his entire right. What share of the boat did he retain? what share did he sell? and what would be the dividend for the last purchaser, if the boat cleared $45,000. *Ans.* $7000.

SECTION II.—*Determinate Fractions, or Compound Numbers.*

THE different methods of increasing and decreasing integers as well as decimal and common fractions, have now, it is believed, been sufficiently exemplified. A class of numbers, however, remains to be noticed, called by some writers DETERMINATE FRACTIONS, from the circumstance of being *limited* in number and variety of expression, while other fractions are unlimited in both. But, by most arithmeticians, this class is called, rather inappropriately, COMPOUND NUMBERS. These relate chiefly to the division and subdivision of weights and measures, coins, and time. The total want of uniformity in these divisions makes them exceedingly complex and perplexing. Old habits, unconnected in their origin, have introduced such a variety, that TABLES of these subdivisions have become absolutely necessary. Strictly speaking, these tables are definitions. They will be found below; and, a knowledge of them being essential to the business of life, they should be thoroughly committed to memory.

In France, during the first revolution, a system of weights and measures was established on the *decimal* scale, which was, of course, exceedingly simple and intelligible; and the government of Great Britain is at present engaged in a similar undertaking. The coins of the United States have also been arranged on this scale. But we have not derived all the advantages we might from this simple system, owing to the tenacity with which the people have clung to their old habits of reckoning by pounds, shillings, and pence,—denominations sufficiently perplexing anywhere, but particularly in the United States, as they are not exactly represented by the coins, and as the same denominations possess a different value in the different states. This inconvenience would probably have disappeared long ago, but for the foreign coins which mingle in our circulation. Many attempts have been made in Congress to simplify the system of weights and measures, but hitherto without effect. Such an enterprise, indeed, does not properly belong to an individual nation. To be effectual and thorough, it should be executed by a commission representing all the commercial powers. The movement now making in Great Britain, it is to be hoped, will be followed up by a general congress of scientific men, for the establishment of a system coëxtensive with the field of trade. Meanwhile our youth must be content to waste their time and burden their memory with these unconnected and unphilosophical divisions of the unit of length and capacity.

TABLES OF COIN, WEIGHT, AND MEASURE.

I. Coin.

1. *Federal Money.*

10 mills (marked *m.*) make .	1 cent.	*c.*
10 cents	1 dime.	*d.*
10 dimes	1 dollar.	$.
10 dollars	1 eagle.	*e.*

m	*c*	*d*	$	*e*
10=	1			
100=	10=	1		
1000=	100=	10=	1	
10000=	1000=	100=	10=	1

2. *English or Sterling Money.*

4 farthings (*q.*) make . .	1 penny.	*d.*
12 pence	1 shilling.	*s.*
20 shillings	1 pound.	£.

q	*d*	*s*	£
4=	1		
48=	12=	1	
960=	240=	20=	1.

☞ Farthings (fourthings) are often written as fractions of a penny. Thus, 1 farthing is written $\frac{1}{4}$; 2 as $\frac{1}{2}$; and 3 as $\frac{3}{4}$.

3. *Provincial Currencies.*

While the United States were British colonies, their currency, like that of the mother country, was sterling. Each colony issued its own money in bills of the denomination of pounds, shillings, and pence. During the revolutionary war, these bills depreciated, and in different degrees in the different colonies, so that a pound or shilling no longer had the same value throughout the land. The federal currency of dollars and cents was adopted soon after the peace. But the people still cling to their old habits of expressing prices in pounds, shillings, and pence; and, as the value of these still differs in different places, it is proper that the student should understand the method of changing a sum of money from one currency into another. This is best done by means of the dollar, which serves as a universal measure, being the uniform standard of value.

The relative value of the dollar, and of the pound, and its subdivisions in the Provincial currencies, is as follows:

a. In English, or Sterling Money.

£1=20*s.*=240*d.* $1=4*s.* 6*d.*=54*d.*; therefore,

$$\pounds 1=\$\tfrac{240}{54}=\$\tfrac{40}{9}.$$
$$\$1=\pounds\tfrac{54}{240}=\pounds\tfrac{9}{40}.$$

b. In Canada, Nova Scotia, and New Brunswick.

£1=20*s.* $1=5*s.*; therefore,

$$\pounds 1=\$\tfrac{20}{5}=\tfrac{4}{1}.$$
$$\$1=\pounds\tfrac{5}{20}=\tfrac{1}{4}.$$

c. In New England, Kentucky, and Tennessee.

£1=20*s.* $1=6*s.*; therefore,

$$\pounds 1=\pounds\tfrac{20}{6}=\tfrac{10}{3}.$$
$$\$1=\pounds\tfrac{6}{20}=\tfrac{3}{10}.$$

d. In New York, North Carolina, and, except Vermont, all the States added to the Union since 1786.

£1=20*s.* $1=8*s.*; therefore,

$$\pounds 1=\$\tfrac{20}{8}=\tfrac{5}{2}.$$
$$\$1=\pounds\tfrac{8}{20}=\tfrac{2}{5}.$$

e. In Pennsylvania, New Jersey, Delaware, and Maryland.

£1=20*s.*=240*d.* $1=7*s.* 6*d.*=90*d.*; therefore,

$$£1=\$\tfrac{240}{90}=\tfrac{8}{3}.$$
$$\$1=£\tfrac{90}{240}=\tfrac{3}{8}.$$

f. In South Carolina and Georgia.

£1=20*s.*=240*d.* $1=4*s.* 8*d.*=56*d.*; therefore,

$$£1=\$\tfrac{240}{56}=\tfrac{30}{7}.$$
$$\$1=£\tfrac{56}{240}=\tfrac{7}{30}.$$

II. Weight.

1. *Troy Weight.*

For weighing gold, silver, jewels, liquors, &c.

24 grains (gr.) make . .	1 pennyweight.	dwt.
20 pennyweights . . .	1 ounce. . .	oz.
12 ounces	1 pound. . .	lb.

gr.	dwt.		
24=	1	oz.	
480=	20=	1	lb.
5760=	240=	12=	1.

2. *Avoirdupois Weight.*

For weighing hay, grain, groceries, and all coarse articles.

16 drams (*dr.*) make	1 ounce. . . .	oz.
16 ounces	1 pound. . . .	lb.
25 pounds	1 quarter. . . .	qr.
4 quarters	1 hundredweight.	cwt.
20 hundredweight . . .	1 ton.	T.

dr.	oz.				
16=	1	lb.			
256=	16=	1	qr.		
6400=	400=	25=	1	cwt.	
25600=	1600=	100=	4=	1	T.
512000=	32000=	2000=	80=	20=	1.

Formerly 28 pounds were reckoned to the quarter, 112 pounds to the hundredweight, and 2240 pounds to the ton. But this practice is fast becoming obsolete.

3. *Apothecaries' Weight.*

Used for mixing, but not for selling, medicines.

20 grains (*gr.*) make	1 scruple.	℈
3 scruples	1 dram.	ʒ
8 drams	1 ounce.	℥
12 ounces	1 pound.	℔

gr.	℈			
20=	1	ʒ		
60=	3=	1	℥	
480=	24=	8=	1	℔
5760=	288=	96=	12=	1

175 ounces Troy are equal to 192 oz. Avoirdupois; 1 lb. Troy to 5760 gr.; and 1 lb. Avoirdupois to 7000 grains. The pound and ounce in Apothecaries' weight are the same as in Troy weight; the only difference is in their divisions and subdivisions.

III. Measures of Capacity.

1. *Dry Measure.*

For dry wares, as grain, seeds, fruit, roots, sugar, salt, coal, lime, &c.

2 pints (*pts.*) make	1 quart. . . .	qt.
4 quarts	1 gallon. . . .	gal.
2 gallons	1 peck. . . .	pk.
4 pecks	1 bushel. . . .	bu.
36 bushels	1 chaldron. . . .	ch.
8 bushels	1 quarter. . . .	qr.
5 quarters	1 wey, load, or ton.	wey.
2 weys	1 last.	last.

The last three measures are hardly, if at all, used in the United States.

pts.	qt.					
2=	1	gal.				
8=	4=	1	pk.			
16=	8=	2=	1	bu.		
64=	32=	8=	4=	1	qr.	
512=	256=	64=	32=	8=	1	last.
5120=	2560=	640=	320=	80=	10=	1.

2. *Wine Measure.*

For spirits, wines, oil, honey, &c.

4 gills (*gi.*) make	1 pint . . .	pt.
2 pints	1 quart . .	qt.
4 quarts	1 gallon . .	gal.
31½ gallons	1 barrel . .	bar.
42 gallons	1 tierce . .	tier.
63 gallons	1 hogshead .	hhd.
84 gallons	1 puncheon .	pun.
2 hogsheads . . .	1 pipe, or butt.	p. or b.
2 pipes	1 tun . . .	T.

gi. pt.
4= 1 qt.
8= 2= 1 gal.
32= 8= 4= 1 bar.
1008= 252= 126= 31½=1 tier.
1344= 336= 168= 42 =1⅓=1 hhd.
2016= 504= 252= 63 =2 =1½=1 pun.
2688= 672= 336= 84 =2⅔=2 =1⅓=1 p.
4032=1008= 504=126 =4 =3 =2 =1½=1 T.
8064=2016=1008=252 =8 =6 =4 =3 =2=1.

3. *Ale or Beer Measure.*

For ale, beer, and milk.

2 pints (pt.) make	1 quart. . .	qt.
4 quarts	1 gallon. . .	gal.
36 gallons	1 barrel. . .	bar.
54 gallons	1 hogshead. .	hhd.

pt. qt.
2= 1 gal.
8= 4= 1 bar.
288=144=36=1 hhd.
432=216=54=1½=1.

The dry gallon contains 268⅘ cubic inches; the wine gallon 231 cubic inches; and the beer gallon 282 cubic inches.

IV. Measures of Length.

1. *Cloth Measure.*

$2\frac{1}{4}$ inches (in.) make		1 nail. . .	na.
4 nails	. . . . .	1 quarter. .	qr.
4 quarters	. . . .	1 yard. . .	yd.
3 quarters	. . . .	1 ell Flemish.	Fl. e.
5 quarters	. . . .	1 ell English.	E. e.
6 quarters	. . . .	1 ell French.	Fr. e.

in.	na.			
$2\frac{1}{4}$=	1	qr.		
9 =	4=	1	yd.	
36 =	16=	4=	1	
27 =	12=	3=	1	Fl. e.
45 =	20=	5=	1	E. e.
54 =	24=	6=	1	Fr. e.

2. *Long Measure.*

For length or distance.

3 barley-corns (b. c.) make		1 inch.	in.
12 inches	. . . . . .	1 foot.	ft.
3 feet	. . . .	1 yard.	yd.
$5\frac{1}{2}$ yards, or $16\frac{1}{2}$ feet	. .	1 rod, perch, or pole.	rd.
40 rods	. . . . . .	1 furlong . . .	fur.
8 furlongs	. . . . . .	1 mile.	m.
3 miles	. . . . . .	1 league. . . .	l.
$69\frac{1}{2}$ miles	. . . . . .	1 degree. .	deg. or °.

b. c.	in.							
3=	1	ft.						
36=	12=	1	yd.					
108=	36=	3=	1	rd.				
594=	198=	$16\frac{1}{2}$=	$5\frac{1}{2}$=	1	fur.			
23760=	7920=	660=	220=	40=	1	m.		
190080=	63360=	5280=	1760=	320=	8=	1	l.	
570240=	190080=	15840=	5280=	960=	24=	3 =	1	deg.
13210560=	4403520=	366960=	122320=	22240=	556=	$69\frac{1}{2}$=	$23\frac{1}{6}$=	1

The point, the line, the hand, and the fathom, also belong to this measure. Six points make 1 line; 12 lines 1 inch, used in measuring the length of pendulums for clocks, and other small measurements. The hand is 4 inches, and is used for measuring the height of horses. The fathom is 6 feet, used principally for measuring depths of water. Sixty geographical miles make a degree.

V. Measure of Surface.

For land or square measure.

144 square inches (sq. in.) [12·12, that is, 12 in. long and 12 broad] make	1 square foot. . . .	sq. ft.
9 square feet [=3·3, 3 long and 3 broad]	1 square yard. . . .	sq. yd.
$30\frac{1}{4}$ square yards . . .	1 square rod, perch, or pole.	sq. rd.
40 square rods . . .	1 rood.	R.
4 roods, or 160 sq. rods,	1 acre.	a.
640 acres	1 square mile. . .	sq. m.

sq. in	sq. ft.				
144=	1	sq. yd.			
1296=	9 =	1	sq. rd.		
39204=	$272\frac{1}{4}$=	$30\frac{1}{4}$=	1	R.	
1568160=	10890 =	1210 =	40=	1	a.
6272640=	43560 =	4840 =	160=	4=	1.

By this measure, land, and husbandman's and gardener's work are measured; also the work of artificers, such as boards, glass, pavements, wainscoting, flooring, and every thing that has the two dimensions of length and breadth. Land is usually measured by Gunter's chain, which is four rods in length, and is divided into 100 equal parts, called links.

Pupils are sometimes at a loss to understand why there are nine square feet in a square yard, and 144 square inches in a square foot. The following figures will make the matter plain:

No. 1.

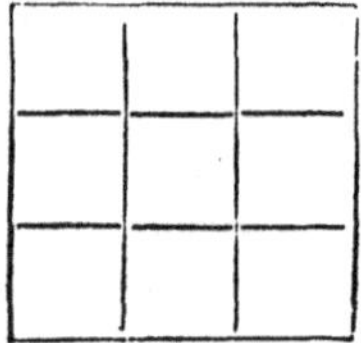

No. 2.

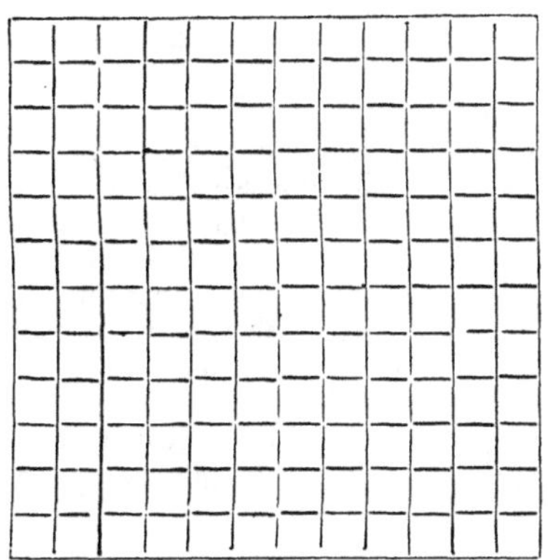

No. 1 represents a square yard; that is, each of its four sides is one yard in length. Each of the sides are divided into three parts, representing feet, by lines running across the figure, which is thus divided into nine equal surfaces, each representing one square foot. Now, if we take the whole length, 3 feet, and one foot in breadth, we shall have $3 \times 1 = 3$ square feet. Taking 2 feet broad, and 3 feet long, we have $2 \times 3 = 6$ square feet. And, taking the whole figure, we have $3 \times 3 = 9$ square feet. By using a similar process in No. 2, it will appear that there are 144 square inches in one square foot. The figures also distinctly show why 3 square feet are only $\frac{1}{3}$ of 3 feet square; and 12 square inches $\frac{1}{12}$ of 12 inches square.

VI. Measure of Solidity.

For solid or cubic Measures.

1728 cubic inches (c. i.)$=12 \times 12 \times 12$, that is, 12 long, 12 broad, and 12 deep, make	1 cubic foot. .	c. f.
27 cubic feet$=3 \times 3 \times 3$	1 cubic yard.	c. yd.
40 cubic feet of round timber, or 50 of square timber,	1 ton. . . .	T.
42 cubic feet of shipping	1 ton. . . .	T.
16 cubic feet	1 cord foot.	c. ft.
8 cord feet, or 128 cubic feet . .	1 cord of wood.	c.

A pile of wood 8 feet long, 4 feet wide, and 4 feet high, contains just one cord, since $8\times4\times4=128$.

By this measure, firewood, timber, stone, and other articles which have the dimensions of length, breadth, and thickness, and are of regular form, are measured.

It has been shown that a *square* yard, or yard of surface, by having *two* dimensions, contains $3\times3=9$ square feet. In like manner, a *cubic*, or solid yard, having *three* dimensions, contains $3\times3\times3=27$ cubic feet, as will evidently appear from an inspection of the figure. The difference between a cube of 3 feet and 3 cubic feet, will also be apparent, the one being only $\frac{1}{9}$ of the other.

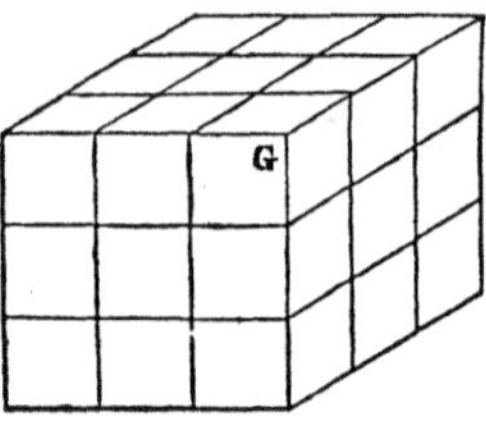

VII. Measure of Circular Motion.

60 seconds (″) make	1 minute. . .	(′).
60 minutes	1 degree. . .	°.
30 degrees	1 sign. . .	S.
12 signs, or 360° . . .	1 circumference.	C.

″	′	°	S.	C.
60=	1			
3600=	60=	1		
108000=	1800=	30=	1	
1296000=	21600=	360=	12=	1.

This measure is used for estimating latitude and longitude, and also in measuring the motions of the heavenly bodies. Every circle, whether great or small, is supposed to be divided into 360 equal parts, called degrees. A degree of a great circle of the earth contains 60 geographical miles, equal to $69\frac{1}{2}$ statute miles.

VIII. Measure of Time.

60 seconds (sec.) make	1 minute. .	min.
60 minutes	1 hour. .	hr.
24 hours	1 day. . .	day.
7 days	1 week. . .	wk.

sec.	min.			
60=	1	hr.		
3600=	60=	1	day.	
86400=	1440=	24=	1	wk.
604800=	10080=	168=	7=	1.

Four weeks are sometimes called a month. In computing interest, 30 days are considered a month, when no particular ones are named. The calendar months are 12 in number. Their length is as follows:

January . .	31	July	31
February . .	28	August . . .	31
March . . .	31	September . .	30
April . . .	30	October . . .	31
May . . .	31	November . .	30
June . . .	30	December . .	31

When the hundreds of any centennial year, and when the tens and units of any other year, are divisible by 4, every such year is called *leap-year*, and then February has 29 days. The number of days in each calendar month will be more easily remembered by committing to memory the following lines:

Thirty days hath September,
April, June, and November;
February hath twenty-eight alone;
All the rest have thirty-one;
Except in leap-year, when, in fine,
February's days are twenty-nine.

The solar, or true year, consists of 365 days, 5 hours, 48 minutes, and 48 seconds. The Julian year consists of 365 days and 6 hours. The calendar year consists of 365 days for three successive years; every fourth year, which is called bissextile, or leap-year, having 366. The calendar year is thus adjusted to the Julian year. By the omission of the odd day of the first year of the century (which would always be leap-year) for three out of four centuries, the calendar year is so nearly adjusted to the true, or solar year, that the only correction it will require will be the suppression of a day and a half in five thousand years.

IX. Books.

A sheet folded in 2 leaves is called a folio.

"	" 4	quarto, or 4to.
"	" 8	octavo, or 8vo.
"	" 12	duodecimo, or 12mo.
"	" 18	octodecimo, or 18mo.
"	" 24	24mo.

X. Miscellaneous.

12 things . . . . . make	1 dozen . . .	doz.
12 dozen	1 gross . . .	gro.
12 gross	1 great gross.	
20 things	1 score.	
24 sheets of paper . . .	1 quire.	
20 quires (19 good, 1 broken)	1 ream.	
200 pounds	1 barrel of pork, beef, shad, [or salmon.	
196 pounds	1 barrel of flour.	
30 pounds	1 barrel of anchovies.	
112 pounds	1 barrel of raisins.	
256 pounds	1 barrel of soap.	
7½ pounds	1 gallon of train oil.	
11 pounds	1 gallon of molasses.	
14 pounds	1 stone of iron or wood.	
8 pounds	1 stone of meat.	
28 pounds	1 tod.	
56 pounds	1 firkin of butter.	
94 pounds	1 firkin of soap.	
112 pounds	1 quintal of fish.	
30 gallons	1 barrel of fish.	
364 pounds	1 sack.	
19½ cwt..	1 fother of lead.	
32 gallons	1 barrel of cider.	
8 bushels of salt . . .	1 hhd. at sea.	
7½ bushels of salt . . .	1 hhd. on shore.	

Specimen of the mode of questioning the classes, after they have recited a table.—1. How many mills make a cent? How many cents make a dollar? How many mills in a dollar, then? How many dollars make an eagle? How many mills in an eagle, then? How many cents in an eagle? 2. How many farthings in a penny? How many pence in a shilling? Then how many farthings in a shilling? How many shillings in a pound? How many farthings in a pound? How many pence in a pound? 3. How many grains in a pennyweight? How many pennyweights in an ounce? Then how many grains in an ounce? and so on throughout the tables, till they are thoroughly committed to memory.

a. Change of Form.

[It has already been shown, when treating of *Common* Fractions, p. 193, 4, that it is sometimes extremely convenient to change their form, without altering their value, and that this is effected by multiplying or by dividing *both terms* by the same number. Such a change of form is equally convenient and necessary in the case of *Determinate* Fractions, and it is effected in precisely the same manner. This we shall readily perceive if we only notice that the sole difference between them is, that the denominations of the one are *limited* in number, and *expressed in words* or *signs*, while those of the other are *unlimited* in number, and expressed by *figures written under them*. Thus, if a pound sterling is considered the unit, 5 *shillings* is the same thing as $\frac{5}{20}$. If we wish to change the sum into pence, by multiplying by 12 (the number of pence in a shilling), we have 60 *pence*, or $\frac{60}{240}$. Here the intimate connection of determinate with common fractions is too evident to escape notice. By multiplying the denominator (shillings) by 12, it is changed to *pence*, *reducing* the value of the determinate fraction twelve fold, just as, by multiplying the denominator of the common fraction, $\frac{5}{20}$, by 12, we change it to $\frac{5}{240}$, *reducing its* value twelve fold. And as, by multiplying the numerator in both fractions by 12, we *increase* their value twelve fold, it is evident that by multiplying both terms in either fraction by the same number, the value of the fraction is unchanged.

Again: if a bushel be considered the unit, 8 quarts is the same as $\frac{8}{32}$. If we wish to change the quarts to gallons, dividing both terms by 4 (the number of quarts in a gallon), the 8 *quarts* become 2 *gallons*, and the common fraction becomes $\frac{2}{8}$. In neither case is there the slightest change of value. For, by dividing the denominator of the determinate fraction by 4, the *quarts* are changed into *gallons*, thus *enhancing* the fraction four fold; and by dividing the numerator by 4, thus *diminish-*

ing the fraction four fold, the one effect completely counterbalances the other, and leaves the value of the fraction unchanged.

It is evident, then, whether these numbers be considered as fractions or as compound numbers, that, when we wish to change their form from one of a *greater* to one of a *less* value, it must be performed by multiplication; because the *greater* number of *less* value will be equivalent to the *less* number of *greater* value. And, on the contrary, when we wish to change their form from a denomination of *less* value to one of *greater*, it must be performed by division, since the *smaller* number of *greater* value will be equivalent to the *greater* number of *less* value. Thus, to change 4 pounds to shillings, the 4 must be multiplied by 20 (the number of shillings in a pound), since 80 shillings=4 pounds. And, to change 80 shillings into pounds, the 80 must be divided by 20, since 4 pounds=80 shillings. Hence, also, it results that questions of this sort may be proved by changing the number back to its original denomination.]

☞ Federal money being arranged on the *decimal* scale, no other operation is necessary, in changing a number from one denomination to another, than a mere change of the separatrix. Thus, to change 5 eagles through all the inferior denominations, and *vice versa*,

e.	*$*	*d.*	*c.*	*m.*	*c.*	*d.*	*$*	*e.*
5=	50'=	500'=	5000'=	50,000'=	5000'0=	500'00=	50'000=	5'0000

Exemplification for the Black-board.

1. Change £3 5*s.* 6*d.* 3*q.* to farthings.

£	*s.*	*d.*	*q.*
3	5	6	3
960	48	4	

2880+240+24+3=3147 farthings.

Suggestive Questions.—How many farthings in one pound? Why, then, are the three pounds multiplied by 960? Of what denomination, then, is the 2880? How many farthings in one shilling? Why, then, are the 5*s.* multiplied by 48? Of what denomination, then, is 240? How many farthings in one penny? Why, then, are the pence multiplied by 4? Of what denomination, then, is 24? Of what denomination is the 3? Of what denomination, then, are all the four numbers? How many farthings, then, in all?

2. Change 3147 farthings to pounds, shillings, and pence.

```
Farthings, 3147(960 Divisor.
Divisor, 48)267   £3 5s. 6d. 3q. total quotient.
Divisor,   4)27
             3
```

Suggestive Questions.— How many farthings make one pound? Why, then, are the whole number of farthings divided by 960? What is the quotient of 3147 divided by 960? Of what denomination, then, is the 3? Of what denomination is the remainder, 267? How many farthings make one shilling? What is the quotient of 267 by 48? Of what denomination, then, is the 5? Of what denomination is the remainder, 27? Why divided by 4? Of what denomination, then, is the quotient, 6? Of what denomination is the remainder, 3? What, then, do 3147 farthings amount to in pounds, shillings, and pence?

3. How many grains in 24 lb. 3 oz. 15 dwt. of silver?

```
   lb.     oz.   dwt.
   24       3     15
  5760     480    24
 ──────────────────────
 138240+1440+360=140040 grains.
```

Suggestive Questions.—How many grains of silver in a pound? In an ounce? In a pennyweight? Why, then, are the pounds, ounces, and pennyweights, severally multiplied by these numbers? Of what denomination, then, are the three products?

4. How many pounds, &c., in 140040 grains of silver?

```
    140040(5760
     24840     24 lb. 3 oz. 15 dwt.
 480)1800
   24)360
      120
```

Suggestive Questions.—Why are the grains of silver divided by 5760? What is the quotient? Of what denomination? What is the remainder? Why divided by 480? Why is the remainder of that division divided by 24?

5. Change £$\frac{2}{9}$ to the fraction of a penny.

$$\frac{2}{9}\times240=\frac{480}{9}$$

6. Change the same sum, namely, £$\frac{2}{9}$, to pence.

$$\frac{2}{9}\times240=\frac{480}{9}=53\tfrac{1}{3}\text{ pence.}$$

or (as $240=3\times80$) $3\times\frac{2}{9}\times80=\frac{160}{3}=53\frac{1}{3}$, as before.

7. Change $\frac{480}{9}$ pence to the fraction of a pound.

$$\text{d.}\frac{480}{9}\div240=\frac{2}{9}£.$$

8. Change the same sum, namely, $53\frac{1}{3}$ pence, to the fraction of a pound.

$$53\tfrac{1}{3}=\frac{160}{3}\div240=\frac{160}{720}=\frac{2}{9};$$

or, by inspection, omitting superfluous steps,

$$\frac{160}{240}=\frac{2}{9}.$$

9. Change $\frac{1}{4}$ of $\frac{3}{5}$ of a cwt. to the fraction of a pound.

$$\frac{1}{4}\times\frac{3}{5}\times100=\frac{300}{20}=\frac{15}{1}\text{ lb.; or }\frac{1\cdot3\cdot100}{4\cdot5\cdot\ 1}=\frac{15}{1}.$$

10. What part of $\frac{3}{5}$ of a cwt. is 15lbs.?

$$15\div100=\frac{15}{100}\div\frac{3}{5}=\frac{5}{20}=\frac{1}{4}\text{; or }\frac{\frac{15}{100}}{\frac{3}{5}}=\frac{1}{4}.$$

Ans. $\frac{1}{4}$ of $\frac{3}{5}$ of 100 lbs. or a cwt.

11. Change £$\frac{3}{4}$ $\frac{5}{6}$*s.* $\frac{3}{5}$*d.* to the fraction of a penny.

Changing to least common denominator,

$$\begin{array}{lll} £\ \frac{45}{60} & \frac{50}{60}s. & \frac{36}{60}d. \\ \ 240 & 12 & \\ \hline \end{array}$$

$$\frac{10800}{60}+\frac{600}{60}+\frac{36}{60}=\frac{11436}{60}=\frac{953}{5}d.$$

Suggestive Questions.—Why are the given numbers changed to least common denominator? *Ans.* Because they are to be —. Why are the pounds multiplied by 240, and the shillings by 12? Of what denomination, then, are the products?

12. Change $\frac{953}{5}d.$ to fractions of pounds, shillings and pence.

$$\begin{array}{l} 5)\underline{953} \\ 12)\underline{190\frac{3}{5}} \\ \quad \underline{15s.\ 10\frac{3}{5}d.}\text{, or} \\ £\tfrac{15}{20}=£\tfrac{3}{4}\ \tfrac{10}{12}s.=\tfrac{5}{6}s.\ \tfrac{3}{5}d. \end{array}$$

13. Change 3*s*. to the fraction of a pound, and then back to shillings.

$$\overset{s.}{3}\div 20=\overset{£}{\tfrac{3}{20}}\times 20=3s.$$

14. Change £4 15*s*. 9*d*. to the fraction of a pound.

$$\begin{array}{ccc} £ & s. & d. \\ 4 & 15 & 9 \\ 240 & 12 & \\ \hline \end{array}$$

$$960+180+9=\overset{d.}{1149}\times\tfrac{1149}{240}=\tfrac{383}{80}$$

15. Change £$\frac{383}{80}$ to determinate fractions; that is, to pounds, shillings, and pence. [This is nothing more than to get rid of the common fraction by performing the division indicated.]

80)383(£4 15*s*. 9*d*.

Change 63 rem'r. to shillings, $\underline{20\times 63}$

1260

Change 60 rem. to pence, $\underline{12\times 60}$

720

16. Change 15*s*. to the decimal of a pound.

$$\tfrac{15}{20}=£0{\cdot}75.$$

17. Change £0·75 to a determinate fraction.

£'75×20=15*s*.

18. Change 12*s*. 6*d*. 3*q*. to the decimal of a pound.

$$\begin{array}{ll} 12 & 6 \quad 3 \\ 48 & 4 \\ \hline \end{array}$$

$$576+24+3=\tfrac{603}{960}=£0{\cdot}628125.$$

19. Change £0'628125 to determinate fractions.

$$\begin{array}{r} £0{\cdot}628125 \\ 20 \\ \hline s.\ 12{\cdot}562500 \\ 12 \\ \hline d.\ 6{\cdot}7500 \\ 4 \\ \hline q.\ 3{\cdot}00 \end{array}$$

Suggestive Questions.—Does the given number amount to more or less than a pound? Why is it multiplied by 20? How many shillings are there in the product? How much remainder? *Ans.* 5625 tens of thousandths of a shilling. Why are the ciphers neglected? Why is the remainder multiplied by 12? What is the integer in the product? What is its denomination? Why is the remainder multiplied by 4? What is the denomination of the product? What, then, is the answer?

Remark.—A more concise method has been devised of changing shillings, pence, and farthings, to the decimal of a pound, and of changing decimals of a pound to shillings, pence, and farthings, sufficiently correct for all practical purposes, which is called *finding the decimal of shillings, pence, and farthings, by inspection.* It will be readily understood by the aid of the following questions: What part of a pound is two shillings? Is every 2*s*. $\frac{1}{10}$ or '1 of a pound, then? If any number of shillings be divided by 2, then, will the quotient be the number of *tenths* of a pound? What is the only possible remainder when a number is divided by 2? What is the half of a tenth? *Ans.* 5 —. What decimal of a pound, then, is 1 shilling? Fill up the blanks, then, in the statements that follow:

Shillings.	Decimals of a pound.
1=	0'..
2=	0'.
3=	0'..
4=	0'.
5=	0'..
6=	0'.
7=	0'..
&c.	&c.

So much for shillings. And now as to pence and farthings. How many farthings in a pound? What part of a pound, then, is a farthing? Which is the larger portion of any thing, $\frac{1}{3}$ or $\frac{1}{4}$? $\frac{1}{960}$ or $\frac{1}{1000}$? Must the number, then, representing a given number of farthings be increased or diminished, if they are considered as 1000ths of a pound in place of 960ths? What part of 960 added to itself will make 1000? ($\frac{1}{24}$) What part of a given number of farthings, then, when added to the number, will reduce its value from 960ths to 1000ths of a pound? On this principle, what part of a pound will 24 farthings be? What portion, then, of any number of farthings must be added to that number to change it from 960ths to 1000ths of a pound?

N. B.—As the remainder will frequently be large when the number of farthings is divided by 24, it is thought best, when that number falls short of 12, to make an increase: from 12 to 36, then, increase it by 1; above 36, increase it by 2. This principle, applied to the 18th and 19th examples, gives us,

Ex. 18.—12*s*. . . . . ='6
6*d*. 3*q*.=27*q*. . . ='028 Why 28?
12*s*. 6*d*. 3*q*. . . . . . . ='628

Ex. 19.—'6 . . . . ='12*s*.
'028=27*q*. . . . = 6*d*. 3*q*.
'628 . . . . . . =12*s*. 6*d*. 3*q*.

20. Change 13*s*. 9*d*. 2*q*. to the decimal of a pound, and rechange to determinate fractions.

12s. . . . . . . =‘6
1s. . . . . . . =‘05
9d. 2q.=38q. =‘040 Why 40?
13s. 9d. 2q. . . . =‘69

‘6 . . . . =12s.
‘05 . . . . = 1s.
‘04=$\frac{4}{100}$=$\frac{40}{1000}$=38q. = 9d. 2q.
‘69 . . . . =13s. 9d. 2q.

21. Change 2 roods and 20 square rods to the decimal of an acre.

2 20
40
80+20=$\frac{100}{160}$=‘625a.

22. Change ‘625 acre to denominate fractions.

‘625
4
2‘500
40
20‘0 *Ans.* 2 roods, 20 square rods.

Exercises for the Slate or Black-board.

1. Change £46 5s. 11d. 3q. to farthings.
2. Change 44447 farthings to pounds, &c.
3. Change £54 to pence.
4. Change 12960 pence to pounds.
5. Change 62 lb. 7 oz. 14 dwt. 18 gr. Troy to grains.
6. Change 360834 grains Troy to pounds, &c.
7. Change 4 cwt. 3 qr. 24 lb. 10 oz. 12 dr. to drams.
8. Change 127916 drams to cwt., &c.
9. Change 3 lb. 1 ℥ 2 ʒ 1 ℈ 1 gr. to grains.
10. Change 17901 grains, apothecaries' weight, to lbs., &c.
11. Change 18 bu. 3 pk. 1 gal. to gills.
12. Change 4832 gills to bushels, &c.
13. Change 5 pipes, 1 hhd. 3 qts. of oil to gills.

14. Change 22200 gills of oil to pipes, &c.
15. Change 12 gal. 2 qts. 1 pt. of milk to pints.
16. Change 101 pints of milk to gallons, &c.
17. Change 15 yds. 3 qu. 3 n. to nails.
18. Change 255 nails to yards, &c.
19. Change 5 miles, 3 fur. 12 rods, 3 yds. 2 ft. 6 in. 2 b. c., to barley corns.
20. Change 1029224 barley corns to miles, &c.
21. Change 3 acres and 3 roods to square inches.
22. Change 23522400 square inches to acres, &c.
23. Change 36 tons of round timber to cubic inches.
24. Change 2488320 cubic inches of round timber to tons.
25. Change 24 tons, 17 feet of square timber to cubic inches.
26. Change 2102976 cubic inches of square timber to tons.
27. Change 1 circumference, 2 signs, 15 degrees, to minutes.
28. Change 26100 minutes to circumferences, &c.
29. Change a solar year, namely, 365 days, 5 hours, 48 minutes, and 48 seconds, into seconds.
30. Change 31556928 seconds into days, &c.
31. Change £$\frac{4}{5}$ to the fraction of a penny.
32. Change $\frac{960}{5}d.$ to the fraction of a pound.
33. Change $\frac{3}{160}$ of an ell English to the fraction of a nail.
34. Change $\frac{3}{8}$ of a nail to the fraction of an ell English.
35. Change 25 yards to ells English. [$25\times\frac{4}{5}$. Why?]
36. Change 20 ells English to yards. [$20\times\frac{5}{4}$. Why?]
37. Change $\frac{16}{17}$ of a minute to the fraction of a day.
38. Change $\frac{1}{1530}$ of a day to the fraction of a minute.
39. Change 2 qr. 8 lb. to the decimal of a cwt.
40. Change 0·58 cwt. to quarters, &c.
41. Change 3 qr. 2 n. to the decimal of a yard.
42. Change 0·875 yd. to quarters, &c.
43. Change $\frac{6}{7}$ of an acre to roods and square rods.
44. Change 3 roods, $17\frac{1}{7}$ square rods, to the fraction of an acre.
45. Change, by inspection, 13*s.* 9*d.* 3*q.* to the decimal of a pound.
46. Find the value of £·691.
47. Change 18*s.* 3*d.* 1*q.* to the decimal of a pound, by inspection.
48. Find the value of £·914.

49. Change, by inspection, the following sums severally into decimals of a pound: 15*s.* 3*d.*; 8*s.* 11*d.* 2*q.*; 10*s.* 6*d.* 1*q.*; 1*s.* 8*d.* 2*q.*; 2*q.*; 2*d.* 3*q.*

50. Find the value of £·762, £·448, £·526, £·085, £·002, £·011.

But, besides these changes from one *denomination* to another, in the several species of determinate fractions, it is frequently necessary to change numbers from one *species* to another. Thus, *longitude* may be changed to *time*, or time to longitude; and a sum of money may be resolved into an equivalent sum in another currency, a process technically called *Exchange*.

1. *To change Longitude into Time, and vice versa.*

The earth performs an entire revolution on its axis in 24 hours;* and, as its circumference is divided into 360°, it follows, that the motion of the earth's surface, from west to east, is $\frac{1}{24}$ part of 360°=15°. Consequently,

15° of motion, or longitude=1 hour of time.
1° of motion, or longitude=$\frac{1}{15}$ hour=4 minutes.
1′ of motion, or longitude=$\frac{4}{60}$ min.=4 seconds.

Hence, when two places on the earth differ in longitude, they will have a corresponding difference in the hour, or the time of day; and, when the time of day differs in two places, there will be a corresponding difference in their longitude, the time being most advanced in the places situated most easterly. Thus, when the sun is setting at any one place, it will be an hour higher at a place 15° directly west, and it will have been set for an hour at a place 15° directly east. Again, when it is noon at any one place, it will be one o'clock, p. m., at all places 15° east, and eleven, a. m., at all places 15° west of the place.

Exemplifications for the Black-board.

1. Change 24° of longitude into time.

* This is not, strictly speaking, correct. The exact time of the earth's revolution on its axis is a trifle less than 24 hours. But this revolution, combined with the earth's motion in its orbit, brings the sun on the meridian, on an average, once every 24 hours, and thus produces exactly the same effect on *time* as if the earth revolved on its axis in 24 hours, with no other motion.

As every 15° of longitude=1 hour of time,

$$15)\underline{24} \quad \text{1h. 36m.} \quad \text{or} \quad 24 \times 4 = 96\text{m.}=\text{1h. 36m.}$$

2. Change 1 hour, 36 minutes, into longitude.

As every 4 min.=1° long. h. min. 1 36=96(4 min. = 24° or h. min. 1 36 × 15 = 24° 0

Exercises for the Slate or Black-board.

1. Change 180° of longitude into time.
2. Change 12 hours of time into longitude.
3. Change 17° 24′ 30″ into time.
4. Change 1 h. 9 min. 38 sec. into longitude.
5. Change 35° 20′ 15″ into time.
6. Change 2 h. 21 min. 21 sec. into longitude.
7. What is the difference of time between Washington and London; the longitude west of the former being 77° 1′ 48″, that of the latter, 0°? Is the time at London faster or slower than that of Washington?
8. If the difference in time between Washington and London is 7 h. 8 m. $7\frac{1}{5}$ sec., what is their difference of longitude?

2. *Exchange.*

Exemplifications for the Black-board.

1. Change £5 12*s.* sterling to federal money.

$$£5\ 12s.=£5{\cdot}6\times\tfrac{40}{9}=\$24\tfrac{8}{9}.$$

2. Change $\$24\frac{8}{9}$ to sterling money.

$$\$24\tfrac{8}{9}\times\tfrac{9}{40}=£5{\cdot}6=£5\ 12s.$$

Suggestive Questions.—No. 1. Why is £5‘6 multiplied by $\frac{40}{9}$. See p. 227. No. 2. Why is $\$24\frac{8}{9}$ multiplied by $\frac{9}{40}$.

Exercises for the Slate or Black-board.

1. How much New York currency is equal to £98 Pennsylvania currency?
2. How much Pennsylvania currency is equal to £104$\frac{8}{15}$ New York currency?
3. What sum in New England currency will pay a debt of £376 10*s.* in New York?
4. What sum in New York currency will pay a debt of £282 7*s.* 6*d.* in Boston?
5. How much New England currency must a Vermont merchant pay to cancel a debt in New York of £144 10*s.*?
6. How large a debt in New York currency will be cancelled by the payment of £108 7*s.* 6*d.* New England currency?
7. How many dollars will pay a debt of £86 in Philadelphia?
8. How many pounds Pennsylvania currency will be paid by \$229$\frac{1}{3}$?
9. How many Bergonia ducats, at 52*d.* each, can be had for 216$\frac{2}{3}$ Saragossa ducats, at 66*d.* each?
10. How many Saragossa ducats, at 66*d.* each, can be had for 275 Bergonia ducats, at 53*d.* each?
11. When the exchange from Antwerp to London is at £1 4*s.* 7*d.* Flemish for £1 sterling, how many pounds sterling must be paid in London to balance £236 Flemish at Antwerp? See "Provincial currencies," p. 227, for the manner of solving this.
12. When the exchange from London to Antwerp is at £1 sterling for £1 4*s.* 7*d.* Flemish, what must a merchant pay in Antwerp for a bill of £192 sterling?
13. In a settlement between C, of Philadelphia, and D, of London, C is found indebted £750 2*s.* sterling. How many dollars should he remit to pay this balance, when exchange is \$4·56 per pound sterling?
14. A merchant of Philadelphia remitted to London property to the amount of \$3420·456. For how much sterling money could he draw at the rate of \$4·56 per pound sterling?

b. Addition, Subtraction, Multiplication, and Division, of Determinate Fractions.

The only peculiarity in these processes consists in *carrying* by the *denominator* of the several fractions, or by the subdivisions of the several units (accordingly as we consider the numbers as determinate fractions or as compound numbers), in place of carrying by *tens*.

Exemplifications for the Black-board.

1. Add together £54 14*s*. 3*d*. 2*q*.; £23 5*s*. 8*d*.; £65 19*s*. 6*d*.; £42 17*s*. 4*d*. 3*q*.; £36 15*s*. 8*d*. 1*q*.; £9 17*s*. 6*d*.; and prove by adding from above downwards.

£	*s*.	*d*.	*q*.
54	14	3	2
23	5	8	
65	19	6	
42	17	4	3
36	15	8	1
9	17	6	
£233	10	0	2

Suggestive Questions.—What is the amount of the column of farthings? Equal to how many pence? To which column should the penny be carried? What is the amount of the column of pence? Equal to how many shillings? To what column, then, does the 3 belong? What is the amount of the column of shillings? Equal to how many pounds? To which column, then, does the 4 belong?

	cwt.	qr.	lb.	oz.	dr.	
2. From	224	0	20	8	12	Minuend.
Take	37	2	16	12	14	Subtrahend.
	186	2	3	11	14	Difference.
Proof,	224	0	20	8	12	Amount of Difference and Subtrahend.

Suggestive Questions.—Can 14 drams be taken from 12 drams? How many drams in an ounce? If, then, we change one of the 8 oz. in the minuend into drams, how many drams will there be from which to take the 14 drams? Since we have changed one of the ounces in the minuend into drams, how many ounces remain from which the 12 oz. may be taken? Is it necessary, then, to change one of the pounds into ounces in the minuend? How many ounces will there then be? Since one of the pounds has been changed into ounces, how many pounds remain from which to take the 16 pounds? Is it necessary to change one of the hundredweights into quarters? How many will then remain from which to take the 37 pounds?

3. How long a period elapsed from May 3, 1852, to Jan. 15, 1854?

	y.	m.	d.
From	1854	1	15
Take	1852	5	3
Leaves	1y.	8m.	12d. *Ans.*

4. Multiply 1 cwt. 2 qr. 14 lb. 12 oz. 3 dr. by 236, and prove it by division.

cwt.	qr.	lb.	oz.	dr.	
1	2	14	12	3	
				10	
16	1	22	9	14	product of 10.
				10	
164	3	1	2	12	prod. of 10×10=100.
				2	
329	2	2	5	8	product of 200.
49	1	17	13	10	product of 3×10=30.
9	3	13	9	2	product of 6.

388 cwt. 3 qr. 8 lb. 12 oz. 4 dr.(236 divisor, 236

Rem'r., 152×4 Proof, 1 cwt. 2 qr. 14 lb. 12 oz. 3 dr., quot.

611

Rem'r., 139×25

3483

1123

Rem'r., 179×16

2876

Rem'r., 44×16

708

...

Suggestive Questions.—The multiplication above will be readily understood. The questions that follow relate to the division. What is the quotient of 388 cwt. divided by 236? Of what denomination is the remainder, 152? How many quarters in a cwt.? Why, then, is the remainder, 152, multiplied by 4? Why is the product 611 in place of 608? What is the quotient of 611 quarters divided by 236? Of what denomination is the remainder, 139? How many pounds in a quarter? Then why multiplied by 25? What is the quotient of 3483 pounds divided by 236? Of what denomination is the remainder, 179? How many ounces in a pound? Why, then, is the remainder, 179, multiplied by 16? Why is the product 2876 in place of 2864? What is the quotient of 2876 divided by 236? Of what denomination is the remainder, 44? How many drams in an ounce? Why, then, is the remainder, 44, multiplied by 16? Why is the product 708 in place of 704? What is the quotient of 708 drams divided by 236?

Exercises for the Black-board or Slate.

1. Add together £14 12*s.* 5*d.* 2*q.*; 6*s.* 0*d.* 1*q.*; £33; £67 4*s.* 0*d.* 1*q.*; £3 15*s.* 6*d.* 2*q.*; £29 19*s.* 9*d.*; £55 9*d.*; £37 17*s.* 6*d.* 1*q.* *Ans.* £241 16*s.* 0*d.* 3*q.*

2. Add together 3 a. 2 r. 29 sq. rd. 6 sq. yd.; 15 a. 3 r. 17 sq. rds. 18 sq. yd.; 5 a. 3 r. 6 sq. rd. 3 sq. yd.; 15 a. 1 r. 18 sq. rd. 2 sq. yd. *Ans.* 40 a. 2 r. 30 sq. rd. 29 sq. yd.

3. What is the difference between 9 cwt. 2 qr. 18 lb., and 6 cwt. 3 qr. 24 lb.? *Ans.* 2 cwt. 2 qr. 19lb.

4. What is the difference between 9 m. 6 fur. 15 rods, and 18 m. 3 fur. 12 rods? *Ans.* 8 m. 4 fur. 37 rods.

5. Find the time elapsed from March 19, 1838, to Feb. 17, 1852. *Ans.* 13 y. 10 m. 28 d.

6. Find the time from Dec. 16, 1853, to Jan. 28, 1854. *Ans.* 1 m. 12 d.

7. Find the time from May 21, 1796, to April 15, 1828. *Ans.* 31 y. 10 m. 24 d.

8. Find the time from Jan. 1 to Oct. 16, 1853. *Ans.* 9 m. 15 d.

9. Multiply 1 lb. 4 ℥ 3 ʒ 1 ℈ 12 gr. by 6, 14, 150, and 325, separately, and prove each product by division.

10. Multiply 42 gal. 2 qt. 1 pt. by 6 and by 72, and prove each separate product by division.

11. Multiply 16 yds. 3 qr. 1 na. by 14, and prove by division.

Besides the multiplication, as above, by integers, there are two other kinds of multiplication of determinate fractions, namely, that of one *lineal* fraction by another, to form surfaces and solids; and that of fractional *coins* by other determinate fractions. The first of these is called *Multiplication of Duodecimals*, or *Cross Multiplication;* the other is called *Practice.*

1. *Multiplication of Duodecimals, or Cross Multiplication.*

It was stated above, p. 248, that the only peculiarity in the computation of determinate fractions was that of carrying by their *varying* denominations, in place of carrying *uniformly* by tens, as in integers. Cross multiplication, however, may seem an exception to this remark. But it is only a seeming exception, as will presently appear. A surface has been defined as the product of the length and breadth of the sides of any substance or space; solids are estimated by the product of their respective lengths, breadths, and depths. Now, it frequently happens that the extent of these sides consists of more than one denomination. For instance, a substance may be 6 feet 5 inches long, and 4 feet 1 inch broad, and in such a case the pupil might be at a loss how to multiply the one number by the other. But this difficulty vanishes, if we consider these dimensions as *mixed numbers* in common fractions, the foot representing the integer, and the inches the fraction. The length will then be $6\frac{5}{12}$, the breadth $4\frac{1}{12}$, and the multiplication can be readily performed as has been already shown. But the mode of computing such numbers by Cross Multiplication is shorter than that by common fractions, all changes from one kind of fraction to another being avoided. A single example will make the subject plain and easy. First, however, it is necessary to remark, that inches, in these operations, are called *primes*, or twelfths of a foot, and that the primes are subdivided into twelfths, called *seconds*, or 144ths of a foot, and those seconds into twelfths, called *thirds*. Hence their name, *duodecimals*, which signifies *twelfths*. They are respectively marked thus: ′, ″, ‴.

Exemplification for the Black-board.

1. What are the solid contents of a block of marble 6 ft. 4′ long, 3 ft. 5′ wide, and 2 ft. 6′ thick?

feet	′		
6	4		Length.
3	5		Width.
19	0		Product of 6 feet 4′ by 3 feet.
2	7	8	Product of 6 feet 4′ by 5′, or $\frac{5}{12}$.
21	7	8	Superficies of one side.
2	6		Thickness to opposite side.
43	3	4	Product of 21 ft. 7′ 8″ by 2 ft.
10	9	10	Product of 21 ft. 7′ 8″ by 6′.
ft. 54	1′	2	Solid contents.

Suggestive Questions.—Recollecting that a foot is the integer, and primes (′), seconds (″), and thirds (‴), fractional parts, what are 3 times 4′? Equal to how many feet? 3 times 6 feet and 1 foot carried? 5′×4′ (or $\frac{5}{12}\times\frac{4}{12}$), 20 of what denomination? Equal to how many primes? 5′×6 ft. (or $\frac{5}{12}\times6$) 30, and 1 carried, of what denomination? Equal to how many feet? [The same questions, varying only in the numbers and denomination, may be applied to the rest of the exemplification.]

The result of such computations as these may be restored to its original elements by division; that is, the thickness of a solid may be found, if its solid contents and the superficies of one of its sides be given; and one side of a superficies may be found, if its superficies and the other side be given. Thus, in order to prove the above computation, let 54 ft 1′ 2″ be the solid contents of a block of marble, and 21 ft. 7′ 8″ the superficies of one of its sides, to find its thickness:

```
                   ft.  ′   ″ ft.  ′  ″                [one side.
Solid contents,    54   1   2(21   7  8 (divisor) superficies of
                   10   9  10      2  6 (quotient) thickness.
                    .   .   .
```

Again, given superficies and length, to find breadth:

```
                         ft.  ′  ″ ft.  ′
Superficies of a side,   21   7  8( 3   5 (divisor) length of side.
                          1   1  8      6  4 (quotient) breadth of
                          .   .  .                           side.
```

Exercises for the Slate or Black-board.

1. Multiply 17 ft. 7′ by 1 ft. 5′, and prove by resolution into its original elements by division.
2. Multiply 4 ft. 6′ by 3 ft. 10′, and prove.
3. How many cubic feet in a block 2 ft. 3′; by 6 ft. 5′; by 8 ft. 4′? Prove.
4. How many cubic feet in a block whose dimensions are 3 ft. 6′, 2 ft. 1′, and 1 ft. 2′? Prove.

2. *Practice.*

Practice will be sufficiently understood from a few illustrations.

1. What will 6 cwt. 2 qr. 12 lb. of sugar cost, at £3 15*s.* 6*d.* per cwt.

£	*s.*	*d.*			
£3	15	6	price of 1 cwt.		
		6		cwt.	
22	13	0	price of	6	qr.
1	17	9	price of		2 = $\frac{1}{2}$ cwt.
	18	$10\frac{1}{2}$	price of		1 = $\frac{1}{2}$ of 2 qr.
	7	$6\frac{3}{5}$	price of		10 lb. = $\frac{1}{10}$ cwt.
	1	$6\frac{3}{25}$	price of		2 lb = $\frac{1}{5}$ of 10 lb.
25	18	$8\frac{11}{50}$	price of	6 cwt. 3 qr. 12 lb.	

2. What will be the cost of 55 bushels 3 pecks 5 quarts of wheat, at 10*s.* 2*d.* 3*q.* per bushel?

£	*s.*	*d.*	*q.*		
	10	2	3	price of 1 bushel.	
			11		
5	12	6	1	price of 11 bushels.	
			5		
28	2	7	1	price of 55 bu.	
	5	1	$1\frac{1}{2}$	price of	2 pk. = $\frac{1}{2}$ bushel.
	2	6	$2\frac{3}{4}$	price of	1 pk. = $\frac{1}{2}$ of 2 pecks.
	1	3	$1\frac{3}{8}$	price of	4 qts. = $\frac{1}{2}$ peck.
		3	$3\frac{11}{32}$	price of	1 qt. = $\frac{1}{4}$ of 4 qts.
£28	11	10	$1\frac{31}{32}$	price of 55 bu.	3pk. 5 qt.

3. What will 24 lb. of sugar cost, at $11'25 per cwt.

[2 divisors] 25 5)11'25
2'25 price of 20 lb. = $\frac{1}{5}$ cwt.
'45 price of 4 lb = $\frac{1}{25}$ cwt.
2'65 price of 24 lb.

To some students the last operation may appear more like division than multiplication. And, in effect, multiplying by $\frac{1}{5}$ or by $\frac{1}{25}$, &c., really is division. For multiplication, it will be remembered, is taking the multiplicand as many times as there are *units* in the multiplier.

Exercises for the Slate and Black-board.

1. What will 7 yds. 3 qr. 2 na. of cloth come to, at £2 2*s.* 6*d.* per yard? *Ans.* £16 14*s.* 8*d.* 1*q.*
2. What is the value of 6 cwt. 3 qr. 12 lb. of sugar, at £3 7*s.* 8*d.* per cwt.? *Ans.* £23 4*s.* 10$\frac{1}{2}\frac{1}{5}$*d.*
3. What would 37 T. 14 cwt. 2 qr. iron cost, at £5 14*s.* 8*d.* per ton? *Ans.* £216 5*s.* 9$\frac{3}{5}$*d.*
4. What will 20 a. 2 r. 25 sq. rd. of land cost, at $29 per acre? *Ans.* $603'5625.
5. What will 75 yd. 2 qr. of broadcloth cost, at $4'75 per yard? *Ans.* 358'625.
6. What is the value of 13 lb. 10 oz. 12 dwt. 16 gr. of silver, at £4 17*s.* 6*d.* per pound? *Ans.* £67 13*s.* 10*d.* 3*q.*
7. What will 4 bu. 2 pk. 3 qt. of beans cost, at $1'12$\frac{1}{2}$ per bushel? *Ans.* $5'166+.
8. What is the cost of 7 hhd. 7 gal. 2 qt. of molasses, at £2 3*s.* 6*d.* per hhd.? *Ans.* £15 9*s.* 8$\frac{1}{7}$*d.*
9. What will 1 cwt. 3 qr. 12 lb. of raisins cost, at £2 11*s.* 8*d.* per cwt.? *Ans.* £4 16*s.* 7$\frac{2}{5}$*d.*
10. What will 57 cwt. 3 qr. 8 lb. of cordage cost, at £3 17*s.* 6*d.* per cwt.? *Ans.* £224 1*s.* 9*d.* 3*q.*
11. What will 14 gal. 2 qt. 1 pt. of milk cost, at 2*s.* 6*d.* per gallon? £1 16*s.* 6*d.* 3*q.*
12. What will 32 bu. 2 qt. 1 pt. of rye cost, at 2*s.* 3*d.* per bushel? *Ans.* £3 12*s.* 2*d.*
13. What will 25 bu. 3 pk. 2 qt. of oats cost, at 1*s.* 6*d.* per bushel? *Ans.* £1 18*s.* 8*d.* 2*q.*
14. What is the value of 3 cwt. 2 qr. 10 lb. of raisins, at £2 14*s.* per cwt.? *Ans.* £9 14*s.* 4*d.* 2*q.*

15. If 1 cwt. of rice cost $9·30, what is the value of 144 cwt. 2 qr. 21 lb.? *Ans.* $1345·8.

16. What is the value of a silver tankard, weighing 1 lb. 7 oz. 14 dwt., at £3 16*s.* per lb.? *Ans.* £6 4*s.* 9*d.*+

CHAPTER IV.

PRACTICAL APPLICATIONS

OF THE METHODS OF INCREASE AND DECREASE, PROMISCUOUSLY ARRANGED.

THE different modes of increasing and decreasing numbers, whether integral or fractional, having now been fully developed and illustrated, it will be proper to furnish the pupil with a variety of questions for practice, promiscuously arranged, to accustom him quickly to decide as to the appropriate mode of solution in every case likely to occur in practical business. The following general principles will aid in forming this decision. Still, however, much must be left to his own judgment in the *application* of the various modes of solving questions with which he has become familiar.

I. All questions in which *quantities of the same kind* are to be *counted together* are solved by ADDITION; it being always remembered that quantities of *different kinds* cannot be numbered or added together, unless, by *changing their denomination*, we bring them to the same name. Thus, although a farmer may enumerate together 2 horses, 18 cows, 2 oxen, 4 calves, 75 sheep, and 8 pigs, their *denomination* must first be changed to some common term, such as *live stock*, &c. The same principle applies to the case of fractions, whether common or determinate. Those of *different denominations* cannot be added together without a change to one common denomination. Thus $\frac{3}{4}$ and $\frac{5}{6}$ cannot be added; but by changing $\frac{3}{4}$ to $\frac{9}{12}$ and $\frac{5}{6}$ to $\frac{10}{12}$, they become capable of union, forming together $\frac{19}{12}$. Again, 5 lb. and 12 oz. cannot be added. But the *de-*

nomination pounds may be changed to ounces, when the 5 lb. being equal to 80 oz., the whole forms 92 oz.

II. When we wish to ascertain the *difference* between two numbers *of the same kind* we have recourse to SUBTRACTION. The same observations apply to this as to ADDITION, namely: that the numbers, whether integer or fractional, must be of the same kind or denomination before their difference can be ascertained.

III. MULTIPLICATION applies to cases where a quantity occurs repeatedly; the number called the multiplier showing *how often* the repetition occurs.

IV. DIVISION applies to cases where a quantity or number is to be *divided equally* among a number of persons, or into a number of equal portions. It is also applied to find the *price* of a single piece of which a number has been purchased for a certain price.

In MULTIPLICATION and DIVISION it is *not* necessary that the separate numbers be of the *same denomination*, either in the case of integers or fractions. In the former case, the question is what is the *amount* of a certain number taken a certain *number of times;* in the latter, *how many times* is one number contained in another.

Where one or more divisors and multipliers enter into a computation, the same result will follow, in whatever order they are taken; and these numbers may be either used separately, or collected into one product. Thus, if 20 is to be multiplied by 4, and by 5, and by 6, and divided by 3 and by 8, these numbers may be used in the order given, or in any other order whatever; or, to shorten the process, each series may be collected into one product. By this last method the 20 will be multiplied by 120, and divided by 24. The process may be still more abridged by using the *quotient* of these products in place of the *products themselves*, considering that quotient as a multiplier or divisor according as the one or the other proves to be the greater. Thus, the product of the multipliers being 120, and that of the divisors 24, the quotient 5 is a multiplier; whereas, had the product of the divisors been 120, and that of the multipliers been 24, the quotient 5 would have been a divisor. All this, however, is but another form of cancellation, as becomes evident when exhibited in a fractional form. Thus,

$$20\times\frac{4\times5\times6}{3\times8}=20\times\frac{120}{24}=20\times5.$$

Similar remarks apply when addition and subtraction enter into a computation. The order in which the numbers are taken is indifferent; and, in place of the respective numbers, their *difference* may be used, adding it when the sum of the additive numbers is the greater, and subtracting it when the sum of the subtractive numbers is the greater.

☞ When the number to be divided or diminished will not run into fractions by the process, it will always save time to divide before multiplying, or to subtract before adding.

Exercises for the Slate or Black-board.

1. A merchant bought, in the spring, goods to the amount of $106,409, and on the first of January following found he had sold to the amount of $74,326; what amount of goods was left unsold? *Ans.* $32,083.

Suggestive Questions.—What do we want to know here? Is it the sum of these two numbers, or their difference? or is either of those numbers to be taken a certain number of times, or to be divided into a number of equal portions?

2. A merchant bought 340 pieces of cotton. In every piece there were 26 yards. How many yards were there in all? *Ans.* 8840 yards.

Suggestive Questions.—What is required here? In *every piece* 26 yards. How many in all? Is it the sum, difference, product, or quotient?

3. A merchant making an inventory of his stock, finds he has cotton cloth to the value of $356, linen $152, broadcloth $575, cassimere $264‘75, silk goods $254‘25, ginghams $125, calicoes $240, and various small articles to the amount of $336‘56. What is the whole value of the stock?

Ans. $2303‘56.

4. A man died leaving an estate worth $21,156 to be equally divided among his six children. How much would they have apiece? *Ans.* $3,526.

5. A merchant, on New Year's day, sent his clerk to collect debts and make some purchases. He received from John Stokes $265, from William Budd $375, from Jacob Jones $526, and from Thomas Strickland $623. The clerk then bought at one of the cotton mills 360 yards of cotton cloth, at 6 cents per yard; 525 yards of calico, at 10 cents per yard; 240 yards of gingham, at $18\frac{3}{4}$ cents. He also bought at a

woollen manufactory 200 yards of broadcloth at $2'25, 140 yards at $2'75, 75 yards at $4, and 63 yards at $5; and before he reached home he also bought 33 yards of carpeting at 60 cents. He paid over the balance to his employer, who divided the cents equally between his two little boys; the dollars he divided equally among his four daughters as a New-Year's present. How much did each of his children receive?

Ans. $50 to each daughter, and 5 cents to each boy.

The above question embraces Addition, Subtraction, Multiplication and Division. The *amount* of money received by the clerk must first be ascertained. Secondly, the *cost* of each article that he purchased, and the total *amount* of the purchases. Thirdly, the *difference* between the receipts and payments will show what money he paid his employer. Lastly, the cents being divided into two parts, and the dollars into four, will show how much each child received. The addition and subtraction may be one operation.

6. A lady who had 19 five dollar bank notes in her pocket-book, and five dollars in silver change in her purse, went out one morning to make some purchases. She bought 7 lbs. of beef, at 6 cents per lb.; a fish for 25 cents; a piece of corned beef for 30 cts.; a cabbage for 4 cts.; a peck of potatoes 8 cts.; a pound of sausages 10 cts.; a bunch of celery 10 cts. Having sent these articles home, she called at a store, where she made the following purchases: 8 yds. of calico, at 18 cts.; 8 yds. of mousseline de laine, at 25 cts.; 2 linen handkerchiefs, at 62½ cts.; a yard of linen, 75 cts.; 8 yds. cotton cloth, at 18 cts.; a pound of tea, 75 cts.; a loaf of sugar, weighing 9 lbs., at 15 cts.; 4 lbs. of raisins, at 10 cts.; a dozen of eggs, 8 cts. At a bookstore she paid $3 for her yearly subscription to a magazine, and bought some new books to the amount of $4. She next called on a poor widow, who had lately lost her only daughter, and presented her with $2. Lastly, she stopped at the tinsmith's, where she bought a watering-pot for $1, 6 tin pans, at 25 cts. each, and a dipper for 25 cents. How much money had she remaining when she got home?

Ans. $77'50.

7. A merchant in Burlington, Vt., received 17 packages of goods from Boston, each weighing 72 lbs. The total freight by railroad was $\frac{1}{3}$ of a cent per lb., the cartage to his store was 25 cts. for the whole. What was the whole expense of carriage from Boston to his store? *Ans.* $4'33.

8. The same merchant bought $17\frac{3}{4}$ cwt. of wool, at $\$40\frac{1}{2}$ per cwt. What was the whole cost? *Ans.* $\$718\frac{7}{8}$.

9. A farmer mowed $78\frac{1}{2}$ acres of meadow land, which yielded on an average $2\frac{2}{3}$ tons per acre. He kept 63 head of cattle through the winter on the hay, and sold, besides, $62\frac{1}{3}$ tons. How much hay did he make in the whole, and how much did each of the cattle consume on an average through the winter? [That is, how much hay did he cut? How much did he sell? How much did all his cattle eat? How much did each of them eat?] *Ans. to the last question,* $2\frac{1}{3}$ tons.

10. What do 9 pieces of cloth, of $28\frac{1}{2}$ yards each, come to, at $\$3{\cdot}37\frac{1}{2}$, or $\$3\frac{3}{8}$, per yard? *Ans.* $\$865\frac{11}{16}$.

11. A captain of a vessel has on board 205 bales, each paying $\$1{\cdot}25$, or $\$1\frac{1}{4}$, freight; 275 packages, each paying $87\frac{1}{2}$ cents, or $\$\frac{7}{8}$; 150 tons of other goods, each ton paying $\$12{\cdot}62\frac{1}{2}$, or $\$12\frac{5}{8}$; and 6 passengers, each paying $75. What does the whole freight and passage money amount to? *Ans.* $\$2840{\cdot}62\frac{1}{2}$.

12. A man who owned $\frac{3}{8}$ of a ship sold $\frac{3}{4}$ of his share. What share of the vessel did he sell, and what share did he keep? *Ans.* to last question, $\frac{3}{32}$.

13. The net profits of the vessel for one year after this sale was $32000. How much was each of the above shareholders entitled to receive? *Ans.* $9000 and $3000.

14. I have 765 pieces of cloth, and am about to put them up into bales of 15 pieces each. How many bales will there be? Prove by trial.

15. How many days are there in 24,480 minutes? *Ans.* 17.

16. A farmer threshed grain 7 days: the first day $12\frac{1}{2}$ bushels: the 2d, $18\frac{1}{3}$; the 3d, $24\frac{1}{8}$; the 4th, $30\frac{3}{5}$; the 5th, $32\frac{3}{4}$; the 6th, $44\frac{2}{3}$; the 7th, $15\frac{1}{2}$. He paid his workmen in grain. To one man he gave $3\frac{1}{2}$ bushels, to another $2\frac{1}{2}$; he returned to his neighbor what he had last borrowed of him to go to mill, which was $7\frac{3}{8}$ bushels; and he sent 10 bushels to mill to be ground for his stock, and $5\frac{1}{10}$ for family use. The next day he sold half of what was left, and the day following sent half of the remainder to the store in payment of the balance of his account. How much was then left in his granary? *Ans.* $37\frac{1}{2}$ bushels.

17. If 18 grains of silver make a thimble, and 12 dwts. make a teaspoon, how many, of each an equal number, can be made from 15 oz. 6 dwts. of silver? *Ans.* 24.

18. A man divided 75 cents among his three sons. As often as he gave the eldest 7 cents, he gave the second 5, and the third 3. How many cents did each receive? Prove by trial.

19. A carpenter bought 16 pieces of timber, each 16 inches square, and 13 feet long. What was the number of cubic feet, and what was the cost, at $12\frac{1}{2}$ cents per cubic foot?

Ans. to last question, \$$46\frac{2}{9}$.

20. A mason undertook to build a bridge for \$300. He hires six laborers to assist him, to each of whom he pays $62\frac{1}{2}$ cents per day, and their board, which he calculates to cost 25 cents each, for every working day. The work was finished in 40 days. How much did he get for his labor and superintendence, calculating his board at the same rate as that of his men? *Ans.* \$80.

21. Bought 16 bales of cotton. Four of them weighed each 1024 lbs.; 6 weighed each 998 lbs.; and the remainder 1054 lbs. each. What was their cost at $6\frac{1}{4}$ cents per lb.?

Ans. \$1025'50.

22. What are the solid contents of a wall 75 feet long, 3 feet 9 inches thick, and 24 feet 2 inches high; and how many bricks did it take, if (including the mortar) they occupied 9 inches in length, $4\frac{1}{2}$ in width, and $2\frac{1}{2}$ in depth?

Ans. to last question, 116,000.

23. There is a room 16 feet long, 14 feet broad, and 11 feet high. A mason engages to plaster it for 20 cents per square yard, the doors and windows being counted the same as an unbroken wall. What will be the amount of his bill, including the ceiling? *Ans.* \$19'$64\frac{4}{9}$.

Suggestion. In order to solve such questions correctly, the pupil should *realize* the statements, by looking around the room in which he is placed. Thus, let him ask himself How many long sides of the room? What are their dimensions? How many short sides? their dimensions? What are the dimensions of the ceiling? What, then, are the dimensions of the whole room? How many square feet in a square yard?

24. What will be the cost of a floor-cloth for an entry which is 34 feet long, and 10 feet wide, in which the stairs occupy a space of 14 feet by 3 feet 9 inches, making an allowance of $\frac{1}{18}$ of a yard for waste in cutting, at \$1'$12\frac{1}{2}$ per square yard? *Ans.* \$36.

25. How many cubic feet of earth will fill a dock 120 feet long, 75 feet broad, and 10 feet deep? A man has engaged

to fill it for 15 cents a load, his cart being 8 feet long, 4 feet broad, and 1 foot 6 inches deep. What will be the expense, supposing the heaping of the loads will balance the settling of the earth? *Ans.* $281'25.

26. What must be the depth of a square vessel 1 foot 6 inches broad, and 2 feet 4 inches long, that shall hold 8 feet cubic measure? *Ans.* 2 feet, $3\frac{3}{7}$ inches.

27. Three brothers emigrate to the western country. The eldest buys 640 acres at $2'25 per acre; the second buys 240 acres for $600; the third bought as much land as both his brothers for $1100. How much had the eldest to pay; how much did the second pay per acre; how much land did the third buy, and at what price per acre; and how much land had the three brothers?

Ans. $1440; $2'50; 880 acres, at $1'24+; 1760 acres.

28. A carpenter hewed 25 pieces of timber 8 inches square and 36 feet long; 16 pieces a foot square and 42 feet long; 23 pieces 18 inches by 20, and 26 feet long; 12 of 10 inches square and 32 feet long; and 15 pieces of 8 inches square and 18 feet long. What will be the amount of his bill at $6\frac{1}{4}$ cents per cubic foot? *Ans.* $200'$43\frac{3}{4}$.

29. What will be the cost of painting two rooms, including walls, ceilings and floors, at 7 cents per square yard? Their height is 12 feet 4 inches; the length of one is 32 feet, and its breadth 24; the length of the other is 24 feet and the breadth 16 feet 6 inches. No allowance to be made for doors or windows. *Ans.* $36'$62\frac{1}{27}$.

30. A workman engaged to plaster a house at 21 cents per square yard, of which the following are the dimensions: on the first floor there are two rooms, each 20 feet by 18, and 11 feet high, and an entry 10 feet by 36, of the same height. In each room, one of the ends, 18 feet wide, is not to be reckoned, as compensation for doors, windows and fireplace; in the entry neither end is to be reckoned, as compensation for doors and windows. On the second floor there are two rooms the same size as those below, on which a similar allowance is to be made; the entry is the same size as below, excepting that a room 10 feet square is taken off one end; neither end of the upper entry is to be counted, as compensation for doors and windows in the entry. What will be the amount of the plasterer's bill? *Ans.* $152'04.

31. The Hoosac Mountains are two degrees west of Boston.

When it is noon at Boston, what is the hour at these mountains? *Ans.* 8 minutes before 12.

32. Easton, Massachusetts, is 6° east longitude from Washington, and Circleville, Ohio, is 6° west from Washington. How many degrees of longitude are these towns apart? what is the hour at Circleville when it is noon at Easton; and what is the hour at Easton when it is 15 minutes past 6 P. M. at Circleville? *Ans. to last question,* 3 minutes past 7, P. M.

33. A meteor was seen for a few moments at Washington, at 14 minutes past 9 o'clock, P. M. It was also seen at Paris, Kentucky, which is 7° west, and at Easton 6° east from Washington. What was the time of the meteor's appearance at Paris and at Easton?
Ans. at Paris 46 m. past 8, P. M.; at Easton 38 m. past 9, P. M.

34. A vessel, named the Hamilton, sailed from New York to China, by way of the Cape of Good Hope, and thence to the Sandwich Islands, where she arrived on a Saturday morning by her reckoning. About an hour after her arrival, the Louisa, another vessel from New York came into the harbor, having come by way of Cape Horn. What day was it by the Louisa's reckoning, supposing both vessels to have kept the days of the week correctly? *Ans.* Friday.

35. What would be the length of the day (i. e., from noon to noon) on board a steamship bound from New York to Liverpool, supposing her to traverse exactly 5° of longitude daily; and what would be the length of the day on board, returning, supposing her to traverse daily the same number of degrees of longitude? *Ans.* 23 h. 40 m.; 24 h. 20 m.

36. A merchant, dying, left a will devising as follows: to his wife $10,000; to each of his three sons $12,000; to each of his two daughters half that amount; to each of his three clerks $500; to his porter $100; the residue of his estate to be equally divided between his wife and daughters. His estate proved to be worth $75,000. How much did the wife and each daughter receive? Prove by trial.

37. A man owning $\frac{1}{4}$ of a ship sold $\frac{1}{4}$ of his share for $1875. How much was the whole ship worth at that rate?
Ans. $30,000.

38. The cargo of a ship was valued at $37,230 and $\frac{1}{4}$ of the ship at $7500. What was the value of both ship and cargo?
Ans. $67,230.

39. A man having $100 purchased 4 cows, which took all

the money he had except \$28. How much did he pay apiece for them? *Ans.* \$18.

40. A farmer hired a boy for a year for \$40 and a suit of clothes. But after he had stayed 5 months they agreed to part, the boy to receive full wages for the time he stayed. He received $6\frac{1}{6}$ dollars and the suit of clothes. How much money is left for 7 months' wages, and what were the clothes valued at? *Ans. to last question* \$18.

41. A mechanic hired a journeyman at 8 shillings and his board, for each day he worked; but if he worked anywhere else, or was idle, the journeyman was to pay 5 shillings a day for his board. They settle at the end of fifty days, and the journeyman receives £10 18*s*. How many days did he work for the mechanic? *Ans.* 36 days.

42. A farmer brought a basket of eggs to market, and offered them for sale at 7 cents a dozen. A person passing by stumbled against the basket and broke 5 dozen. Being paid for these, he resolved to sell the rest for 8 cents a dozen, by which he would receive as much as if he had sold the full number for 7 cents, besides the money for the broken eggs, which would then be clear profit. How much did he gain by raising the price? How many dozen must he have thus sold, then? How many dozen did he bring to market? Prove by trial with the answer found.

43. A man brings eggs to market, and first sells 4 more than the half of them, then he goes further, and sells half the remainder and 2 over. Now 6 eggs more than half the remainder are stolen from him, and, dissatisfied with the loss, he returns home with the 2 eggs which remained in his basket. How many eggs had he at first? Prove by trial with the answer found.

44. I take a certain number, multiply it by $3\frac{3}{7}$, take 60 from the product, multiply the remainder by $2\frac{1}{2}$, and subtract 30, when nothing remains. What is the number? Prove by trial with the result.

45. Required two numbers, whose sum is 70, and whose difference is 16. If the sum is first divided equally, what process will make the one 16 larger than the other? Try.

46. Two purses together contain \$300. If you take 30 out of the first and put them into the second, then there is the same in each. How many dollars does each contain? Prove by trial with the result.

Suggestive Questions.—If one purse contains $6 and another $4, what is the difference between the purses? *What part* of the difference must be taken from the one and put into the other to make them equal? Then what must be the difference between two purses, when $30 taken from one and put into the other makes them equal?

47. A says to B, give me $100, and I shall have as much as you. No, says B, give *me* rather $100, and then I shall have twice as much as you. How many dollars has each? Prove by trial with the result.

Suggestive Questions.—What is the difference between B's money and A's? (See the principle developed in the last exercise.) How much will that difference be increased if A gives B $100? How much will the whole difference then be? If this difference makes B's money double that of A, how much, must A then have? How much, then, had A before he gave $100 to B? If A originally had ——, and the difference was then ——, how much had B originally?

48. A copyist was asked how many sheets he wrote weekly, and answered, "I only work 4 hours a day, and cannot finish 70 sheets, which I wish to do; but if I could work 10 hours a day then I should write exactly as many above 70 sheets as I now write less than that number." How many sheets did he write weekly? Prove by trial.

Suggestive Question.—If working 10 hours produces as great a surplus as working 4 hours does a deficiency, how many hours must he work to produce the exact result, or how many sheets would he finish in 4 *and* 10 hours?

49. There are three brothers; the eldest 30, the 2d 20, and the 3d 6. When will the ages of the two younger amount to that of the elder? The sum of the ages of the *two* younger must, of course, increase with double the rapidity of the age of the elder. Prove by trial.

50. A baker works 313 days in a year (being the number of working days, except in leap-year and in those years which commence on a Sunday), and bakes 9 barrels of flour each day, at 196 lbs. a barrel. The bread weighs $\frac{1}{3}$ more than the flour (on account of the water used). If he sell the bread at 4 cents a pound, how much will be his profit for the year, when the flour costs $5 per barrel? *Ans.* $15,362·04.

51. A man who has $750 a year saves $145 annually. His income being raised to $1200, how much can he spend daily,

counting 365 days to the year, if he save double as much as he did before? *Ans.* \$2‘49$\frac{23}{73}$.

52. I once had an untold sum of money lying before me. From this I took away the third part, and put in its stead \$50. A short time after I took from the sum thus augmented, the 4th part, and put again in its stead \$70. I then counted the money, and found \$120. What was the original sum? Prove by trial.

Suggestive Questions.—What portion of the money is left when $\frac{1}{3}$ is taken away? How much, then, after \$50 is added? How much is $\frac{1}{4}$ of $\frac{2}{3}+50$? How much, then, remains after that is deducted? How much, then, after \$70 more is added? Now, if $\frac{1}{2}+$\$107‘50 is equal to \$120, how much is $\frac{1}{2}$ equal to? Then how much is the whole?

53. A cistern which is filled by two pipes in 12 minutes, can be filled by one of these pipes in 20 minutes; in what time can it be filled by the other?

Suggestive Questions.—What portion of the cistern is filled by *both* pipes in one minute? What portion is filled by *one* of them in one minute? The difference between those two fractions is equal to what?

54. A cistern has two pipes. By one it can be filled in 20 minutes, and by the other in 30 minutes. What time will it require to fill it when both pipes are running?

55. If 9 barrels of flour cost \$45, what will 15 barrels cost? See Practical Exercises on Fractional Quantities, p. 96, examples 5 to 16.

56. If 15 barrels of flour cost \$75, what will 9 barrels cost?

57. If a horse travel 30 miles in 6 hours, how many miles will he travel in 11 hours?

58. How many miles will a horse travel in 6 hours, if he can accomplish 55 miles in 11 hours?

59. If 4 yards of cloth cost \$12, what will 11 yards cost?

60. What will be the price of 4 yards of cloth, when 11 yards cost \$33?

61. If 3 men can mow a meadow in 10 days, in how many days can it be mowed by 6 men?

62. When 6 men can mow a meadow in 5 days, how many days will be necessary for 3 to mow it?

63. If 3 qrs. of a yard of velvet cost 99 cts., how many yards can I buy for \$37‘62?

64. If $28\frac{1}{2}$ yards of velvet cost \$37·62, what will be the cost of 3 quarters of a yard?

65. What is the value of a silver tankard, weighing 1 lb. 7 oz. 14 dwt., at 6*s.* 8*d.* per ounce?

☞ The pounds may be brought to ounces, and the pennyweights to the decimal of an ounce.

66. If a silver tankard, weighing 1 lb. 7 oz. 14 pwts., cost £6 11*s.* 4*d.*, how much is that per ounce?

67. If a staff, 4 feet long, cast a shade on level ground 7 feet long, what is the height of a steeple whose shade at the same time is 198 feet?

68. When a steeple, whose height is 198 feet, casts a shadow on level ground of the length of $346\frac{1}{2}$ feet, how long would be the shadow of a staff of 4 feet?

69. A journeyman pays \$12 for 4 weeks' board. What will be his bill for 24 weeks at the same rate? *Ans.* \$72.

70. A farmer exchanged 30 bushels of rye for 120 bushels of potatoes. How much rye must he give at that rate for 600 bushels of potatoes? *Ans.* 150 bush.

71. If 10 men can build a wall, 360 rods long, in 9 days, how many rods can 1 man build in one day? 2 men in 1 day? 2 men in 2 days? 24 men in 75 days?

Ans. to last question, 7200.

72. If a man travel 100 miles in 5 days, travelling 4 hours each day, how far could he travel in 12 days, going 10 hours each day? *Ans.* 600 miles.

73. If 6 men build a wall 20 feet long, 6 feet high, and 4 thick, in 16 days, in what time will 24 men build one 200 feet long, 8 feet high, and 6 feet thick? *Ans.* In 80 days.

74. If the freight of 8 hhds. of sugar, each weighing $12\frac{1}{2}$ cwt., 20 leagues, cost £16, what must be paid for the freight of 50 tierces, each weighing $2\frac{1}{2}$ cwt., 100 leagues?

Ans. £100.

75. A journeyman paid \$72 for 24 weeks' board; how much would be the charge for 4 weeks, at the same rate?

Ans. \$12.

76. How many yards of matting, 1·5 feet wide, will cover a room 18 feet wide, and 30 feet long? *Ans.* 120 yards.

Suggestive Questions.—How many *square* feet in the room? How many of these feet will be covered by 1 foot of matting? How many feet of matting, then, will cover 540 square feet? How many yards, then?

77. If the interest of $100 for 1 year be $6, what will be the interest of the same sum for 4 years? For 6 years? For 6 months (or ½ a year)? For 1 month (or 30 days)? For 2 days? For 15 days? *Ans. to the last*, $0'25.

78. If the interest of $100 for 1 year be $6, what will be the interest of $1 for the same time? Of $145? Of $27'50? Of $1472? Of $562? Of $25'25? Of $304?
Ans. to the last, $18'24.

79. If the interest of $109 be $7 for 1 year, what will be the interest of $325 for the same time? Of $62'50? Of $235? *Ans. to the last*, $16'45.

80. If the interest of $350 for 1 year be $17'50, in what time will the interest on the same sum be $87'50? Be $78-'75? Be $91'875? *Ans. to the last*, 5¼ yrs.

81. What is the interest of $100 for 1 year, when the interest of $1750 for 4 years is $350? Is $280? Is $560?
Ans. to the last, $8.

Partnership.

82. Three men, A, B, and C, trade in company, and agree to share the profits in proportion to the amount of property each furnishes to the common stock? A puts in $2000, B $4000, and C $6000. They gain $3000. What is each man's share of the gain? Observe that $3000 is gained by trading with $12,000. What is the gain for every dollar put in? Prove by the sum of the gains.

83. A, B, C, and D, are concerned in a joint stock of $1000, of which A's part is $150, B's $250, C's 275, and D's 325. On the adjustment of their accounts, they find they have lost $337'50. What is the loss of each partner?

Suggestive Questions.—What is the loss on each dollar of stock? Then what is each man's several loss? Prove by addition of the several losses.

84. A ship, worth $3000, being lost at sea, of which $\frac{1}{6}$ belonged to A, ½ to B, and the rest to C, what loss will each sustain, supposing $450 to have been insured on her?

Suggestive Questions.—What is the whole loss? What part of the vessel belonged to C? What was each man's loss? Prove by taking the sum of the insurance and of the several losses.

85. A, B, and C, freighted a ship with 68,900 feet of boards. A put in 16,520 feet, B 28,720, and C the rest. In

a storm, 13,780 feet were thrown overboard to lighten the vessel. How many feet had C? What loss would be sustained on every foot of boards? How many feet would each individual lose? Prove as before.

86. A and B, venturing equal sums of money, cleared by joint trade $273. By agreement, as A executed the business, he was to have 8 per cent., and B 5 per cent. How much was A allowed for his trouble?

Suggestive Questions.—Out of every $8 that A had, how much was for his trouble in doing the business? How many 13s in 273? How *many times*, then, had he 3 dollars for his trouble? Prove as before.

87. In a profitable transaction, A and B were partners. A put in $45, and took $\frac{3}{5}$ of the gain? What did B put in?

Suggestive Questions.—If A took $\frac{3}{5}$ of the gain, what *proportion* of the capital must he have advanced? Then what proportion must B have advanced? If $45 was $\frac{3}{5}$ of the capital, what is $\frac{1}{5}$? $\frac{2}{5}$? Prove.

88. A, B, and C, put $720 in a partnership concern, and gained $540, of which, as often as A took $3, B took $5, and C $7. What did each advance, and what did each gain?

Suggestive Questions.—As their profits were divided according to their advances, for every $15 advanced A paid $3. Then how many 15s in $720? Prove.

89. A bankrupt is indebted to A $120, to B $230, to C $340, to D $450; and his whole estate amounts only to $560. How must it be divided among his creditors, or, which is the same thing, how much will each receive for every dollar due? Prove as before.

90. Divide 360 into four such parts as shall be to each other in the proportion of 3, 4, 5, 6; that is, every time the first has 3, the second shall have 4, &c. Prove by addition.

91. Divide $540 into four parts, bearing to each other the proportion of $\frac{2}{5}$, $\frac{1}{6}$, $\frac{1}{4}$, and $\frac{11}{60}$, and prove by addition.

92. The taxes in a certain town amount to $4000. It contains taxable property to the amount of $125,000, and 500 polls (that is, individuals who are personally taxed), who pay 50 cents each. What part of the tax was laid on property? What was the tax on every dollar? What was the amount of A's tax, whose list is $1400 and 1 poll; B's, whose list is $1200 and 2 polls; and C's, whose list is $400 and 1 poll?

and how much did the remainder of the town contribute for polls and for property? Prove by addition.

93. Four farmers hire a pasture for $75. A puts in 75 sheep for 16 weeks; B 50 for 24 weeks; C 120 for 20 weeks; and D 50 for 4 weeks. How much does it cost them per week for each sheep, and how much has each to pay towards the rent? Prove by addition.

94. Three farmers hire a pasture. A puts in 80 sheep for 4 months; B 60 sheep for 2 months; C pays $21·60 towards the rent, and puts in 72 sheep for 5 months. What was the whole rent of the pasture, and what share did A and B severally pay? Prove.

95. Four merchants traded in company. A put in $400 for 5 months; B $600 for 7 months; C $960 for 8 months; and D 1200 for 9 months. By misfortunes at sea they lost $617. What must each man sustain of the loss?

Suggestion.—$400 for 5 months is equal to $100 for how many months? Prove.

96. A, with a capital of £100, began trade Jan. 1, 1850. He took in B as a partner on the first day of March following, with a capital of £150; and, three months afterwards, they admitted C as a third partner, who brought into the business £180. After trading together till the first of Jan., 1851, they found there had been gained since A's commencing business £264. How must this be divided among the partners? Prove.

97. Suppose the gain mentioned in the last question had been from the time of C's entering into partnership, how should it be divided then? Prove.

98. A merchant, A, commences business on the first of Jan., with a capital of $3000. He takes B into partnership on April 1, and C on July 1. Their profits were $2400, which were to be divided in proportion to the amount of capital furnished by each, and the length of time it was used. It so happened that each received $\frac{1}{3}$, or $800. How much capital did B and C severally furnish?

Exchange.

99. How much New York currency is equal to £150 of Canada currency? See table of Provincial Currencies, p. 227, and examples, p. 248. See, also, p. viii.

100. How much Canada currency will pay a debt of £240 in New York?

101. How many dollars will pay a debt of £77 16*s.* 8*d.* in Charleston, South Carolina?

102. How much of South Carolina currency can be cancelled by \$333$\frac{4}{7}$?

103. What sum in New England currency will cancel a debt of £93 18*s.* 6*d.* in Fayetteville, North Carolina?

104. What sum in North Carolina or New York currency will cancel a debt of £70 8*s.* 10$\frac{1}{2}$*d.* in New England?

105. A traveller wishes to change £233 16*s.* 8*d.* sterling for Venice ducats, at 4*s.* 9$\frac{1}{2}$*d.* per ducat. How many ducats must he have? *Ans.* 976.

Mean Numbers and Values, commonly called Alligation.

106. A merchant mixes 8 lbs. of tea worth 35 cts. per pound, with 24 lbs. worth 44 cts. per pound. What is the value of the whole mixture? and what is the mean value, or value of the mixture per pound?

Ans. to the last question, 41$\frac{3}{4}$ cts.

107. A farmer mixes 18 bushels of corn, worth 75 cts. per bushel, with 9 bushels of oats, at 33$\frac{1}{3}$ cts. per bushel, and 4 bushels of bran, at 15 cts. per bushel. What is the value of the whole mixture, and how much its mean value per bushel? Prove as below.

18 bushels, at 75 cents =
9 " at 33$\frac{1}{3}$ " =
4 " at 15 " =
} =31 bushels, at — =

108. The same man makes another mixture of 4 bushels of rye, at 60 cents per bushel; 15 bushels of buckwheat, at 40 cts.; and 8 bushels of corn, at 75 cts. per bushel. What is the mean value of the mixture per bushel? Prove as above.

109. A grocer bought 3 barrels of sugar; one at 5 cts. per lb., one at 6, and one at 7 cts. per lb. What is the average or mean value of the whole, supposing each barrel to contain the same weight of sugar? Prove.

110. A goldsmith melts together 4 lbs. of gold, of 22 carats fine; 1 lb., of 20 carats fine; and 1 lb., of 16 carats fine. What is the fineness of the mixture; that is, the mean number of carats of fineness? Prove.

☞ The fineness of gold is estimated by carats. Pure gold

is 24 carats fine. The smaller the number of carats, the less pure is the metal.

111. The average or mean height of the thermometer at a certain place for the month of December was 30 degrees; the average for January was 27°; and the average for February was 24°. What was the average height for the three months?

Ans. 27°.

112. A merchant wishes to mix coffee, at 10 cts. and 14 cts. per pound, so that the compound shall be worth 12 cts. per pound. Should the quantities of the two sorts be equal or unequal?

Suggestive Questions.—How much does the price of the *least* costly fall short of the price of the mixture? How much does the price of the *most* costly exceed that of the mixture? Will the gain on 1 lb. of the former, then, exactly balance the loss on 1 lb. of the latter? Should the quantities of the two sorts, then, be equal or unequal?

113. What proportions of coffee at the following prices, namely, at 6 cts., 7 cts., 10 cts., and 12 cts., per lb., should be mixed so as to make a compound worth 9 cts. per lb.?

	Prices of ingredients.	Difference from price of mixture.	Quantities of ingredients.	PROOF. lb. cts.
	6 . . .	—3 . .		1 at 6= 6
Price of mixture, 9 cts.	7 . . .	—2 . .		3 at 7=21
	10 . . .	+1 . .		3 at 10=30
	12 . . .	+3 . .		2 at 12=24
				9 at 9=81

Suggestive Questions.—If the mixture is to be sold at 9 cts. per lb., what will be the gain on each pound of the 6 cent? What will be the loss on each lb. at 10 cents? Then how many lbs. should go into the mixture of that losing 1 cent, to balance the 1 lb. which gains 3? Will 1 lb. at 6 cents, then, exactly balance 3 lbs. at 10? Mark the quantities, then, 1 and 3, in the blank column, opposite 6 and 10. Compare the two quantities you have just written, and say why the *quantities* are in inverse order to the *differences* in price from the mixture. Will a *greater difference* always require a *less quantity*, and a *less difference* require a *greater* quantity? Balance the prices, 7 and 12, in the same manner, avoiding fractions by doubling, or trebling, &c., both numbers. One lb. at 12 will be balanced by what quantity at 7? How can this fraction be got rid of? Why must there be 1 lb. of the 6 cent coffee? *Ans.* Because it is balanced by — of —? Why 3 lbs. of 10 cent coffee? *Ans.* Because they are balanced by — of —. Why should there be 3 of the 7 cent? Why 2 of the 12?

114. How many pounds did the mixture of the last example contain? If a bag, holding 36 lbs., then, were required to be filled of such a mixture, how many times would the mixing be repeated? Then how many pounds would be required of each kind of coffee? Prove as in last example.

115. If a merchant, wishing a mixture like that named in Ex. 113, should put in 16 lbs. of the 6 cent coffee, how many lbs. would be required of each of the others? Prove.

116. Again: if he wished to use 16 lbs. of the 12 cent coffee, how many times must the proportions of the others be repeated to make a similar mixture? How many pounds would such a mixture contain? Prove.

117. A farmer had several kinds of provender which he wished to mix for his cattle and horses, so as to form a compound worth 50 cts. a bushel, namely, corn meal, worth 75 cts. a bushel; rye, worth 55; oats, 40; shorts, 25; and wheat bran, 15 cents a bushel. What proportions of each sort should be used, and how many bushels would the smallest quantity of the mixture contain, excluding fractions of a bushel?

Ans. to the last question, 13 bushels.

	Prices.	Diff.	Quantities.
Compound, 50.	75	+25	
	55	+ 5	
	40	—10	
	25	—25	
	15	—35	

Suggestive Questions.—In what proportions may we mix the meal at 75 cts. and that at 25 cts. to form a compound worth 50 cts.? Mark the quantities 1 each, then. In what proportions can those of 55 and 40 be so mixed? Enter them, then. In what proportions those of 55 and 15? Enter them, then, separating the two numbers at 55 by the sign of addition, +. Why? Prove as in example 113 above.

118. If 208 bushels of the above compound were wanted, how many times would the proportion of each sort be repeated, and how many bushels would there be of each? Prove.

119. If $5\frac{1}{2}$ bushels of the corn meal were used, how much would be required of each of the others? Prove.

120. A grocer has four sorts of sugar, worth 4 cts., 5 cts., 7 cts., and 8 cts. a pound. He would make a mixture of 200 lbs., worth 6 cts. a pound? What quantity must be taken of each sort? Prove.

121. A goldsmith has four sorts of gold, namely, of 22 carats fine, of 20 carats fine, of 18 carats fine, and of 15 carats fine. He would make a mixture of 48 oz. of 17 carats fine. How many oz. of each sort must he take? Prove.

122. Afterwards, of the same material, he wished to make a mixture of the same fineness, containing 4 oz. of 20 carats fine. How many ounces must he take of each of the other sorts? Prove.

---

123. A rectangular field was 16 rods long and 12 wide. How many square rods did it contain?

124. What is the width of a rectangular field containing 192 square rods, whose length is 16 rods?

125. There are 192 rods in a rectangular field, whose width is 12 rods. What is its length?

126. There is a square field whose sides are 16 rods long. How many square rods does it contain?

127. What is the length of a square field containing 256 square rods?

128. The sides of one of the square fields of a farm is 40 rods long, and those of another 80. How many times is the one larger than the other?

129. There are two square fields in a farm, one of which is 40 rods long; the other is 4 times the size. What is the length of its sides?

130. The inside of a box is 2 feet every way. How many cubical feet does it contain?

131. The contents of a box with equal sides are 8 cubical feet. What are its length, width, and depth inside?

132. A farmer erected a stable 50 feet long by 25 feet wide. The height of the gable was 8 feet. The eaves projected a foot over each side of the building, and the roof was 2 feet longer than the frame, so as to project a foot over each gable. How many thousand shingles would be required for the roof, if one thousand shingles cover 10 feet square?
Ans. 18 nearly.

133. One man exchanged with a broker £4 10*s.* 10*d.* sterling for 11 crowns and 7 dollars; and another man, at the same rate, £1 15*s.* for 4 crowns and 3 dollars. How much were the crown and dollar severally valued at? Prove by trial.

Suggestive Questions.—What is 3 times the amount of the first exchange? 7 times the amount of the 2d? What is the *difference* between the exchanges thus increased? What, then, is the value of a crown? Of a dollar? In what respects does this operation differ from bringing fractions to the same denomination?

134. Required two such numbers that if $\frac{1}{3}$ of the first be added to $\frac{5}{6}$ of the second, the sum shall be 66; and if $\frac{5}{6}$ of the first be added to $\frac{1}{3}$ of the second, the sum shall be 60. Prove by trial.

135. If the greater of two numbers be divided by the less, the quotient is 6, and the sum of the two numbers is 252. What are the numbers? Prove by trial.

136. A gentleman gave $4350 for a house-lot, the land being valued at $2 per foot. If it had been 6 feet wider, it would have cost $5394. What were the length and breadth of the lot? Prove by trial.

137. A boy bought at one time 5 apples, 6 pears, and 4 oranges, for 48 cents; at another time, 3 apples, 4 pears, and 5 oranges, for 43 cents; and again, 2 apples, 3 pears, and 6 oranges, for 43 cents, all at the same rate. What did he pay for each kind of fruit? Prove by trial.

138. There were 5 Sundays in the month of February in 1852. In what year will this occur again; that is, when will the first day of February fall on a Sunday in a bissextile or leap year? *Ans.* In 1880.

Suggestive Questions.—How many days of the week does the year advance from one bissextile to another? What is the smallest number of *fives* exactly divisible by 7? Then how many bissextiles must elapse till 5 Sundays again occur in February?

Equation of Payments.

139. A man bought a farm for $2000, one half of which was to be paid in two years, and the remainder in 4 years; that is, the purchaser was to have the use of $1000 of the purchase money for 2 years, and the use of the remaining

$1000 for 4 years. At what time may the whole be paid at once without loss to either of the parties?

	$		years		$
Use of	1000	for	2	=	2000 for one year.
	1000	"	4	=	4000 for one year.
paid	2000			=	$6000 for one year.

Suggestive Questions.—How long must he keep the $2000 so as to balance the use of $6000 for one year? By what process can this be ascertained? By addition, subtraction, multiplication, or division?

140. A man owed his neighbor $300, which he engaged to pay as follows: $50 in 2 months, 100 in 4 months, and $150 in 6 months. When may the whole be paid at once without loss to either party?

	$		ms.		$		
Use of	50	for	2	=	100	for 1	month.
	100	for	4	=	400	for 1	"
	150	for	6	=	900	for 1	"
	$300			=	$1400	for 1	"

Suggestive Question.—$1400 for 1 month=$300 for how many months?

141. A friend lent me $400 for three months. How long should I lend him $100 to balance the favor?

Ans. 12 months.

142. A man bought a piece of property for $600, and agreed to pay $100 in 2 months, 200 in 5 months, and the rest in 8 months? What would be the proper time to make one payment of the whole? *Ans.* 6 months.

143. What is the mean time for the settlement of a debt of $800, contracted to be paid as follows: $200 in 3 months, $\frac{1}{3}$ of the remainder in 4 months, $\frac{1}{2}$ of what then remains in 5 months, and the rest in 6 months? *Ans.* $4\frac{1}{2}$ months.

144. One merchant owes another $800, payable in 6 months, but wishes to pay him $200 of the debt in 2 months. How long should the time of payment of the remainder be suspended to balance the favor? *Ans.* $1\frac{1}{3}$ months.

145. A country merchant makes purchases from a merchant in Boston, on a credit of 6 months, as follows: $1500

on May 1, \$400 on June 1, \$500 on July 1, and \$300 on Aug. 1. What is the mean time for the whole from Aug. 1?

Ans. 4 months.

---

146. What is the mean length of the following pieces of cloth: No. 1, 30 yds.; No. 2, 28 yds.; No. 3, 27 yds.; No. 4, 29 yds.; No. 5, 32 yds.; No. 6, 25 yds.; No. 7, 25 yds.

Ans. 28 yards.

147. A man on horseback travelled the following distances: the first day, 30 miles; the second day, 34 miles; the third day, 36 miles; and the fourth day, 42 miles. How many miles did he average a day? *Ans.* $35\frac{1}{2}$ miles.

148. Four men are engaged in building a wall measuring 820 cubic feet. The first can build 9 cubic feet in 4 days; the second, 10 cubic feet in 4 days; the third, 8 cubic feet in 6 days; and the fourth, 7 cubic feet in 3 days. How many days will be necessary to complete the whole wall, when all work together?

149. A merchant had 32 tons of plaster for sale. On examining his sale books at the end of a week, he found that there remained 8 tons more than he had sold. How many tons were sold? Prove.

150. Three farmers bought a pasture jointly, consisting of 140 acres. On dividing it, it was agreed that A's share should be to B's as 6 to 11, and that C should have 4 acres more than A and B together. What is the share of each? Prove.

151. A man being asked how much money he had in his pocket, answered that $\frac{1}{3}$ and $\frac{1}{5}$ of it amounted to \$3·20. How much had he? Prove.

152. A traveller, being asked what o'clock it was, replied that it was between 3 and 4. But a more particular answer being requested, said that the hour and minute hands were exactly together. What was the time? *Ans.* $16\frac{4}{11}$ min. past 3.

Suggestive Questions.—How far are the hands apart at 3 o'clock? In what time will the minute overtake the hour hand?

153. John sets out on a journey, and travels at the rate of 5 miles an hour. He travels 8 hours the first day, and the next morning a friend sets out after him at the rate of 7 miles an hour. If both start at the same hour in the morning, and travel the same number of hours in a day, how far must the friend travel before he overtakes John?

154. How many miles an hour must William travel for 20 hours, in order to overtake John, who is 40 miles ahead, and is travelling at the rate of 5 miles an hour?

155. A company sets out for California by land, and travels at the rate of 30 miles a day. Three days afterwards a man undertook to overtake them, and accomplished his purpose in 10 days. At what rate per day did the man travel?

156. A man bound for California by land, on his arrival at the usual starting point, learns that a company was 3 days ahead of him, with the intention of travelling 30 miles a day. He set out after them, and travels at the rate of 39 miles a day. In how many days will he overtake them?

157. The velocity of sound through the air is found to be 1142 feet per second of time; and, in a healthy person, the number of pulsations is, say 70 in a minute. Now, between the time of observing a flash of lightning and hearing the explosion of the thunder I counted 20 pulsations. Required the number of feet which sound moves in a minute; the number in one pulsation; and the distance of the cloud in miles.

158. A gentleman counted 20 pulsations between the sight of a flash of lightning and the arrival of the sound. By calculation he found the distance of the cloud to be 3 miles, 226 rods, and $8\frac{1}{7}$ feet; allowing the pulsations to be at the rate of 70 in a minute, how many feet did the sound travel in a second?

159. A lunar month, in which the moon makes a complete apparent revolution round the earth, is, say $29\frac{1}{2}$ days. Supposing the moon's motion to be uniform, what is the apparent distance of the sun and moon in degrees, &c., in one day after the change? in 12 days 6 hours after the change? in 19 days? [Take notice that the moon again begins to approach the sun after $14\frac{3}{4}$ days. Why?] *Ans. to the last question*, $128\frac{8}{59}°$.

160. The earth, being 360° in circumference, turns on its axis, say in 24 hours. How many miles are the inhabitants at the equator carried by that movement in one minute, a degree there being $69\frac{1}{2}$ miles? *Ans.* $17\frac{3}{8}$ miles.

161. If an ingot of gold, weighing 9 lb. 9 oz. 12 dwt., be worth $1128·96, what is it worth per grain?

162. If gold be worth 2 cents per grain, what is the value of 9 lb. 9 oz. 12 dwt.?

163. A saves $\frac{1}{3}$ of his income; but B, who has the same

salary, by living at double the expense, runs behind $120 a year. How much has each per annum? *Ans.* $360.

164. Required the sum of which $\frac{1}{3}$, $\frac{1}{4}$, and $\frac{1}{5}$ make $94. *Ans.* $120.

165. A is 20 years of age. B's age equals A's and half of C's; and C's equals them both. What are their several ages? If C's age is considered (1); B's will be ($\frac{1}{2}$)+20; and A's 20; but C's equals A's and B's; how much, then, is the other half of C's age equal to? Prove by trial.

166. The head of a fish is 9 inches long, and its tail is as long as its head and half the body, and the length of the body equals those of the head and tail. What is its whole length? Prove by trial.

167. A laborer hired for 42 days upon this condition, that he should receive 80 cents for every day he worked, and forfeit 40 cents for every day he was idle. On settlement it was found that nothing was due him. How many days did he work? Perform this mentally by division. *Ans.* 14 days.

168. Two foot travellers sat down to dinner, one having 5 small loaves of bread, the other 3 of the same size. A third traveller coming along proposed to join them on condition of paying for his share of the bread. The whole was eaten, and the third traveller paid 8 cents. How should this money be shared between the two owners of the bread?

Ans. 7 to the first, 1 to the second.

169. Tell the difference between 365 times 421 and 375 times the same number, by a mere glance of the eye, and prove by multiplication and subtraction.

170. Tell the difference between 235 times 364 and 240 times the same number, by halving a certain number mentally, and prove as in last exercise.

171. Tell the difference between 328 times 465 and 333 times the same number, by a similar process, and prove.

172. Tell the difference between 425 times 326 and 475 times the same number, by halving a certain number mentally, and prove.

173. Tell the difference between 354 times 248 and 379 times the same number, by dividing a certain number mentally by —, and prove.

174. Tell the difference between 432 times 368 and 457$\frac{1}{4}$ times the same number, by a similar mental process, and prove.

175. Tell the difference between 216 times 344 and 241 times the same number, by a mental process, and prove.

176. Tell the difference between 284 times 275 and 334 times the same number, by a mental process, and prove.

177. Tell the difference between 341 times 432 and 367 times the same number, by a mental process, and prove.

178. Tell the difference between 286 times 76 and 310 times the same number, by a mental process, and prove.

179. Tell the difference between 525 times 340 and 552 times the same number, by a mental process, and prove.

180. Find, by inspection, the product of 2·44 by 24·9, by division and subtraction, and prove by multiplication on the slate.

Suggestive Questions.—If 2·44 be taken 25 times in place of 24·9 times, how many times is it taken too much? What is $\frac{1}{10}$ of 2·44?

181. Find, by inspection, the product of ·5 by 23·84, by division, and prove by multiplication.

Suggestive Question.—Why are there only two decimal places in the product, when there are three in the two factors?

182. Find, by inspection, the square of 49 by division and subtraction, and prove by multiplication.

183. Find, by inspection, the square of 26, by division and addition, and prove by multiplication.

184. Find, by inspection, the squares of 24, 16, 14, by processes similar to those in the last two problems, and prove by multiplication.

185. Find, by inspection, the square of 75, by division and subtraction of the quotient, and prove by multiplication.

186. Find, by inspection, the square of 3·1, by multiplication and addition, and prove by multiplication only.

187. Find, by inspection, the product of 324 by 49, by division and subtraction, and prove by multiplication.

188. Find, by inspection, the product of 64 by $4\frac{3}{4}$, by division and subtraction, and prove by multiplication.

189. Find, by inspection, the product of 72 by $5\frac{1}{4}$, by division and addition, and prove by multiplication.

190. Find the product of 16 by $12\frac{1}{2}$ by division.

Solution. 8)1600=16×100.
200

Suggestive Questions.—How many times $12\frac{1}{2}$ in 25? How many times 25 in 100? Then how many times $12\frac{1}{2}$ in 100? If, then, we multiply by 100 in place of $12\frac{1}{2}$, how many times will the product be too large? Why, then, should we divide by 8?

191. Find the product of 24 by $37\frac{1}{2}$, by division and subtraction.

Solution. 2)2400
24×50= 1200—300(=$\frac{1}{4}$ of 1200)=900.

Abbreviated. 2)24
12—3=900.

192. Find the product of 64 by $62\frac{1}{2}$ (=50+$12\frac{1}{2}$) by division and addition.

Solution.
2)6400=64×100
4)3200=64× 50
800=64× $12\frac{1}{2}$
4000=64× $62\frac{1}{2}$

193. Find the product of the same numbers by division and multiplication.

Solution.
8)6400=64×100
800=64×$12\frac{1}{2}$
5
4000=64×$62\frac{1}{2}$

194. Find the product of 328 by $87\frac{1}{2}$ by division and subtraction.

Solution.
8)32800=328×100
4100=328× $12\frac{1}{2}$
28700=328× $87\frac{1}{2}$

195. Find the product of 176 by $112\frac{1}{2}$ by division and addition.

Solution.
8)17600=176×100
2200=176× $12\frac{1}{2}$
19800=176×$112\frac{1}{2}$

196. Find the product of 3724 by 125, by division and addition.

Solution. $4)372400 = 3724 \times 100$
$93100 = 3724 \times 25$
$\overline{465500} = 3724 \times \overline{125}$

197. Find the product of 3426 by 175.

Solution. $3426 \times 200 = 685200$
Less $\frac{1}{8}$ 85650
$3426 \times 175 = \overline{599550}$

198. Find the product of 4273 by 225.

Solution. $4273 \times 200 = 854600$
More $\frac{1}{8}$ 106825
$4273 \times 225 = \overline{961425}$

199. Find the above 9 products by inspection, without the aid of pen or pencil.

200. Find the following products severally by inspection, and prove each by calculation: 4572 by $12\frac{1}{2}$; 3968 by 25; 8128 by $37\frac{1}{2}$; 1437 by 50; 2456 by $62\frac{1}{2}$; 7415 by 75; 9284 by $87\frac{1}{2}$; 667·8 by $112\frac{1}{2}$; 1529 by 125; 7924 by $137\frac{1}{2}$; 6414 by $162\frac{1}{2}$; 2256 by 175; 6328 by $187\frac{1}{2}$; 4428 by $212\frac{1}{2}$; 7732 by 225; 3648 by $487\frac{1}{2}$.

201. The property of John and Thomas was estimated by the assessors at $18,600, and John was twice as rich as Thomas. What was the capital of each?

Ans. John's $12,400, Thomas' $6200.

Suggestive Questions.—If Thomas's property was considered one share, how many shares had John? Into how many parts, then, should the property be divided?

202. The sum of $2500 is to be divided between two persons, the one to have $4 as often as the other $1. How much would each receive? *Ans.* $2000 and $500.

203. A man sold a farm for $3960, and received payment in bank notes and gold, there being $4\frac{1}{2}$ times as much in paper as in gold. How much was there of each?

Ans. $720 in paper, $3240 in gold.

Suggestion.—Get rid of the fraction by multiplication.

204. Divide 1394 into two such parts that the one may be contained in the other $3\frac{1}{4}$ times. *Ans.* 328 and 1066.

205. Divide $144 between two persons so that their respective shares will be as 7 to 5. *Ans.* $84 and $60.

206. If the 3d and 4th part of a field together contain 14 acres, how many acres are there in the field? *Ans.* 24 acres.

207. Two brothers purchased a house, of which the one paid the 4th and the other the 8th part, amounting together to $300. What was the whole price of the house?

Ans. $800.

208. In a certain congregation of 1300 souls, there were 9 times as many women and 3 times as many men as children. How many were there of each?

Ans. 100 children, 300 men, and 900 women.

209. A man bought a house for $3040, and paid $3\frac{1}{2}$ times as much in bank notes as in gold, and $2\frac{1}{3}$ times as much in silver as in bank notes. How much did he pay of each sort?

Ans. $240 in gold, $840 in bank notes, and $1960 in silver.

210. Divide $1120 among three persons so that the first one's part may be to that of the second as 2 to 3, and the part of the third as great as that of the other two.

Ans. $224, $336, and $560.

211. Divide $2600 among three persons, A, B, and C, so that B may have $\frac{1}{3}$ more than A, and C twice as much as A.

Ans. A $600, B $800, and C $1200.

212. A tax of $594 is to be raised in three towns, A, B and C, in proportion to their population. Now the population of A to B is as 3 to 5, and that of B to C 8 to 7. How much money must each town raise?

Ans. A $144, B $240, and C $210.

213. The estate of a bankrupt, amounting to $21,000, is to be divided among four creditors, A, B, C, D, in proportion to their claims. Now A's claim to B's is as 2 to 3, B's claim to C's as 4 to 5, and C's claim to D's as 6 to 7. How much will each creditor receive?

Ans. A $3200, B $4800, C $6000, D $7000.

214. A young man spent $\frac{1}{3}$ of his income in board and lodging, $\frac{1}{8}$ in clothes and washing, and $\frac{1}{10}$ in incidental expenses, and saved annually $318. What was his yearly income?

Ans. $720.

215. At the close of a partnership, two merchants divided their profits, to the amount of $12,000, so that one partner received half as much as the other, exclusive, however, of

$500 to the latter for his management of the business. How much did each receive?

Ans. The one $7666⅔, the other $4333⅓.

216. Divide $1925 into three portions, so that the second may be $150 greater than the first, and the third $125 greater than the second. *Ans.* $500, $650, $775.

217. A man died leaving an estate of $7500 to his widow, two sons, and three daughters, in such a manner that each son was to receive twice as much as each daughter, and the widow $500 more than all the children together. How much is the share of each person?

Ans. The widow $4000, each son $1000, each daughter $500.

218. On New Year's day a father presented $100 to his five children, dividing it so that each received $2 more than the next younger child. What was the share of the youngest?

Ans. $16.

219. A man left his property by will in such a manner that his widow was to receive one half of it, less $3000, his son one third, less $1000, and his daughter one fourth, more $800. What was the amount of the estate, and what the share of each legatee? *Ans.* $38,400, $16,200, $11,800, $10,400.

		Less.	More.
Widow,	$\frac{1}{2}$	3000	
Son,	$\frac{1}{3}$	1000	
Daughter,	$\frac{1}{4}$		800
	$\frac{13}{12}$	4000	800

Suggestive Question.—How much is one twelfth of the estate?

220. A certain property valued at $5600 is to be divided among 5 persons as follows:—B is to receive twice as much as A, and $200 more; C three times as much as A, less $400; D half of what B and C receive together, and $150 more; and E one fourth of what all the others receive, and $475 more. What is the share of each?

Ans. A $500, B $1200, C $1100, D $1300, and E $1500.

221. If I have a certain number of dollars in my purse, and by *adding* 24 to it the sum becomes $80, how many dollars were in it at first? Prove. By what process was it found?

222. I have a certain number of dollars in my purse, and, after having *subtracted* $24 from it for the payment of expenses, there remain $56. How many dollars were in at first? Prove. What was the process?

223. A certain number, *multiplied* by 24, gives 1800 for product. What was the original number? Prove. What was the process?

224. A certain number *divided* by 24, gives 75 for quotient. What was the original number? Prove. What was the process?

225. How many dollars did you pay for that? says William to John. If you multiply the number by 7, replied John, add 3 to the product, divide this by 2, subtract 4 from the quotient, the remainder will be 15. How much was paid? Prove. See the last 4 examples.

226. A man whose age is 30 years has a son aged 10. In how many years will this man, who is now three times, be only twice as old? Prove.

227. The boy mentioned in the preceding exercise has a brother aged 6. In how many years will the ages of both the boys together equal that of their father? Prove.

228. A cistern has three pipes. By the first it can be filled in 2 hours, by the second in 3 hours, and by the third in 4 hours. In what time can it be filled when all three run together? *Ans.* in $55\frac{5}{13}$ minutes.

229. A cistern of $365\frac{1}{4}$ cubic feet has three pipes. The first discharges 52 cubic feet in $3\frac{1}{4}$ minutes; the second 51 cubic feet in $3\frac{2}{5}$ minutes; and the third 27 cubic feet in $2\frac{1}{4}$ minutes. In what time can the cistern be filled if they all run at once? *Ans.* In $8\frac{85}{172}$ minutes.

230. A contribution being necessary for a purpose in which a number of persons were equally interested, 80 cents each was first proposed, which was found to produce $10 too much, and then 60 cents each was tried, which was found to produce $10 too little. What was the amount wanted, and how many were the contributors? Prove by trial.

231. A merchant, having a piece of unsalable silk, disposed of it to a lady at prime cost. Having sent it home to her without measurement, and being unable to find the original bill, he could only ascertain the length and wholesale price by recollecting that if he had sold it for $1·25 per yard his profit would have been $12, whereas, at $1 per yard, it would have netted him only $6. What was the number of yards, and their prime cost? Prove by trial.

232. The income of two brothers taken together is $1000 per annum. If the income of the elder was increased sixfold

and that of the younger fourfold, their joint income would be $4800. What was the income of each?

Ans. Elder's income $400, younger's $600.

	Elder.		Younger.		$
	1	+	1	=	1000
	6	+	4	=	4800
Therefore	4	+	4	=	4000
Difference	2	+	0	=	800

233. Find two numbers such that if the first be taken 4 times and the second 3 times, their sum is 2760; and if the first be taken 5 times and the second 7 times, their sum is 4490. Prove by trial.

234. A person has a certain number of pieces of money which he wishes to arrange in the form of a square. At the first trial there were 190 over; but, when the side of the square was enlarged by 4 more pieces there only remained 14. How many pieces of money had he? Draw a square on the slate, and enlarge it by 4 additions. Prove by trial.

235. A person has 330 coins, consisting of eagles and dollars. Their value amounts to $1500. How many are there of each? Prove by trial.

Suggestive Questions.—How many coins would there be if the money was all in eagles? How many more should there be? Every eagle changed into dollars gives how many additional coins? Then how many eagles must be changed into dollars to make 330 in all?

236. What number is that, which, if you multiply it by $4\frac{3}{5}$, take 19 from the product, multiply it again by $3\frac{1}{4}$, and take away 13, nothing will remain? Prove by trial.

237. A boy having a basket of apples, sold half of them and 3 more to one of his neighbors, and half the remainder and 5 more at the next house. Finally, he sold half of what still remained and three more, and then found he had only three left. How many had he at first? Prove by trial.

238. A man bequeathed, by will, to his widow $250 and half the remainder of his property; to his son a fifth of the residue and $525 more; and the remainder, amounting to $1375, to his daughter. What was the whole amount of the property? Prove by trial.

239. A person being asked how many dollars he had, replied, if you add to them their third part and 176 more, and then multiply by $2\frac{1}{2}$, the sum will as much exceed \$1000 as it now falls short of it. Prove by trial.

CHAPTER V.

PRACTICAL APPLICATION

OF THE ELEMENTS OF ARITHMETIC BY MEANS OF RATIO AND PROPORTION.

DEFINITIONS.

I. *Ratio* is the relation which one quantity has to another *of the same kind*, as expressed by the quotient of the one divided by the other. Thus, the ratio of 4 to 2 is $\frac{4}{2}$ or 2, and the ratio of 5 to 6 is $\frac{5}{6}$. A ratio is sometimes written 4 : 2, or $4 \div 2$. But the fractional form, as $\frac{4}{2}$, is the most convenient for arithmetical computations. The two numbers which constitute a ratio are called its *terms*.

II. A ratio, like a common fraction, remains unchanged in value: (1) when both its terms are *multiplied* by the same number; (2) when both terms are *divided* by the same number; (3) when the same proportional part of each is *added* to or *subtracted* from both. Thus, $\frac{12}{6}$, $\frac{24}{12}$, $\frac{8}{4}$, $\frac{10}{5}$, $\frac{8}{4}$, are merely different forms of the same ratio (all being equal to 2), although both terms in the second have been multiplied by 2, in the third divided by 3, in the fourth increased by $\frac{1}{4}$th, in the fifth diminished by $\frac{1}{5}$th.

III. *Proportion* is the *equality* of two ratios. Thus, $\frac{4}{2} = \frac{12}{6}$ is a *proportion*, both the ratios being 2. Although both terms of a *ratio* must relate to *things of the same kind*, it is not necessarily so with the two ratios of a *proportion*. These may relate either to things of the same kind, or to things of different kinds. Thus, the proportions horses. dollars. days. days.

$$\frac{4}{2}=\frac{12}{6}, \text{ and } \frac{4}{2}=\frac{12}{6}$$

are both correct.

IV. Proportion is used in the resolution of arithmetical problems in cases where *one* of the ratios is *incomplete* from one of its terms being unknown. This unknown term is readily supplied by a mere change of form in the complete ratio; that is, by changing its denomination to that of the incomplete one. Thus, in the proportion $\frac{\ }{16}=\frac{4}{8}$, the unknown term is found to be 8 by changing the complete ratio to 16ths. Again, in the proportion $\frac{\ }{4}=\frac{9}{12}$, the unknown term is found to be 3, by changing $\frac{9}{12}$ to 4ths.

V. Proportion is either *simple* or *compound*. It is *simple* when both ratios are simple; *compound* when one of them is compound. Thus, the ratios $\frac{1\,5}{4\,5}=\frac{4}{12}$, form a simple, and $\frac{3}{5}$ of $\frac{5}{9}=\frac{4}{12}$ a compound proportion. The latter may be more concisely written $\frac{3\cdot 5}{5\cdot 9}=\frac{4}{12}$. See p. 58, l. 5; and 83, l. 16.

VI. In every problem in proportion there is an affirmation and a question. Thus, in the problem, "Bought 9 yards of cloth for £5 12*s.*; what will be the cost of 72 yards?" the affirmation is that "9 yards were bought for £5 12*s.*;" and the question is, "what will be the cost of 72 yards?" The affirmation, however, is more commonly put in the form of a supposition, as, "*If* 9 yards cost £5 12*s.*" Again, in the problem, "If a man travel 90 miles in 3 days of 12 hours long, how far will he travel in 8 days of 10 hours long?" the first clause, to the comma, is the affirmation; the last clause, to the note of interrogation, is the question.

VII. The number belonging to the *imperfect* ratio may always be ascertained from the words asking the question. Thus, in the first of the above problems, it is "£5 12*s.*," since the question is "What will be the *cost?*" In the second problem it is "90 miles," the question being, "How *far* will he travel?" The arrangement of the terms of the *perfect* ratio depends on the answer to a question which should be put to every perfect ratio, and to every part of it if it be compound, namely, LESS OR MORE? that is, "What effect will the numbers in the perfect ratio produce on the unknown term; increase or decrease it?" The larger term is placed above or below, according to the reply. Thus, in the first of the following exercises, 5 is known to be the imperfect ratio from the question "How many *barrels?*" and, in order to know whether the $75 or $25 is the upper term of the perfect ratio, the question "How many barrels, LESS or MORE, for $75?" the reply, "MORE," shows that the larger number should be placed above, in order that the unknown number may be larger than 5. Again, in the fourth exercise below, 8 is known to be the imperfect ratio, because the question is "in how many *days?*" 120 is placed above 70, because 70 men will take "MORE" days to build the wall than 120; 24 is placed above 30, because 24 feet can be built in "LESS," time than 30; and 50 above 40, because 50 will take "MORE" than 40.

VIII. In the resolution of problems, four cases occur:

1. The lower term of the perfect ratio may be a factor of the number in the imperfect ratio, as $\frac{\quad}{20}=\frac{3}{4}$. Here the problem is solved by multiplication alone, which, by changing $\frac{3}{4}$ to $\frac{15}{20}$, shows the unknown number to be 15.

2. The lower term of the perfect ratio may be a multiple of that of the imperfect ratio, as $\frac{\quad}{8}=\frac{6}{24}$. Here the problem is solved by division alone, which, by changing $\frac{6}{24}$ to $\frac{2}{8}$, shows the unknown number to be 2.

3. The lower term of the perfect ratio may be neither a multiple nor a factor of the imperfect ratio, as $\frac{\quad}{10}=\frac{4}{7}$. Here both multiplication and division are required, namely, the number in the *imperfect* ratio is first *supplied* by multiplication, thus, $\frac{4\cdot10}{7\cdot10}=\frac{4}{7}$, and then the lower term of the *perfect* ratio (7) is *removed* by division, thus, $\frac{4\cdot10}{7\cdot10}=\frac{5\frac{5}{7}}{10}$, giving $5\frac{5}{7}$ for the unknown term. See these three cases exemplified, pp. 204—211.

4. In compound proportion, the lower term of the perfect ratio may contain a factor, or a multiplier, or both, of the imperfect ratio, as $\frac{\quad}{24}=\frac{6\cdot3}{9\cdot4}$. Here, a practised eye would detect a factor in 4, and a multiple in 9; and, therefore, in transferring the ratio to the slate for solution, would have divided the $\frac{6}{9}$ by 3, and multiplied the $\frac{3}{4}$ by 2, making $\frac{2\cdot6}{3\cdot8}$, showing at

a glance that the unknown number was 12. Again, if the proportion $\frac{\quad}{20}=\frac{4\cdot6}{15\cdot8}$, were given, observing the 5 and the 4 in 15•8, a change might be made in transferring the perfect ratio to the slate, writing $\frac{1\frac{1}{3}\cdot3}{5\cdot4}$, or, still better, $\frac{2\cdot2}{5\cdot4}$, in place of $\frac{4\cdot6}{15\cdot8}$, which would show at a glance that the unknown number was 4. As there is a greater variety of factors and multiples in compound than in simple proportion, cases in the former are frequently more easily solved than in the latter, as will presently more distinctly appear.

Questions to be put by the teacher.—What is a ratio? Give an example. Show the different *forms* in which a ratio may be expressed. What does the character placed between the terms of a ratio signify? Do the numbers of a ratio always relate to things of the same kind? What operations may be performed on a ratio without changing its value? What is proportion? Must both ratios of a proportion relate to things of the same kind? Must they always relate to things of different kinds? In what cases is proportion used in arithmetic? How may the unknown term be found? How many kinds of proportion are there? How is simple, distinguished from compound, proportion? How many kinds of problems are there in proportion? What is the first? *Ans.* When the lower term of the imperfect ratio is, &c. How are problems of this kind solved? What is the second? How is this kind solved? The third? How solved? The fourth? How solved? Try to give an instance of the 4th kind, with a multiple and factor of the imperfect ratio in the lower term of the perfect ratio.

Exemplifications for the Black-board.

1. If 5 barrels of flour can be bought for $25, how many barrels can be bought for $75?

Barrels. $

Statement. $\frac{\quad}{5}=\frac{75}{25}$ *Ans.* 15.

Suggestive Questions.—What do we want to know? *Ans.* How many ——? Then the imperfect ratio consists of barrels, namely, 5. What is the perfect ratio? 25 and 75 dollars.

Will the number of barrels be "LESS OR MORE" than 5? In what order, then, should the 75 and 25 be placed in the perfect ratio? How can the perfect ratio be changed to 5ths? To what case does this question belong?

2. If 3 men can build a wall in 12 days, in what time can 4 men build it?

Time. Men.

Statement. $\frac{\cdot}{12}=\frac{3}{4}$ *Ans.* 9.

Suggestive Questions.—What are the 3 words asking the question? To what, then, does the imperfect ratio relate? To what does the perfect ratio relate? Will 4 men take "LESS OR MORE" time? In what order, then, should the terms of the perfect ratio be placed? How can the perfect ratio be changed to 12ths? To what case does the question belong?

3. If 8 oz. of silver are worth \$9, what are 3 oz. worth?

\$ oz.

Statement, $\frac{\ }{9}=\frac{3}{8}$. Change to $\frac{\ }{9}=\frac{3 \cdot 9}{8 \cdot 9}$. *Ans.* \$$3\frac{3}{8}$.

Suggestive Questions.—What words ask the question? Should the answer be in money or in ounces? Will the 3 oz. be worth "LESS OR MORE" than the 8 oz.? In what order, then, should the terms of the perfect ratio be placed? How shall the perfect ratio, $\frac{3}{8}$, be changed to 9ths? To what case does this question belong?

4. If 120 men can build a wall 30 feet high, and 40 feet long, in 8 days, in how many days can 70 men build a wall 24 feet high, and 50 feet long, the breadth in both cases being the same?

Statement in full. Abbreviated.

$\frac{\ }{8}=\frac{120 \cdot 24 \cdot 50}{70 \cdot 30 \cdot 40}$ $\frac{\ }{8}=\frac{12 \cdot 4 \cdot 10}{7 \cdot 5 \cdot 8}$ *Ans.* $13\frac{5}{7}$.

Suggestive Questions.—What is the question here? Time. What number, then, forms the imperfect ratio? Does the answer depend on the number of men, or on the height of the wall, or on its length, or on all these circumstances taken together? Is this a simple or a compound proportion, then? How was the first part of the perfect ratio abbreviated? *Ans.* By dividing both terms by —. How was the second? the third? Why was the third divided by 5 rather than by 10? How was the abbreviated statement changed to the simple ratio of 8ths? *Ans.* By dividing by — and —, or rather by their product. To what case does this question belong?

Exercises for the Slate or Black-board.

5. At $54 for 9 barrels of flour, how many barrels may be purchased for $186? *Ans.* 31 barrels.

Suggestive Questions.—Does the imperfect ratio relate to barrels or money? Will the number of barrels be "LESS or MORE than 9? Which term, then, of the perfect ratio should be placed above? Make the statement, and then give the answer by inspection.

6. A house was built in a year by 20 workmen, but, being totally destroyed by fire, it was necessary that it should be rebuilt in 5 months. How many workmen must be employed, LESS OR MORE? Make the statement, and then give the answer by inspection.

7. If a house can be built by 48 workmen in 5 months, how long a time would be necessary for 20 workmen to build such another?

8. If 60 lbs. at Boston make 56 lbs. at Amsterdam, how many lbs. at Boston will be equal to 350 at Amsterdam? Add $\frac{1}{14}$th to perfect ratio. Why?

9. If 375 lbs. at Boston make 350 lbs. at Amsterdam, how many lbs. at Amsterdam would make 60 lbs. at Boston? Deduct $\frac{1}{15}$th from each term of the perfect ratio. Why?

10. If 95 lbs. Flemish make 100 lbs. American, how many American lbs. are equal to 550 lbs. Flemish? Add $\frac{1}{19}$th. Why?

11. If it take $578\frac{18}{19}$ lbs. American to make 550 lbs. Flemish, how many lbs. American will make 95 lbs. Flemish?

12. How many yards of matting, 2 ft. 6 inches broad, will cover a floor 27 ft. long and 20 feet wide? In stating, double the perfect ratio. Why?

13. If it take 72 yards of matting, 2 ft. 6 in. broad, to cover a room 27 feet long, what is the width of the room?

14. If 24 yards at Boston make 16 ells at Paris, how many ells at Paris will make 129 yards at Boston? Make the statement without changing the figures in the ratio, and ascertain the answer by inspection.

15. How many ells at Paris will make 24 yards at Boston, if 129 yards at Boston make 86 ells at Paris?

16. How many yards of cloth, 3 qrs. wide, are equal in measure to 30 yds., 5 qrs. wide? Ascertain the answer by inspection without any statement.

17. What is the width of a piece of cloth, 50 yards long, equal in measurement to 30 yards, 5 qrs. wide?

18. If 30 men can perform a piece of work in 11 days, how many men can accomplish a piece of work of the same kind, 4 times as large, and in $\frac{1}{5}$ part of the time? 1. In stating, multiply the time by 5 (Why?), then ascertain the answer by inspection. 2. Ascertain the answer without statement.

19. A wall that is to be built to the height of 27 feet, was raised 9 feet by 12 men in 6 days. How many men must be employed to finish the wall in 4 days, at the same rate of working? How many feet in height remain to be done? In stating, divide the ratio of the height by 3 (why not by 9?), and ascertain the answer by inspection.

20. If 18 feet of a wall was built by 36 men in 4 days, how many days will 12 men take to finish it to the height of 27 feet?

21. A borrowed of his friend B $250 for 7 months, promising to do him the like kindness. Some time after, B had occasion for $300. How long may he keep it to receive full amends for the favor? Longer or shorter? How do you ascertain the answer to be $\frac{5}{6}$ of 7 by inspection, without any statement?

22. If 6 pairs of gloves cost $7, what will be the cost of 50 pairs? What proportional part must be added to the perfect ratio? Get the answer without statement. *Ans.* 58\frac{1}{3}$.

23. If 6 men build a wall 20 feet long, 6 feet high, and 4 feet thick, in 16 days, in what time will 24 men build one 200 feet long, 8 feet high, and 6 feet thick? How often is the factor 2 repeated in the imperfect ratio? In writing the statement, then, preserve as many 2s in the consequent of the perfect ratio, endeavoring to dispense with as many of the other factors as may be. If properly stated, the answer will appear at a very slight inspection. If it does not, try to state in another form.

24. If 24 men can build a wall 200 feet long, 8 feet high, and 6 feet thick, in 80 days, how many men will be required

to build one 20 feet long, 6 feet high, and 4 feet thick, in 16 days?

25. How many men may be hired for 16 days for $103·04, if the wages of 4 men for 3 days be $11·04? State, and divide by 44·16. Why?

26. If the wages of 7 men for 16 days amount to $103·04, how many days will 4 men work for $11·04 at the same rate? State, and divide by 25·76. Why?

27. If 8 men mow 36 acres of grass in 9 days, by working 9 hours each day, how many men will be required to mow 48 acres in 12 days, by working 12 hours each day? If each part of the perfect ratio be written in its lowest terms, the answer can be found by inspection.

28. If 6 men can mow 48 acres in 12 days, working 12 hours a day; how many acres can 8 men mow in 9 days, working 9 hours a day?

29. If 10 cows eat 12 tons of hay in 9 weeks, how many cows will eat 56 tons in 21 weeks? State in lowest terms, and find the answer by inspection.

30. If 20 cows eat 56 tons of hay in 21 weeks, how long will 12 tons last 10 cows at the same rate? Write the perfect ratio in lowest terms, and increase it one half. Why?

31. If 4 cows eat $4\frac{4}{5}$ tons of hay in $8\frac{2}{3}$ weeks, how many cows will eat $9\frac{3}{5}$ tons in $4\frac{1}{3}$ weeks? In stating, multiply the first part of the perfect ratio by 5, the latter part by 3? Why?

32. If 16 cows eat $9\frac{3}{5}$ tons of hay in $4\frac{1}{3}$ weeks, how long will $4\frac{4}{5}$ tons last 4 cows?

33. If $\frac{4}{5}$ of a bushel of wheat cost $\$\frac{15}{16}$, how much may be bought for $\$\frac{3}{4}$? To what does the imperfect ratio relate? How much of wheat?

$$\frac{\quad}{\frac{4}{5}}=\frac{\frac{12}{16}\cdot\frac{4}{5}}{\frac{15}{16}\cdot\frac{4}{5}}=\frac{12\cdot\frac{4}{5}}{15\cdot\frac{4}{5}}=\frac{\frac{48}{75}}{\frac{4}{5}}.$$ *Ans.* $\frac{16}{25}$ of a bushel.

Suggestive Questions.—Whence comes this $\frac{12}{16}$ in the first form of the perfect ratio? By what number is the second multiplied? Whence comes the $\frac{48}{75}$ in the third? The $\frac{16}{25}$ of the answer? ☞ The first and second form of the perfect ratio are superfluous, except to a mere beginner. Solve the question three times: 1st, as above; 2d, by omitting the first form of the perfect ratio; 3d, omitting the first two forms; that is, performing them mentally.

34. If £$\frac{3}{5}$ purchase 24 doz. steel pens, what will $\frac{5}{6}$ of a penny purchase? Whence comes the $\frac{720}{5}$?

$$\frac{\quad}{288}=\frac{\frac{5}{6}\cdot 288}{\frac{720}{5}\cdot 288}=\frac{\frac{5}{720}\cdot\frac{5}{6}\cdot 288}{288}.$$ *Ans.* $1\frac{2}{3}$ pens.

35. If 7 times $\frac{3}{9}$ of $\frac{7}{8}$ of an estate be worth \$15,000, what is $\frac{2}{9}$ of $\frac{3}{8}$ of it worth? The perfect ratio is $\frac{2}{49}$ by a mental process. Try it. *Ans.* \$$612\frac{12}{49}$.

36. Bought $\frac{3}{8}$ of a yard of cloth for \$$1\frac{1}{4}$. What will $\frac{4}{5}$ of a yard cost at the same rate? State the imperfect ratio in the form of an improper fraction, and find the answer \$$2\frac{2}{3}$ by inspection.

37. If $\frac{3}{8}$ of a yard of cloth cost \$$\frac{5}{7}$, what will $\frac{2}{5}$ of a yard cost?

$$\frac{\quad}{\frac{5}{7}}=\frac{\frac{2}{5}\cdot\frac{5}{7}}{\frac{3}{8}\cdot\frac{5}{7}}.$$ *Ans.* \$$\frac{16}{21}$ by inspection.

38. If $\frac{3}{5}$ yd. cost \$$\frac{7}{8}$, what will $27\frac{3}{4}$ yds. cost?

39. If $27\frac{3}{4}$ yards of cloth cost \$$40\frac{15}{32}$, what will $\frac{3}{5}$ yd. cost?

40. A merchant, owning $\frac{3}{8}$ of a vessel, sold $\frac{2}{3}$ of his share for \$750. How much of the vessel did he own, and what was the value of the whole vessel at that rate? Value, \$3000.

41. The purchaser mentioned in the last exercise sold $\frac{1}{4}$ of his share for \$200. How much did he gain by the sale?

Ans. \$12'50.

42. A man bought $\frac{3}{8}$ of a barrel of flour, and sold $\frac{1}{2}$ of it to one of his neighbors at the same rate for \$1'$12\frac{1}{2}$. What would a barrel come to at that rate?

43. When flour is \$6 a barrel, what is the value of $\frac{1}{2}$ of $\frac{3}{8}$ of a barrel?

44. If $\frac{4}{5}$ of $\frac{3}{8}$ of a yard of cloth cost \$$1\frac{4}{5}$, what will $\frac{2}{3}$ of $\frac{3}{5}$ yd. cost?

45. If $\frac{2}{3}$ of $\frac{3}{5}$ of a yard of cloth cost \$2'40, what will $\frac{4}{5}$ of $\frac{3}{8}$ yd. cost?

46. If a train of cars move at the rate of 20 miles an hour, what portion of a mile do they travel in a second of time?

47. If a train of cars move uniformly $\frac{1}{180}$ of a mile per second, what is their rate per hour?

48. What would a pile of building stone cost, measuring 30 feet long, 26 feet broad, and $4\frac{1}{2}$ feet high, at \$1·25 per perch of $16\frac{1}{2}$ feet long, 1 foot high, and $1\frac{1}{2}$ feet broad? Use no fractions in the perfect ratio.

49. A pile of building stone, measuring 30 feet long, 26 feet broad, and $4\frac{1}{2}$ feet high, was sold for \$$177\frac{3}{11}$; how much is that per perch of $16\frac{1}{2}$ feet long, 1 foot high, and $1\frac{1}{2}$ feet broad?

50. How many cords of wood are contained in a pile 200 feet long, 10 feet high, and 36 feet broad? The imperfect ratio here is 1. *Ans.* $562\frac{1}{2}$ cords.

51. A justice of the peace has an income of \$1500 per annum, and the perquisites of his office average \$7 per week. How much will he save per annum, if his expenses average \$15 per week, counting 52 weeks to the year? Imperfect ratio, 8. Why? Solve this first by the aid of a statement; afterwards mentally, without the use of a slate or black-board.

52. The perquisites of the office of a justice of the peace are found to average \$7 per week. His expenses average \$15 per week, and yet he finds his savings at the end of each year of 52 weeks to amount to \$1084. What must be the amount of his annual income from other sources?

53. A contractor employed 400 workmen on a railroad for 4 weeks, paying them a ration of provisions *every* day, and 50 cents per *working*-day; that is, for 6 days in the week. At the end of the 4 weeks, he employed 200 additional hands at the same rate. The work was finished at the close of 12 weeks from its commencement. What was the whole expense, supposing the rations to cost 25 cents each? The rations and daily pay should be calculated separately. Why?

Ans. \$44,800.

54. Another railroad contractor employed 360 workmen for 12 weeks, and paid 85 cents per day, without rations; but, owing to rainy days and sickness, it was found that they only worked 18 days in four weeks on an average. How much had he to pay? *Ans.* \$16,524.

55. A man left by will $\frac{3}{4}$ of his estate to his widow, $\frac{3}{4}$ of the remainder to a son, and the rest, amounting to \$100, to his daughter. How much did he leave in all? This may be done mentally. *Ans.* \$1600.

56. If $\frac{3}{8}$ of a pole stands in the mud, 3 feet in the water, and $\frac{1}{4}$ of its length above the water, what is the length of the pole? *Ans.* 8 feet.

Interest, Discount, Commission, Insurance, and Percentage generally.

DEFINITIONS.

I. Interest is the sum of money given for the loan or use of another sum of money.

II. Three elements enter into all calculations of interest, namely: the *principal*, or money lent; the *interest* paid for the *use* of the principal, sometimes called *use;* and the *time* for which the principal is lent.

III. The interest will evidently be proportional to the *time;* for, whatever interest is paid for the use of $100 for one year, twice that interest will be required for the use of the same sum for two years, and half that interest for half a year. The interest is also proportional to the *principal;* since, if $6 be required for the use of $100 for any given time, $3 will evidently be required for the use of $50, and $12 for the use of $200 for the same time.

IV. The whole sum paid back to the lender, which of course includes both principal and interest, is called the *amount*.

V. In questions of interest, the Latin terms *per cent.* and *per annum* are very frequently used. The former means *for a hundred;* the latter, *for a year.* Thus, "at $6 per cent. per annum," means "at $6 for a hundred for a year." Sometimes one, at other times both these terms are omitted. But they must never be left out of the calculation. The interest paid *per cent.*, or for a hundred, is frequently called the *rate*.

VI. Interest is either *simple* or *compound*. *Simple interest* is that which is reckoned and allowed upon the *principal only* during the whole time of the loan; but *compound interest* is reckoned, not only on the principal, or sum lent, but also on the interest, if it remains unpaid after it becomes due. Thus, reckoning by *simple* interest, if $6 be the interest of any sum for one year, $12 will be the interest for two years; whereas by *compound* interest it will be $12·36; for, in the former case, the same interest is charged in both years, whereas, in the latter, the interest is charged on $100 the first year, and on $106 (the *amount* of principal and unpaid interest) the second year. In order to discourage protracted settlements, the law does not allow compound interest on money lent; yet, in purchasing annuities, reversions, leases, &c., it is always allowed.

VII. In calculating interest, a month is reckoned as 30 days, unless the name of the month is specified; and a year is reckoned as 360 days.

VIII. A *note* is a written promise to pay a certain sum of money, or its value in goods, on demand (that is, when demanded), or at some future day mentioned. Hence all notes are called *promissory* notes. Some notes are drawn payable to *bearer.* But a *negotiable* note is one payable to some person, or *order. Indorsement* on a note means writing on the *back* of it. Indorsements are of two kinds: 1. When a person to whose order a negotiable note is made payable writes his name on its back, he becomes responsible for its payment, if properly notified that it is due and unpaid. 2. Indorsements are also records of partial payments of principal or interest on a note, written on the back of it. The sum or debt for which a note is given is called the *principal*, or *face of the note;* the person who gives it is called the *signer*, or *drawer;* and, when the note is indorsed by the person in whose favor it is drawn, the signer is called the *principal,* because the holder must first look to him for payment; the person indorsing it is called the *indorser*, and the person to whom it is indorsed when sold, the *indorsee*, or *assignee.*

IX. *Discount* is a deduction made on the payment of a debt before it becomes due. It only differs from interest by being *deducted* from the principal, whereas interest is *added* to it. In some of the states, banks and private individuals lend money on notes, by advancing the full amount of the note to the borrower, and charging interest thereon. In other states, it is customary to *discount* notes; that is, to advance the amount of the note, less the discount, and to charge no interest. The difference between the two methods will be best exhibited by an example. In the former case, the borrower draws a note, say for $100, payable with interest at 6 per cent. in one year. For this he receives $100, and at the end of the year pays $106. In the latter case, the note says nothing of interest; the borrower receives $94, and pays $100 at the end of the year. Thus, the one pays an interest of $6 for the use of $100 for a year, and the other pays a discount of $6 for the use of $94 for a year, making a difference of $\frac{18}{47}$ of 1 per cent.

Present worth of any sum implies that it is payable at a future time without interest. The present worth, then, is such a sum as would at interest amount to the debt when due. Thus, the present worth of $106, due a year hence, is $100; because $100 at interest for that time amounts to $106.

Commission is an allowance for buying or selling goods for another person, generally so much *per cent.* as may be agreed on.

Profit or *loss* on the purchase or sale of any kind of property, is also frequently reckoned at so much per cent.

Questions by the teacher.—What is interest? How many elements enter into all calculations of interest? Name them. What is the principal? The interest? The amount? What is the meaning of *per cent.*? of *per annum?* Are these terms ever omitted? Can they be omitted in the calculation? State the difference between simple and compound interest. How many days are reckoned to a month when no particular month is specified? How many days to a year?

Exercises for the Slate or Black-board.

1. What is the interest of $400 for 6 years, at 5 per cent. per annum?

$$\frac{\ }{5}=\frac{400\cdot6}{100\cdot1}\quad \textit{Ans.}\ \$120,\ \text{by inspection.}$$

Suggestive Questions.—What four words ask the question? What, then, is the imperfect ratio? What is the ratio of the principals? How is the 100 expressed in the question? Why is $400 placed above in the perfect ratio? What is the ratio of time? How is one year expressed in the question? Why is 6 years placed above?

2. In what time will the interest of $400 come to $120, at 5 per cent.?

$$\frac{\ }{1}=\frac{100\cdot120}{400\cdot\ 5}\quad \textit{Ans.}\ 6\ \text{years, by inspection.}$$

Suggestive Questions.—What three words ask the question? What, then, is the imperfect ratio? Where do you find the 1 year in the question? How are the two principals expressed in the question? Will it take less or more time for $400 than for $100 to produce $120 of interest? Should the larger or smaller principal, then, be placed above? Of what numbers does the ratio of the interest consist? Why should the larger number be placed above?

3. What principal will produce $120 in 6 years, at 5 per cent. per annum?

$$\frac{\quad}{100}=\frac{120\cdot1}{5\cdot6}$$ *Ans.* $400, by inspection.

Suggestive Questions.—What two words ask the question? In what two words is the given principal expressed? What two sums of interest are given? Why is the larger placed above? How are the two numbers forming the ratio of time expressed? Why is the smaller placed above?

4. What will be the amount of $400 in 6 years at 5 per cent.?

$$\frac{\quad}{5}=\frac{400\cdot6}{100\cdot1}$$ *Ans.* Interest $120, amount $520, by inspection.

Suggestive Questions.—What is meant by "amount"? How, then, can the amount be found?

5. What is the interest of $54 for 4 months, at 6 per cent. per annum?

$$\frac{\quad}{6}=\frac{54\cdot4}{100\cdot12} \text{ or } \frac{‘54\cdot2}{1\cdot6}$$ *Ans.* $1‘08, by inspection.

Suggestive Questions.—How is the 12 in the perfect ratio expressed in the question? In the abbreviated perfect ratio, why is the ratio of the months expressed by $\frac{2}{6}$ rather than by $\frac{1}{3}$? (See the imperfect ratio.)

6. What is the interest of $36 for 7 months, at 6 per cent. per annum?

$$\frac{\quad}{6}=\frac{‘36\cdot3‘5}{1\ \ 6}$$ *Ans.* $1‘26, by inspection.

Suggestive Questions.—The first part of the perfect ratio, $\frac{‘36}{1}$, is the ratio of the principals. By what number is it divided? By what number is the remaining part of the perfect ratio, i. e., the ratio of the months, divided?

7. What is the interest of $240 for 12 days, at 6 per cent.?

$$\frac{\quad}{6}=\frac{240\cdot12}{100\cdot360} \text{ or } \frac{2‘4\cdot‘2}{1\cdot6}$$ *Ans.* ‘48, by inspection.

Suggestive Questions.—How is the first part of the perfect ratio abbreviated? How is the second part? Is the 360 in the perfect ratio expressed or understood in the question?

NOTE. The first or complete form of the perfect ratio should be dispensed with as soon as possible, passing at once to the abridged form by a mental process.

8. What is the interest of $54 for 37 days, at 6 per cent.?

$$\frac{\quad}{6}=\frac{‘54\cdot3‘7}{1\cdot36}$$ *Ans.* $33\frac{1}{3}$ cents, by inspection.

Suggestive Questions.—Whence comes 36 in the perfect ratio? Why is the 54 changed to ‘54? the 37 to 3‘7? The answer is the product of ‘09 and 3‘7; whence comes the ‘09?

9. What is the interest of $48 for 25 days, at 7 per cent.?

$$\frac{\quad}{7}=\frac{‘48\cdot2‘5}{1\cdot36}=\frac{‘01\frac{1}{3}\cdot2‘5\cdot7}{1\cdot1\cdot7}$$ *Ans.* $23\frac{1}{3}$ cents, by inspection.

Suggestive Questions.—Whence comes ‘$01\frac{1}{3}$ in the second form of the perfect ratio? (Examine it in connection with the first form.) Why are both terms multiplied by 7?

10. What is the interest of $350 for 3 years, 4 months, and 6 days, at 5 per cent.? The time may be considered as 1206 days, or as $3\frac{21}{60}$ years (Why?), or as 3 years and 126 days, or as stated above. In the two latter use separate calculations for the different periods. Perform the calculation in each of the four different methods. A slate or black-board will be necessary, as the question cannot readily be solved by inspection. *Ans.* $58‘62$\frac{1}{2}$.

11. What is the interest of $75 for 5 months and 17 days, at 7 per cent.? Bring the time into days for first calculation, and prove by calculating for months and days separately. *Ans.* $2‘435+.

12. Calculate the interest of the following sums, at 6 per cent. per annum.

$100 for 60 days . .	
50 for 30 " . .	
25 for 90 " . .	
27‘50 for 75 " . .	
72‘75 for 25 " . .	
Carried over,	$2‘271

Brought over,	$2‘271
$36 for 2 years . . .	
75 for 18 months . .	
32 for 3 " . .	
27 for 4 " . .	
89 for $1\frac{1}{2}$ " . .	
Total,	$15‘0285

13. A merchant in New York, who allowed his customers a credit of three months, charging interest at the end of that period, until paid, at 6 per cent. per annum, found the following unpaid charges on his books on the first day of January, 1853, viz.

			Interest.	Amount.
Against	A, $500	Feb. 1, 1852.		
"	B, $260	Feb. 14, "		
"	C, $380	June 1, "		
"	D, $137'50	July 1, "		
"	E, $550	Aug. 1, "		
"	F, $225	Sept. 15, "		
	$2052'50		$45'47½	$2097'97½

Calculate the interest and several amounts due by the debtors on said first of January, allowing half a month for the latter halves of Feb. and Sept.

14. What is the amount of $500 for 1 year and 6 months, at 6 per cent.?

15. What sum will amount to $545 in 1 year and 6 months, at 6 per cent.?

16. At what rate per cent. will $500 amount to $545 in 1 year and 6 months?

17. In what time will $500 amount to $545, at 6 per cent. per annum?

18. What is the interest of £20 15*s.* 6*d.* for 3 months, at 5 per cent. [Bring shillings and pence to decimal of a pound. See p. 242.]

19. At what rate per cent. will £20 15*s.* 6*d.* amount to £21 0*s.* 8¼*d.* in 3 months?

20. What is the amount of £20 in 2 years and 6 months, at 6 per cent. per annum?

21. In what time will £20 amount to £23, at 6 per cent.?

22. What is the interest of £540 7*s.* 6*d.* for 1 year, at 5 per cent.?

23. If £540 7*s.* 6*d.* amount to £567 7*s.* 10½*d.* in 1 year, what is the rate per cent.?

24. Find the interest of £25 8*s.* 4*d.* at 6 per cent. for 7 months and 20 days.

25. If the interest of £25 8*s.* 4*d.* be 19*s.* 5¾*d.* at 6 per cent., what is the time?

26. What is the interest of £100 for a month, at 5 per cent. per annum?

27. What principal will amount to £100 8*s.* 4*d.* in a month, at 5 per cent.?

28. Calculate the amount due on the following note, on the first day of January, 1853.

$100. Burlington, Vt., March 1, 1852.

On demand I promise to pay to the order of Jonathan Wheeler, one hundred dollars, with interest at 6 per cent. per annum, for value received. Tobias Cheney.

The following indorsements were on the above note: May 1, 1852, paid $20. Aug. 1, 1852, paid $30. *Ans.* $53·45.

☞ Various rules have been established by different courts of law to prevent the compounding of interest where partial payments are made on notes. Probably, the most simple and exactly correct one is the following: "Find the amount of the note from the time it first began to bear interest till its final settlement; then find the amount of each several payment at the date of settlement, and subtract their sum from the amount of the note. The balance will show how much is due on the day of settlement."

29. Find the amount due on each of the following notes on the 1st day of January, 1853, after deducting the amount of partial payments agreeably to the above rule, and then find the sum of the whole.

$500. Boston, Jan. 1, 1849.

For value received, I promise to pay to John Smith, or order, on demand, five hundred dollars, with interest at 6 per cent.

A. B.

Indorsements: Jan. 1, 1850, received one hundred and twenty-five dollars. Jan. 1, 1851, received two hundred dollars. Jan. 1, 1852, received one hundred and fifty dollars.

$125. Boston, March 3, 1852.

I promise to pay on demand to the order of John Smith, one hundred and twenty-five dollars, with interest at 6 per cent., for value received. C. D.

Indorsements: June 14, 1852, received forty dollars. Oct. 1, 1852, received twenty-five dollars.

$352. Boston, April 1, 1852.

We promise to pay on demand to the order of John Smith, three hundred and fifty-two dollars, with interest at 6 per cent., for value received. E. F. & Co.

Indorsements: May 1, 1852, received one hundred dollars. May 15, 1852, received fifty dollars. Oct. 1, 1852, received twenty dollars.

$750. Boston, Feb. 14, 1852.

On demand I promise to pay to John Smith or order, seven hundred and fifty dollars, with interest at 6 per cent., for value received. G. H.

Indorsements, July 4, 1852, received two hundred dollars. Sept. 1, 1852, received three hundred dollars.

$150. Boston, Jan. 1, 1852.

On demand I promise to pay to John Smith or order, one hundred and fifty dollars, with interest at 6 per cent., for value received. J. K.

Indorsement: May 1, 1852, received fifty dollars.

$250. Boston, May 4, 1852.

On demand, I promise to pay to the order of John Smith, two hundred and fifty dollars, with interest, at 6 per cent., for value received. L. M.

Indorsements: Oct. 1, 1852, received thirty-seven dollars and fifty cents. Nov. 1, 1852, received twenty-five dollars.

Total amount of the above six notes, $926'736.

30. What is the discount on a note for $500, drawn this day, payable 60 days after date, at 6 per cent. per annum?

☞ Three days of grace are allowed by law for the payment of notes; that is, a note drawn at 60 days is not considered payable for 63 days, and banks charge discount for one day more; that is, they charge for the day on which the note is payable as well as for that on which it is drawn.

31. How much does a bank allow for a note for $200, payable in 60 days, at 6 per cent.? *Ans.* $197'86.

32. How much is the bank discount for a note of $650, payable in 90 days, at 6 per cent.? *Ans.* $10'18.

33. A note for $2500 was discounted in bank at 6 per cent. payable in 90 days with grace. How much money did the owner receive for it? *Ans.* $2460'83.

34. Find the interest of $240 for 5 years, at $4\frac{1}{2}$ per cent.

35. In what time will \$240 amount to \$294 at $4\frac{1}{2}$ per cent. per annum?

36. At what rate will \$240 amount to \$294 in 5 years?

37. What sum of money will amount to \$294 in 5 years, at $4\frac{1}{2}$ per cent. per annum?

38. A father, at the birth of his son, lent a brother \$100, to be paid to his boy with simple interest at 6 per cent., on the boy's attaining his majority. What would be the amount?

39. What sum would amount to \$226 in 21 years, at 6 per cent. per annum?

40. What is the interest of 125 francs for a year, at $4\frac{1}{2}$ per cent. per annum?

41. In what time will 125 francs amount to $130\frac{3}{8}$ francs, at $4\frac{1}{2}$ per cent. per annum?

42. What is the interest of 225 ducats for 6 months, at 4 per cent. per annum?

43. At what rate per cent. will 225 ducats amount to $229\frac{1}{2}$ ducats in 6 months?

44. Sold to John Thomas the following goods at cash prices, under an agreement that he is to pay interest on all sums due from the delivery of the goods until paid, viz., Jan. 1, \$1275; March 1, he paid \$600, and bought \$100 worth of goods; April 1, he paid \$500; May 1, he paid \$1000, and bought to the amount of \$800; July 1, he paid \$400; Sept. 1, he bought goods amounting to \$1500, and paid \$800. On the first day of January following he called to settle. How much was then due, charging interest at 6 per cent.? *Ans.* \$398.

45. In how many years will a sum of money double itself (that is, \$100 produce \$100 of interest), at 6 per cent. per annum? *Ans.* $16\frac{2}{3}$ years.

46. A commission merchant sold goods to the amount of \$5650, for which he charged a commission of $2\frac{1}{2}$ per cent. What was the amount of his profit on the sale?

$$\frac{\quad}{2\frac{1}{2}} = \frac{5650}{100} \quad \textit{Ans. } \$141{\cdot}25.$$

47. How much is the commission of \$725, at 3 per cent.?

48. What amount of goods must be sold to produce a commission of \$21·75 at 3 per cent.?

49. How much per cent. does a commission merchant charge, if his commission amounts to \$21·75 on \$725?

50. A lady, who had \$360 in a savings-bank wished to draw

out 5 per cent. of her deposit. What would be the amount of her draft?

51. If 5 per cent. of a deposit in a savings-bank was $18, what was the whole amount?

52. A lady, who had $360 in a savings-bank, drew out $18. How much was that per cent.?

53. What is $6\frac{2}{3}$ per cent. of $963?

54. $64'20 is how much per cent. of $963?

55. $64'20 is $6\frac{2}{3}$ per cent. of how much?

56. How much is 15 per cent. of $730'24?

57. $109'536 is how much per cent. of $730'24?

58. $109'536 is 15 per cent. of how much?

59. A man insured some property in a mutual insurance office to the amount of $6000, for which he gave a note for 4 per cent. of the amount insured. What was the amount of the note?

60. A man insured his property for 4 per cent., for which he gave a note for $240. What was the amount insured?

61. If the insurance on my household furniture, at $\frac{3}{8}$ of 1 per cent. for a year, amounts to $5'62$\frac{1}{2}$, what is the furniture valued at?

62. How much is the annual insurance on my household furniture, valued at $1500, at $\frac{3}{8}$ of 1 per cent. per annum?

63. How much is the annual insurance on property to the amount of $7500 at $\frac{1}{2}$ of 1 per cent. per annum?

64. How much property would be covered by an insurance for which $37'50 was paid, at the rate of $\frac{1}{2}$ of 1 per cent.?

65. A merchant failing, found that he owed $40,000, and that he had goods to the value of $10,000, a house valued at $4000, cash $2500, and good debts $3500. How much per cent. could he pay to his creditors?

66. On taking an account of his property a merchant found it amounted to $20,000, which was only 50 per cent. of what he owed. How much did he owe?

67. A bankrupt owes A $204'50, B $65, C $150, D $427'50, E $1500, and numerous small debts to the amount of $1653. His whole property only amounted to $3000, which was distributed among his creditors in proportion to their demands. How much per cent. did he pay, and how much did each creditor receive?

68. The estate of a bankrupt, when divided among his creditors, amounted to 75 per cent. of their demands, which

amounted in the whole to $4000. How much did the estate prove to be worth?

69. A farmer in Vermont insured his property in a mutual insurance company, as follows, viz., on his dwelling-house and woodshed attached $2000; on household furniture and clothing therein $350; plate and books $50; provisions and produce in his house $80; piano-forte $150; new barn and shed $200; produce therein $150; old barn and cider-house $90; produce and cider-mill in same $60; corn-barn $30; produce therein $50; farm-house and woodshed $200. To effect this insurance the farmer gave a note without interest at the rate of $4\frac{1}{2}$ per cent. on the amount insured, and paid 3 per cent. of the amount of the note in cash, together with 50 cents for the policy. Three years afterwards he was called on for an assessment of 4 per cent. on the face of the note (without reference to what had been paid). Three months afterwards the whole property was destroyed by fire, and the farmer was paid the full amount for which it was insured. How much did he save by effecting this insurance, allowing interest at 6 per cent. per annum on his payments? *Ans.* $3397'68.

70. I bought 10 shares of stock, at $50 per share, for which I paid 6 per cent. advance ($53 per share). Afterwards the stock fell to 10 per cent. below par. How much did I lose by the fall? *Ans.* $80.

71. A man subscribed for 50 shares of bank stock, at $50 per share. He sold half of them at 6 per cent. advance, and some time afterwards sold the rest at 10 per cent. below par. Did he lose or gain by the transaction, and how much?

Ans. He lost $50.

72. A merchant bought goods to the amount of $3500, and sold them at a profit of 15 per cent. What was his whole gain?

73. Upon the sale of goods to the amount of $3500, a merchant made a profit of $525. How much did he gain per cent.?

74. A merchant purchased 150 barrels of flour at $5 per barrel, and paid 25 cents a barrel for transportation. An accident happened to the flour, which caused him to lose 5 per cent. on the transaction. What was the amount of his loss?

75. A dealer in flour sold 150 barrels for $748'12$\frac{1}{2}$, by which he lost $39'37$\frac{1}{2}$. How much did he lose per cent.?

76. A flour dealer bought in one day the following lots of

flour: 25 barrels at $\$5\frac{1}{8}$; 50 at $\$5\frac{1}{4}$; 90 at $5; 100 at $\$4\frac{7}{8}$; 400 at $5. His store rent was $10 per week; clerk hire, $12 per week; insurance on his flour at the rate of $\frac{1}{30}$ of 1 per cent. per week. What price per barrel will cover all expenses, and afford 10 per cent. profit on the outlay, if sold within the week? *Ans.* $5·54+.

77. A tax of $2000 is assessed upon a certain town, of which $400 is raised by poll-tax, that is, by a tax raised on the citizens by the head, without regard to property, and the remainder on the inhabitants, in proportion to the amount of their real and personal property. The number of taxable citizens is 800, and the whole amount of taxable property in the town is valued at $400,000. How much is the tax per cent. on the property, and what has a farmer to pay on 2 polls in his family, and property to the amount of $1500?

Ans. $\frac{2}{5}$ of 1 per cent. and taxes $7.

SHORT PROCESSES FOR THE CALCULATION OF SIMPLE INTEREST.

I. *When the interest is for one or more years.*

Exemplification for the Black-board.

1. What is the interest of $56 for 2 years, at 6 per cent. per annum?

$$\frac{}{6}=\frac{56\cdot 2}{100\cdot 1}=\frac{56\cdot 2\cdot 6}{100\cdot 1\cdot 6}=\frac{\text{‘}56\cdot 2\cdot 6}{6}$$

Suggestive Questions.—What does 56 represent in the second perfect ratio? What part of the principal do you find in the third perfect ratio? What does the 2 represent? the 6? By what, then, must the hundredth part of the principal be multiplied to give the interest for any number of years? Form a rule, then, to find interest for one or more years:

Multiply the — part of the — by the — and the —, or by their product.

Exercises for the Slate or Black-board.

1. Find the interest of $24 for 6 years at 5 per cent., by inspection, without any statement. *Ans.* $7·20.

2. Find the interest of $64 for 7 years at 6 per cent., by inspection, without statement. *Ans.* $26'88.

3. Interest of $345 for 3 years at 5 per cent., by inspection, without statement. *Ans.* $51'75.

II. *When the interest is for one or more months.*

Exemplification for the Black-board.

1. What is the interest of $61 for 8 months at 6 per cent. per annum?

$$\frac{}{6}=\frac{64\cdot 8}{100\cdot 12}=\frac{'64\cdot 8\cdot 6}{1\cdot 12\cdot 6}$$

Suggestive Questions.—What does the '64 represent in the second perfect ratio? *Ans.* The — part of the —. What does the 8 represent? the 6? Form a rule, then, by which to find interest for months:

Multiply the — part of the — by a fraction in its lowest terms, of which the product of the — and — of — forms the numerator, and 12 the denominator.

Exercises for the Slate or Black-board.

1. Find, by inspection, without statement, the interest of $245 for 9 months, at 4 per cent. per annum. *Ans.* $7'35.

Suggestive Questions.—What is the numerator of the multiplying fraction in this problem? the denominator? To what integer is the fraction equal?

2. Find, by inspection, without statement, the interest of $336 for 8 months, at 6 per cent. *Ans.* $13'44.

3. Find the interest of $27 for 11 months, at 7 per cent. *Ans.* $1'73.

4. Find the interest of £12 10*s.* 5*d.* for 7 months, at 6 per cent. *Ans.* £0 8 9+

III. *When the interest is for one or more days.*

Exemplification for the Black-board.

1. What is the interest of $384 for 16 days, at 6 per cent. per annum?

$$\frac{}{6}=\frac{384\cdot 16}{100\cdot 360}=\frac{'384\cdot 16\cdot 6}{1\cdot 36\cdot 6}$$

Suggestive Questions.—What does '384 represent in the

second perfect ratio? the 16? the 6? the 36 in the denominator? Why not 360? See the principal in the numerator. What is the fraction in its lowest terms by which the thousandth part of the principal is to be multiplied? Give a rule for finding interest for days.

Multiply the — part of the — by a fraction in its lowest terms, whose numerator is the product of — and the —, and whose denominator is 36.

Exercises for the Slate or Black-board.

1. What is the interest of $450 for 18 days, at 5 per cent.? *Ans.* $1‘12½.
2. Find the interest of $220 for 25 days, at 6 per cent. *Ans.* $‘916.
3. Find the interest of $324‘50 for 24 days, at 6 per cent. *Ans.* $1‘298.
4. Find the interest of £365 for 16 days at 5 per cent. *Ans.* 16*s.* 2½*d.*

IV. *When the Time consists of two Determinate Fractions, or of an Integer and one or two Determinate Fractions.*

Exemplification for the Black-board.

1. What is the interest of $420 for 2 years, 4 months, and 18 days, at 6 per cent. per annum?

$4‘20•12= =50‘40 Int. for 2 y. by Case I.
4‘20•$\frac{24}{12}$=2= 8‘40 Int. for 4 m. by Case II.
‘420•$\frac{108}{36}$=3= 1‘26 Int. for 18 d. by Case III.
$60‘06 Int. for 2 y. 4 m. 18 d.

Exercises for the Slate or Black-board.

1. What is the interest of $3475 for 9 months and 12 days, at 6 per cent. per annum? *Ans.* $163‘325.
2. Find the interest for the same sum and time at 7 per cent. *Ans.* $190‘545.
3. Find the interest for the same sum and time at 5 per cent. *Ans.* $136‘104.
4. Find the interest of $0‘73 for 1 year and 8 months, at 6 per cent. *Ans.* ‘073.
5. Find the interest of $7342 for 1 year, 4 months, and 15 days, at 6 per cent. *Ans.* $605‘715.

6. Find the amount of a bond for $875·49 for 5 years, 8 months, and 18 days, at 6 per cent.; also at 7 and at 5 per cent.

Ans. { $1176·78 at 6 per cent.
$1226·82 at 7 per cent.
$1126·74 at 5 per cent. }

☞ The three rules that have been developed above may all be comprehended in one, as follows:

The interest for any given sum may be found by multiplying its — part by a fraction in its lowest terms, whose numerator is the product of the — and —, and whose denominator is 1 for years, 12 for months, and 360 for days.

1. Find the simplest form of the fraction for multiplying the principal when the time is 8 months and the rate 6 per cent. *Ans.* $\frac{4}{1}$.
2. Find the simplest form when the time is 5 months and the rate 7. *Ans.* $\frac{35}{12}$.
3. When the time is 3 months and the rate 8. *Ans.* $\frac{2}{1}$.
4. When the time is 36 days and the rate 6. *Ans.* $\frac{3}{5}$.
5. When the time is 20 days and the rate 7. *Ans.* $\frac{7}{18}$.
6. When the time is 6 days and the rate 6. *Ans.* $\frac{1}{10}$.

☞ The rate of interest is generally fixed by law in the several States of the Union. As in most of these it is six per cent., it may be well to seek for a still more simple rule for calculating interest for months and days when the rate is uniformly of that amount; as follows:

Exemplifications for the Black-board.

For Months.

1. Find the interest of $500 at 6 per cent. for 4 months.

$$\frac{\quad}{6}=\frac{500\cdot 4\cdot 6}{100\cdot 12\cdot 6}.$$

Or, cancelling only the principal and the permanent terms of the ratio,

$$\frac{\quad}{6}=\frac{5{\cdot}00\cdot 4\cdot 1}{1{\cdot}00\cdot 2\cdot 6}.$$

Suggestive Questions.—What is the multiplying fraction in the last formula? What part of the 4 months is $\frac{4}{2}$? If the number of months was 6 what would be the fraction? Will the fraction, then, for any number of months be always equal to half the number of months? Give, then, a rule for finding the interest for months at 6 per cent.

Multiply the — part of the principal by — the number of months.

For Days.

2. Find the interest of \$568 at 6 per cent. for 24 days.

$$\frac{\quad}{6}=\frac{568\cdot 24\cdot 6}{100\cdot 360\cdot 6}.$$

Or, cancelling the principal and the permanent terms of the ratio,

$$\frac{\quad}{6}=\frac{5{\cdot}68\cdot 24\cdot 1}{1{\cdot}00\cdot 60\cdot 6}.$$

Suggestive Questions.—What is the multiplying fraction in the last formula? What part of it would be different if the number of days were altered? Would the fraction, then, always be $\frac{1}{60}$ of the number of days? Give, then, a rule for finding the interest both for months and days, at 6 per cent.?

Multiply the — part of the principal by — the time when it consists of months, and by — of the time when it consists of days.

Exercises for the Slate or Black-board.

1. Calculate the interest of the following sums, as above, at 6 per cent. per annum.

		Brought over,	\$2·271
\$100 for 60 days . .		\$36 for 2 years . . .	
50 for 30 " . .		75 for 18 months . .	
25 for 90 " . .		32 for 3 " . .	
27·50 for 75 " . .		27 for 4 " . .	
72·75 for 25 " . .		89 for 1½ " . .	
Carried over,	\$2·271	Total,	\$15·028

2. Find the interest of £300 for 2 years, 8 months and 15 days, at 6 per cent. *Ans.* £48 15*s.*

3. Find the amount of \$221·75 for 3 years, 7 months and 6 days, at 6 per cent. *Ans.* \$269·647.

☞ There is another method of computing interest when the time consists partly or wholly of determinate fractions, which will probably be found as short and simple as either of those already given, namely, by the method called "Practice," as developed p. 254. Or, the time may be resolved into its lowest denomination, and the whole calculated by one operation. Two examples by each of these methods will make both sufficiently clear.

1. Find the interest of \$265 for 2 years and 5 months, at 7 per cent.

I. *By Practice.*

	\$2‘65	$=\frac{1}{100}$ of the principal.
	14	=product of years and rate.
	37‘10	=Int. for 2 y.
4 m.=$\frac{1}{6}$ of 2 y.	6‘183	=Int. for 4 m.
1 m.=$\frac{1}{4}$ of 4 m.	1‘545	=Int. for 1 m.
	\$44‘828	=Int. for 2 y. 5 m.

II. *By changing the time to its lowest denomination.*

2 y. 5 m.=29 months.

$\frac{1}{100}$ of principal 2‘65×29×7÷12=$\frac{203}{12}$. (m. over 29, r. over 7)

203

12)537‘95

\$44‘829 Int. for 29 m.=2 y. 5 m.

2. Find the interest of \$224‘75 for 3 years, 4 months, and 16 days, at 5 per cent.

I. *By Practice.*

	\$2‘2475	$=\frac{1}{100}$ of principal.
	15	=product of years and rate.
	33‘7125	=Int. for 3 y.
4 m. $\frac{1}{9}$ of 3 y.	3‘7458	=Int. for 4 m.
15 d. $\frac{1}{8}$ of 4 m.	‘4683	=Int. for 15 d.
1 d. $\frac{1}{15}$ of 15 d.	‘0312	=Int. for 1 d.
	\$37‘9578	=Int. for 3 y. 4 m. 16 d.

II. *By changing the Time to its lowest denomination.*

3 y. 4 m. 16 days $=1216\times5$ rate$=6080$.

$\frac{6080}{360}=\frac{152}{9}$

$2·2475$=\frac{1}{100}$ of principal.

```
        152
  9)341·6200
     37·9577+=Int. for 3 y. 4 m. 16 d.=1216 d.
```

☞ Perform all the exercises under the head of "Short Processes," &c., where the time is of two or three denominations, first by "Practice," and then by "Changing the Time," as above.

RECAPITULATION

OF THE ABRIDGED PROCESSES FOR THE CALCULATION OF SIMPLE INTEREST.

I. *For one or more years.*

RULE.—Multiply the — part of the — by the — and the —, or by their product.

II. *For fractional parts of years.*

1. *When the rate is* 6 *per cent.*

RULE.—Multiply the — part of the —, by — of the time when it consists of months, and by — of the time when it consists of days.

2. *When the rate is less or more than* 6 *per cent.*

RULE.—Multiply the — part of the —, by the lowest term of a fraction, whose numerator consists of the product of the — and the —, and whose denominator is 12 for months and 360 for days; or, by "Practice," or by "Changing the Time."

COMPOUND INTEREST.

Definition.—Compound Interest is that which arises from the principal increased by the interest as it becomes due at the end of each year, or other stipulated time of payment. Though, no doubt, strictly just, it is illegal in most countries, most probably on the principle that it is well to encourage frequent settlements, and to discourage the accumulation by constant increase of unpaid interest, which might finally be injurious both to debtor and creditor.

1. Find the amount and interest of $500 at compound interest for 3 years, at 5 per cent. per annum.

	$	
	500	principal lent.
	25	interest for 1 year.
	525	principal at the end of 1 year.
	26'25	interest on $525 for 1 year.
	551'25	principal at the end of 2 years.
	27'56	interest on $551'25 for 1 year.
Amount	578'81	principal at the end of 3 years.
	500	original principal.
	78'81	Compound Int. on $500 for 3 years.

2. Find the amount of $324 at compound interest, for 4 years, at 6 per cent. per annum. *Ans.* $409'04.

3. Find the amount of $532'24 at compound interest for $5\frac{1}{4}$ years, at 4 per cent. per annum. *Ans.* $654'02.

4. Find the amount of the same sum at compound interest for $2\frac{1}{2}$ years, at 6 per cent. *Ans.* $615'96.

☞ Another method for computing compound interest will be developed in the Supplement, under the head of "Progression by Ratios."

PARTNERSHIP.

1. Six villagers hired a pasture for $75. A put in 5 cows for the season, B 2, C 3, D 8, and E and F 1 each. How much was the pasturage for each cow? *Ans.* $3'75.

2. The same men the following year hired the same pasture for the same price. A put in 5 cows for 6 months; B 2 cows for 5 months; C 5 for 4 months; D 8 for 5 months; and E and F 2 each for 5 months. How much had each person to pay?

☞ Five cows for 4 months=1 cow for 20 months.

Ans. A $18'75; B $6'25; C $12'50; D $25; E and F $6'25 each.

3. Four men enter into partnership. A puts in $2500, B

$3000, C $2500, and D $2000. They gain $1500. What is the share of each?

Ans. A $375, B $450, C $375, and D $300.

4. Three men enter into partnership. A puts in $500 for 10 months, B $600 for 6 months, and C $800 for 4 months. They lose $300. How is this loss to be apportioned?

☞ 600 for 6 months=3600 for 1 month.

Ans. A's loss 127\frac{7}{59}$; B's 91\frac{31}{59}$; C's 81$\frac{21}{59}$.

5. Two merchants traded in company. Each put in $500. But A kept his in for 12 months, while B was only in for 6 months. Their gain was $600. How should it be divided?

Ans. A should have $400, B $200.

6. A, B, and C, put money into a joint stock. A put in $40, B and C together $170. They gained $126, of which B took $42. What did A and C gain, and B and C put in respectively?

Ans. A gained $24 and C $60; B put in $70 and C $100.

7. Two merchants entered into partnership for 18 months. A, at first, put into stock $400, and at the end of 8 months put in $200 more. B, at first, put in $1100, and at the end of 4 months took out $280. At the expiration of the 18 months they found the gains to amount to $1052. What is each man's share? *Ans.* A's 385\frac{2260}{2508}$, B's 666\frac{248}{2508}$.

8. A and B went into partnership. A put in on the first of January £150, but B could not put in any till the first of May. What did he then put in to have an equal share at the year's end? *Ans.* £225.

9. Three merchants traded in company. On the first of January they reckoned their gains, of which A and B took £228; B and C £215; and A and C £187 10*s.* What was the whole gain, and the gain of each?

Ans. Whole gain £315 5*s.* Gain of A, £100 5*s.*; of B, £127 15*s*; of C, £87 5*s.*

10. Three merchants, A, B, C, enter into partnership. A advances $1200; B $800, and C $600. A leaves his money 8 months, B 10 months, and C 14 months, in the business. They gain $500. What is the share of each?

Ans. A receives 184\frac{8}{13}$, B 153\frac{11}{13}$, C 161\frac{7}{13}$.

EXCHANGE.

1. Change £5 12*s.* sterling to federal money.

$$\frac{\$}{1}=\frac{5{\cdot}6 \text{ decimal of } £5\ 12s.}{{\cdot}225 \text{ decimal of } \$1 \text{ in pounds.}}=\$24\tfrac{8}{9},\ \textit{Ans.}$$

2. Change the same sum from New England currency, and also from New York currency, to dollars.

3. Change 18\frac{2}{3}$ to New England currency, and $14 to New York currency.

4. Change $36·50 to New York currency.

5. Change £14 12*s.* New York currency to dollars.

6. A merchant sends cotton to England, which is sold there for £2000, besides paying all expenses. To a friend, who wishes to purchase goods in England, he sells a draft for that amount at 6 per cent. advance. How many dollars does he receive for his draft?

$$\frac{}{\$\,1}=\frac{£2000{\cdot}106}{{\cdot}225{\cdot}100}=\frac{2120}{{\cdot}225}=\$9422\tfrac{2}{9},\ \textit{Ans.}$$

7. A merchant of Newbern, N. C., bought goods in New York to the amount of $1000. He directed the seller to draw on him through the Planter's Bank of Charleston, S. C., in the currency of that state. What must be the amount of the New York merchant's draft, when South Carolina money was at 2 per cent. discount in New York, and how much must the Newbern merchant pay in North Carolina currency, exchange with Charleston being 1 per cent. discount only?

Ans. £238·09+ in S. Carolina, and £412·35+ in N. Carolina currency.

8. A merchant in New York ships a quantity of cotton to Liverpool, which sells for £500, besides paying freight, commission, and all other expenses. For how many dollars should he sell his bill of exchange on Liverpool, exchange being 7 per cent. advance? *Ans.* $2377·77.

9. Another merchant sends cotton by the same vessel, which brings the same sum. But, instead of selling his bill in New York, he forwards it to a merchant in New Orleans, in payment for a debt which he owes him. How much should he be credited in New Orleans, if Louisiana funds be at a discount of 2 per cent.? *Ans.* $2426·30.

10. A merchant in Boston has effects at Amsterdam, Holland, to the amount of $3530, which he can remit by way of Lisbon at 840 rees per dollar, and thence to Boston at 8*s.* 1*d.* per milree (=1000 rees); or, by way of Nantz, at $5\frac{2}{5}$ livres per dollar, and thence to Boston at 6*s.* 8*d.* per crown of 6 livres. Which is the most advantageous way of remittance, and what is the difference between them?

Ans. By Lisbon £1198 8*s.* $8\frac{4}{10}$*d.* By Nantz £1059.

11. If 140 braces at Venice be equal to 150 braces at Leghorn, and 7 braces at Leghorn be equal to 4 American yards: how many American yards are equal to $52\frac{4}{15}$ Venetian braces?

Ans. 32 yards.

12. If 40 lb. at Newburyport make 36 at Amsterdam, and 90 lb. at Amsterdam make 116 at Dantzick, how many lb. at Dantzick are equal to 244 at Newburyport?

Ans. $283\frac{1}{25}$ lbs.

13. A merchant in Mississippi purchases goods to the value of $1500 from a merchant in New York. He sells the goods in his store at Jackson, and receives his pay in cotton, which he sends to his correspondent in New Orleans, who forwards it to Liverpool, where it is sold, and the net proceeds remitted to a banker in London, and placed to the credit of the Mississippi merchant. For how much sterling money must the latter draw a bill of exchange on London in favor of the New York merchant, allowing him 3 per cent. interest for the credit on his goods, exchange on London being 7 per cent. advance?

Ans. £324 17*s.* $7\frac{3}{4}$*d.*

14. A banker received 759 ducats at 7*s.* 6*d.* per ducat, and 579 dollars, at 4*s.* 8*d.* per dollar, which he exchanges for Flemish marks, at 14*s.* 3*d.* each. How many ought he to receive? *Ans.* $589\frac{15}{171}$.

15. A bill of exchange for £400 sterling was accepted at London, for an equal value delivered at Amsterdam, at £1 13*s.* 6*d.* Flemish, for £1 sterling. How much money was delivered at Amsterdam? *Ans.* £670 Flemish.

16. A merchant delivered at London £120 sterling, to receive £147 Flemish in Amsterdam. How much was £1 sterling valued at in Flemish money? *Ans.* £1 4*s.* 6*d.*

17. A factor sold goods at Cadiz for 1468 pieces of eight, valued at 4*s.* $6\frac{1}{2}$*d.* sterling each. How much sterling money do those pieces of eight amount to? *Ans.* £333 7*s.* 2*d.*

18. A traveller wished to have an equal number of crowns,

at 5*s.* 6*d.* per crown, and dollars, at 4*s.* 5*d.* each. How many of each sort may he have for £309 8*s.*? *Ans.* 624 of each.

19. A man wished to exchange £527 17*s.* 6*d.* for dollars at 4*s.* 6*d.* each, ducats at 5*s.* 8*d.* each, and crowns at 6*s.* 1*d.* each; and wanted 2 dollars for every ducat, and 3 dollars for every 2 crowns. How many of each should he receive?

Ans. 927 dollars, 463½ ducats, and 618 crowns.

20. A banker is to receive £500. He is offered crowns, at 6*s.* 1½*d.* per crown, which are worth but 6*s.*, or he may have dollars at 4*s.* 5*d.* each, which are worth but 4*s.* 4*d.* Which of these should he receive to have the least loss? and how much will he lose in the payment?

Ans. The smallest loss will amount to £9 8*s.* $8\frac{8}{53}$*d.*

CONJOINED PROPORTION.

When questions are of a complicated nature, which frequently happens in mercantile exchange, where the circulating medium of several foreign countries enter into the computation, they may be solved, perhaps, more simply by what is called Conjoined Proportion than by the usual method, as follows:

Exemplification for the Black-board.

1. If the exchange of London on Genoa be at 47 pence sterling per pezza, and that of Amsterdam on Genoa at 86 grotes per pezza; what is the proportional exchange between London and Amsterdam, through Genoa; that is, how many shillings and grotes Flemish (that is Amsterdam money, 12 grotes to a shilling) are equal to one pound sterling?

1£ Sterling.
£1=240 pence.
*d.*47= 1 pezza.
pezza 1= 86 grotes.

sh. gr.
47)20640($439\frac{7}{47}$÷12=36 $7\frac{7}{47}$, *Ans.*
184
430
7

Suggestive Questions.—In the above statement of 4 lines we have 3 sums of money on the left, given equal to 3 on the right, and if we knew how many grotes were equal to £1 sterling (the first line) all the 4 would be equal. Now, supposing the deficiency on the left to be supplied, would the products of these equal values be also equal? But the product may be found complete on the right, while one factor is wanting on the left; how, then, may that factor be found?

2. If the exchange of London on Madrid be at 42 pence sterling per dollar, or 272 maravedis, and that of Amsterdam on Madrid at 96 grotes Flemish per ducat=375 maravedis, what is the exchange between London and Amsterdam, through Madrid, in shillings and grotes, per pound sterling, allowing 12 grotes for a shilling?

1£ Sterling.
£1=240 pence.
d. 42=272 maravedis.
m. 375= 96 grotes.
Left hand prod. 1575|0)626688|0(397$\frac{1413}{1575}$=$\frac{157}{175}$
15418 — 33*s.* 1$\frac{157}{175}$ gr. *Ans.*
12438
1413

Exercises for the Slate or Black-board.

1. If the exchange from Philadelphia to London was 4 per cent. above par (104 per 100) and from London to Paris 23 liv. 8 sous per pound sterling, what would be the proportional exchange from Philadelphia to Paris, through the medium of London? and how many dollars would purchase a bill on Paris for 1100 livres 15 sous, allowing 20 sous to be equal to 1 livre, and £1 sterling to be equal to \$$\frac{40}{9}$?
Ans. 5 liv. 2 sous per dollar. \$217·43.

2. If, at New York, bills on London are at 5 per cent. above par; the exchange of London on Amsterdam 34*s.* 4*gr.* per pound sterling; and Amsterdam on Paris 54*gr.* for 3 livres; what is the proportional exchange between New York and Paris in francs per dollar, 80 francs being equal to 81 livres? *Ans.* 4·882 fr. per dollar.

3. If the rate of exchange were, Boston on Paris 5·30 francs per dollar, Paris on Lisbon 464 rees per ecu of 3 livres,

what would be the proportional exchange between Boston and Lisbon, viz., how many rees per dollar? *Ans.* 830 rees.

4. If the exchange of London on Lisbon be at 68 pence sterling per milree (=1000 rees), and that of Genoa on Lisbon at 718 rees per pezza; what is the proportional exchange between London and Genoa, through Lisbon, in pence sterling per pezza? *Ans.* $48\frac{103}{125}$.

SUPPLEMENT.

CONTRACTED MULTIPLICATION AND DIVISION OF DECIMAL FRACTIONS.

1. Contracted Multiplication.

It frequently happens, when one decimal fraction is multiplied by another, that the fractional part of the product extends to numbers altogether insignificant. Thus, if it were required to multiply 4·233 by 6·287, the product would extend to six decimal places, the last figure to the right being one-millionth part of 1, a number devoid of worth, even if it related to gold. To save the tedious labor of producing such worthless numbers, then, is frequently a matter of some consequence, especially where the computations are numerous, as in some of the articles in this Appendix. This may easily be effected by proceeding as follows:

1. Place the multiplier under the multiplicand in an *inverted order*, putting the unit's place of the multiplier under that decimal place in the multiplicand, which is the lowest meant to be retained in the product.

2. In multiplying, begin each line of partial products with that figure in the multiplicand which stands directly over the multiplying figure, increasing it by the tens that would have been produced (if any) by multiplying another figure to the right; and also increasing it by *one*, if the right hand figure would have been 5 or upwards; and let the first figures on the right of all the partial products stand directly under each other.

3. When it is desirable to be absolutely *certain* that the last figure retained is that *nearest* to the truth, the work should be extended to one place more than is wished to be retained.

4. The local value of the total product should be ascertained by an inspection of the two factors.

In general, when a decimal fraction is abbreviated by striking off, or omitting, some of the places on the right hand, in order that the last figure retained may be the *nearest* to the truth, whether too great or too little, it should be increased by *one* when the right hand figure is 5 or upwards. Thus, ·1246, abridged to three decimal places, would be ·125, while ·1244 would only be ·124.

Exemplifications for the Black-board.

1. Multiply 4'127643 by 6'25135, retaining only four decimal places in the product.

In full.	Contracted.
4'127643	4'127643
6'25135	531526
2|0638215	247659
12|382929	8255
41|27643	2064
2063|8215	41
8255|286	12
247658|58	2
25'8033|4106805	25'8033

Suggestive Questions on the Contracted Method.—In the first partial product, how many tens are carried to the first figure on the right? Would, or would not, the adjoining omitted figure have been 5 or upwards? By how much, then, has the first figure standing on the right been increased? By how much has the second partial product been increased? Why? By how much has the third? Why? [The answer to these two "whys" is different.] By how much has the fifth been increased? Why? By how much the sixth? Why?

2. Multiply '36425 by '724325, retaining 5 decimal places in the product, so that 4 places may be absolutely certain of being nearest to correctness.

In full.	Contracted.
'724325	'724325
'36425	52463
3|621625	21730
14|48650	4346
289|7300	290
4345|950	14
21729|75	4
'26383|538125	'26384, or '2638 nearest 4 places.

Suggestive Questions on the Contracted Method.—Why is the first partial product increased by *one*? Why the second by two? Why the third by two? Why the fifth by one?

☞ To those who may not perceive why the figures of the multiplicand are placed in an inverted *order*, and in a rather unusual *place*, it may be remarked, that both form a mere mechanical contrivance to save time and labor, by enabling the student instantly to decide where the multiplication by each several figure of the multiplier is to begin. The order in which those figures are taken is of no moment, as has been shown, p. 257.

Exercise for the Slate or Black-board.

1. Multiply 34'265 by 4'396, true to three decimal places, and prove by multiplication in the usual manner.

2. Multiply ‘008 by 3‘796, true to three decimal places, and prove as above.

3. Multiply ‘5264 by ‘0428, true to three decimal places, and prove.

4. Multiply 1‘729 by 7‘218, true to four decimal places, and prove.

5. Multiply 26‘45 by 39‘46, true to two decimal places, and prove.

2. Contracted Division.

When it is desirable in division to limit the number of decimal places in the quotient, it may be done as follows:

1. Take as many figures only, on the left hand side of the divisor, as the whole number of figures required to be in the quotient, and cut off the rest.

2. Let each remainder successively be a new dividend, without bringing down any figure from the original dividend, but, instead thereof, let another figure be continually cut off from the divisor for each quotient figure, till the whole is exhausted, observing, however, as in contracted multiplication, to increase each particular product by the nearest number of tens in the product of the quotient figure, into the figure last cut off in the divisor.

3. When the whole divisor does not contain as many figures as are required to be in the quotient, no figure should be cut off till the figures in the divisor shall be equal to the remaining figures required to be in the quotient, when the cutting off should commence as above directed.

Exemplifications for the Black-board.

1. Divide 74‘33373 by 1‘346787, retaining three decimal places only, or five places in the quotient altogether.

```
74‘33373(1,‘3,4,6,7|87
67339     55‘193 Ans.
 6994
 6734
  260
  135
  125
  121
    4
    4
```

Suggestive Questions.—How many tens are carried into the first partial product? How many into the second? the third? the fourth? the fifth?

☞ The *pupils* should not write the partial product, but make the subtraction, as usual, mentally.

2. Divide ‘07567 by 2‘32467, true to four decimal places; or three significant figures, the first being a cipher.

```
‘07567(2‘3,2|467
   59   ‘0326 Ans.
   13
```

Suggestive Questions.—Are any tens carried into the first partial product? into the second? the third?

3. Divide 5'37341 by 3'74, true to four decimal places.

```
5'37341(3,'7,4
 1633   1'4367
  1374
   252
    28
     2
```

Suggestive Questions.—Are any figures of the original dividend brought down to the partial dividends? If so, say how many, and why? Are any tens carried to the first partial product? to the second? the third? the fourth? the fifth?

4. Divide 1 by '3475, true to three decimal places.

```
1'0000('3,4,7,5
 3050  2'878 Ans.
  270
   27
```

Suggestive Questions.—How many tens are carried to the second partial product? to the third? to the fourth?

Exercises for the Slate or Black-board.

1. Divide 3'467 by '73367, true to four decimal places, and prove by dividing it in the usual manner.
2. Divide 1 by '462849, true to three decimal places, and prove.
3. Divide 1'264531 by '92145, true to three decimal places, and prove.
4. Divide 36'4776 by '9834, true to the units' place, and prove.
5. Divide 2'12457 by '3268, true to three decimal places, and prove.

PROGRESSION.

Progression, in mathematics, signifies a regular or a proportional advance in increase or decrease in numbers. It is of two kinds:

1. *Progression by differences,* commonly, though improperly, called *Arithmetical Progression.*
2. *Progression by quotients (or by ratios),* quite as improperly called *Geometrical Progression.*

I. Progression by Differences.

Definitions.

1. If we take any number, and increase or diminish it continually by another number, we shall form a regular series of numbers, called a progression by differences. Thus:

2	4	6	8	10	12
12	10	8	6	4	2

form two series of progression by differences, the first ascending, the second descending, with the common difference 2.

2. The several numbers are called *terms* of the progression: the *first* and *last* terms are called the *extremes*, and the *intermediate* terms the *means*.

3. Five things are to be considered in every progression by differences, any three of which being known, the remaining two can be found, namely:

1. The first term,	marked a.
2. The last term,	z.
3. The common difference,	d.
4. The number of terms,	n.
5. The sum of the series,	s.

Case I.

Where the first term, the common difference, and the number of terms, are given, to find the last, or any intermediate term (a, d, n, to find z).

Exemplification for the Black-board.

1. Form a progression of 6 terms, with 4 for the first term, and 3 for the common difference. [Let one of the class form it on the black-board, without copying it from the book.]

1st.	2d.	3d.	4th.	5th.	6th.
4	7	10	13	16	19

Suggestive Questions.—How often is d, the common difference, found in the 2d term? What else does the 2d term contain? Then write on your slate as follows:

$$\text{Second}=a+d.$$

How often is d found in the 3d term? Then write that term on your slate, under the 2d, and in a similar manner, namely, Third$=a+2d$, and so on with all the remaining terms. What does the 4th term contain besides a (the first term)? What the 5th? the 3d? Is the common difference, in any term, then, *always* repeated once less than its number? May not the following, then, be considered the first principle in Progression by Differences?

I. Every term in an increasing series of Progression by Differences consists of a (the first term), added to d (the common difference), taken once less than n, the *number* of that term.

Exercises for the Slate or Black-board.

1. When the first term is 6, and the common difference 2, what is the 4th term? the 6th term? the 3d term?

2. A man bought 20 yards of cloth: he engaged to pay 6 cents for the first yard, 8 cents for the 2d, 10 for the 3d, and so on, increasing by the common difference 2: how much did he pay for the last yard?

Ans. 44 cents.

3. What is the 16th term of a progression by differences, whose first term is 0, and the common difference 1?

4. What is the 18th term of a progression, whose first term is 3, and its common difference 4?

5. What is the 25th term of the progression 4, 9, 14, 19, 24, &c.?

Case II.

Where a, z, and n are given, to find d (first, last, and number of terms given, to find the common difference).

Exercises for the Slate or Black-board.

1. What is the common difference in a series whose first term is 4, last term 42, and number of terms 20? Prove by forming the series. If the 20th term proves to be 42, the process is correct.

Suggestive Questions.—What does the 20th term contain besides a? How, then, can d, the common difference, be found?

2. What is the common difference in a series whose first term is 8, last term 15, and number of terms 8? Prove by forming the series. If the 8th term prove to be 15, the process is correct.

3. What is the common difference in a series whose first term is 3, last term 11, and number of terms 10? Prove as above.

4. What is the common difference in a series whose first term is 2, last term 14, and number of terms 8? Prove as above.

5. There are 21 persons whose ages are equally distant from each other. The youngest is 20 years old, and the eldest 60. What is the common difference of their ages? Prove by finding the age of each.

Case III.

Where a, z, and n, are given, to find s (first, last, and number of terms given, to find the sum of all the terms).

Exemplification for the Black-board.

1. Find the sum of a series whose first term is 2, last term 22, and number of terms 11? *Ans.* 264.

Suggestive Questions.—What does the 11th term contain besides a? How, then, can d, the common difference, be found? [Let one of the class write the series on the black-board, and another write the same inverted immediately below, and also the sum of each pair of terms taken vertically.] What is the sum of the first and last terms? of the 2d and 10th? of the 3d and 9th? and so on to the end of the double series.

Are all these sums equal? How many times does each sum contain the common difference? What else? Will this be the case in every series of the kind thus arranged? How many sums are there? How many terms? Will the number of sums and the number of terms always be the same in series thus arranged? What will be the product of one of these sums by the number of terms? To what, then, will half that product be equal? If the first, last, and number of terms be given, then, how can you find the sum of all the terms? May not the following, then, be considered as the second Principle in Progression by Differences?

II. The sum of all the terms in a Progression by Differences, is equal to half the product of the number of terms by the sum of the first and last terms.

Exercises for the Slate or Black-board.

1. What is the sum of a progression by differences, whose first term is 2, last term 100, and number of terms 50? *Ans.* 2550.

2. A debt is to be discharged at 16 several payments, with equal differences. The first payment is to be \$12, the last \$100. What is the amount of the debt, neglecting all consideration of interest? *Ans.* \$896.

3. A man engaged to travel to a certain place in 19 days, and to go but 6 miles the first day, increasing every day by an equal excess, so that the last day's journey may be 60 miles. What is the distance of the journey? *Ans.* 627 miles.

4. How many strokes will the hammer of a common clock make on the bell during the space of 12 hours? *Ans.* 78.

5. Required the sum of all the numbers contained in a multiplication table, extending to 12 times 12. *Ans.* 6084.

Case IV.

Where z, n, and d are given, to find a (the last term, the number of terms, and common difference given to find the first term).

Exercises for the Slate or Black-board.

1. The last term of a progression by differences is 119, the number of terms 24, and the common difference 5. What is the first term? *Ans.* 4.

Suggestive Questions.—What does the last term consist of besides the first term? If that product, then, be taken away, what will remain?

2. A note is to be paid in 10 annual instalments, the several payments being in a progression whose common difference is \$30. The last payment is to be \$400. What is to be the first payment? *Ans.* \$130.

3. A man performs a journey in 19 days, travelling but a short distance the first day, and increasing his daily journey by 3 miles, until the last day, when he travelled 60 miles. How far did he travel the first day? *Ans.* 6 miles.

4. The last term of a progression by differences is $36\frac{3}{4}$, the number of terms is 49, and the common difference $\frac{3}{4}$. What is the first term? *Ans.* $\frac{3}{4}$.

CASE V.

Where a, z, and d are given, to find n (the first and last term, and the common difference, given to find the number of terms).

Exercises for the Slate or Black-board.

1. The first term of a progression by differences is 4, the last term 119, and the common difference 5. What is the number of terms? *Ans.* 24.

Suggestive Questions.—What does the last term consist of besides the first term? If the first term, then, be subtracted from the last, how shall the number of terms less 1 be found?

2. The first term in a progression by differences is $\frac{3}{4}$, the last term $36\frac{3}{4}$, and the common difference $\frac{3}{4}$. What is the number of terms? *Ans.* 49.

3. A note was paid by annual instalments, whose common difference was $30. The first payment was $130, and the last $400. How many instalments were there? *Ans.* 10.

4. In a triangular field of maize, the number of hills in the successive rows forms a progression by differences. In the first row there is but one hill, in the last 81, and the common difference in the number of hills in a row is 2. What is the number of rows? *Ans.* 41.

[The number of cases might be very much enlarged; but this is hardly necessary, as each pupil can readily arrange them for himself. The following equations, with the cases already given, comprehend all that are most common and useful. The exercises given above will answer for the new cases. But it would be more profitable for the student to furnish examples himself. The demonstration of the cases given below, by suggestive questions, would form one of the best exercises that a class could engage in.]

$$z=\frac{2s}{n-a}$$

$$a=\frac{2s}{n-z}$$

$$d=\frac{2s-(2n\times a)}{n\times(n-1)}$$

$$n=\frac{2s}{z+a}$$

$$s=\frac{(z-a+d)\times(a+z)}{2d}$$

II. PROGRESSION BY RATIOS.

Definitions.

1. A series of numbers, which succeed each other regularly by a constant factor, which is called a *ratio*, is called a *Progression by Ratios*. Thus:

2 4 8 16 32 64
64 32 16 8 4 2

form two progressions by ratios, the ratio of the first being 2, and that of the second $\frac{1}{2}$.

2. When the ratio is more than a unit, the series is called an *ascending progression;* when it is less, the series is called a *descending progression.*

3. As in the progression by differences, so in the progression by ratios, five things are to be considered, any three of which being known, the remaining two can be found, namely:

1. The first term,	marked a.
2. The last term,	z.
3. The common ratio,	r.
4. The number of terms,	n.
5. The sum of the series,	s.

Case I.

Where the first term, the ratio, and the number of terms are given, to find the last term.

Exemplification for the Black-board.

1. The first term of a progression by ratios is 2, the ratio is 3, and the number of terms is 6. What is the last term?

2 6 18 54 162 486

Suggestive Questions.—How often is the ratio a factor in the 2d term? By what power of the ratio, then, is the first term multiplied to form the 2d term? How often is the ratio a factor in the 3d term? By what power of the ratio, then, is the first term multiplied to form the 3d? By what power to form the 4th? By what to form the 5th? the 6th? Does every term, then, consist of the first term multiplied by a power of the ratio one less than its number? Find, then, the last term of the above progression, without finding the intermediate ones.

Exercises for the Slate or Black-board.

1. The first term of a progression by ratios is 4, the ratio is 2, and the number of terms 7. What is the last term? [It is to be found without finding the intermediate terms.] *Ans.* 256.

2. A farmer sold 7 cows to a neighbor, for which he was to receive $5 for the first, $10 for the second, $20 for the third, and so on, increasing in progression by ratios. What would be the price of the last cow at this rate? *Ans.* $320.

3. A man once offered to sell a horse on condition that he should have a sufficiency of wheat to pay for the last nail in the animal's shoes, reckoning the first nail at 1 gill, the second at 2 gills, the third at 4, and so on, increasing in progression by ratios. It was found that the horse had 8 nails in each of his four shoes. What would he get for the horse at that rate, reckoning wheat at $1·25 per bushel? *Ans.* $10,485,760.

Compound Interest by Progression.

If the manner in which compound interest is computed, p. 314, be carefully examined, it will be evident that the successive amounts which are considered as new principals, form the terms of a series of progression by ratios, whose first term is the original principal, and the ratio the amount of $1, for one year, at the given rate per cent. The number of terms is equal to *one more* than the number of years, a circumstance which will be readily understood by a glance at the example referred to, the number of principals in the computation being 4, while the number of years is only three. Thus, to find the amount of a given principal at compound interest, for a given number of years, at a given rate per cent., is merely to find the last term of a progression by ratios, when the first term, the ratio, and the number of terms are given.

Exercises for the Slate or Black-board.

1. Find the compound interest of $100 for 4 years at six per cent. *Ans.* $26'247.

2. Find the amount at compound interest of $100 for 17 years at six per cent. *Ans.* $269'277.

3. Find the compound interest of $600 for 20 years at 5 per cent. *Ans.* $1591'978.

4. Find the amount of $100 for 3 years at 6 per cent. per annum, when the interest is to be added at the end of every six months.

Suggestive Questions.—If a new principal be formed at the end of every year, of how many terms would the series consist, reckoning the original principal of $100 as one? At what per cent., then, should the interest be reckoned? But, if the new principal be formed at the end of every half year, of how many terms would the series then consist? At what per cent., then, should the interest be reckoned for the half year? But, if the new principal were formed by adding in the interest every 3 months, how many terms would there then be in the 4th exercise above? How much per cent. must be added in for the 3 months, when the interest is 6 per cent. per annum?

5. Find the amount of $3705 at compound interest, in 3 years and 3 months, at 12 per cent. per annum, the interest being added every 3 months. *Ans.* $5440'918.

6. What will $1000 amount to in 15 years, at 8 per cent. per annum, the interest being compounded half-yearly? *Ans.* $3243'398.

Case II.

Where the last term, the ratio, and the number of terms, are given, to find the first term.

Exercises for the Slate or Black-board.

1. The last term of a progression by ratios is 256, the ratio 2, and the number of terms 7; what is the first term? *Ans.* 4.

Suggestive Questions.—How many terms are there in the above progression? How many times, then, is the ratio a factor in the last term? What other number is a factor in the last term? How, then, can the first term be found?

2. A man bought 7 cows, for which he engaged to pay a certain sum for the first, double the sum for the second, and so on in regular progression to the 7th, the price of which was found to amount to $320. What was the price of the first? *Ans.* $5.

3. A horse was once offered for sale, on condition that the purchaser should give a small quantity of wheat for the first of the 32 nails in his shoes, doubling the wheat for the second, and so on in regular progression to the last. No one offering to take the horse on these conditions, the owner said he would sell it for the price of the last nail. On calculation, it was found that the horse would cost, on this last condition, $10,-485,760, reckoning the wheat at $1'25 per bushel. How much wheat was to be given for the first nail, agreeably to the first offer?

Ans. 1 gill.

Compound Discount.

Definitions.

1. *Compound Discount* is an allowance made for the payment of money before it is due, on the supposition that the money draws compound interest.

2. When compound interest is reckoned, the *present worth* of a debt payable at some future time without interest, is such a sum, as being put out at compound interest, will, in the given time, at the given rate per cent., amount to the debt.

Finding the present worth of a debt, then, resolves itself as follows: given the amount at compound interest, the amount of $1 at the given rate, and the number of terms, to find the principal, or, which is the same thing, given the last term of a progression by ratios, the ratio, and the number of terms, to find the first term.

☞ Observe that the number of terms is always one more than the number of years. Why?

1. What is the present worth of a debt of $126'247, due 4 years hence, at 6 per cent. compound interest? *Ans.* $100.

Suggestive Questions.—How often is the amount of $1 a factor in the above amount? What other number is a factor in it? If the first be known, then, how shall the latter be found?

2. What is the present worth of $269'277, due 17 years hence, at 6 per cent. compound interest? *Ans.* $100.

3. What is the present worth of $1000, due 20 years hence, at 5 per cent. compound interest? *Ans.* $376'889.

4. What is the present worth of $1593'30, due 20 years hence, at 5 per cent. compound interest? *Ans.* $600'50.

Case III.

Where the first term, the last term, and the ratio are given, to find the sum of all the terms.

Exemplification for the Black-board.

1. What is the sum of a series of progression by ratios, whose first term is 2, the ratio 3, and the number of terms 6? Write the series at length, with its proper signs of addition, and then form a new series, by multiplying each term of the old series by the ratio, placing one over the other, so that each term shall be removed one step to the right of that by which it was produced. Thus:

From new series 6+18+54+162+486+1458=3s (or s×r)
Take old series 2+6+18+54+162+486 = s
2 . . . . from 1458=2s (or s×(r–1))

Suggestive Questions.—How was the new series formed? *Ans.* By — each term of the old series by the —. What *is* the ratio in this problem? Is the sum of the new series, then, 3 times the amount of the sum of the old series? Had the ratio been 5, or 7 (or any other number), would the old series have been 5 or 7 (or any other number) times the amount of the new? How many times the sum of the old series is the difference between the sums of the old and new series? Does that depend on the number of the ratio? Had the ratio been 5, or 7, how many times the sum of the old series would then have been the difference between the two series? Would it always be s×(r–1)? What is the difference between the first term of the old series, and the last of the new, ascertained at a glance? Is this number the difference between the two series? Is it equal to twice the sum of the old series? What, then, is the sum of the old series? How, then, can you ascertain the sum of any series of progression by ratios, from the first term, the last term, and the ratio, without forming the intermediate terms? May not the following, then, be considered the first principle in Progression by Ratios?

I. The sum of any series in a Progression by Ratios is equal to the last term, multiplied by the ratio, diminished by the first term, and divided by the ratio less one.

1. What is the sum of the series, 1, 3, 9, 27, &c., to 12 terms? *Ans.* 265720.

2. What is the sum of 10 terms of the series, whose first term is 2, and the ratio 2? *Ans.* 2046.

3. A man bought a horse, agreeing to give 1 dollar for the first nail in one of his shoes, 3 for the second, 9 for the third, and so on. The shoe contained 8 nails. What was the cost of the horse? *Ans.* $3280.

4. A man agreed to buy 30 bushels of wheat: the first bushel for 2 cents, the second 4 cents, the third 8 cents, doubling the price of each preceding bushel for that of the next. What would the 30 bushels cost, and what would be the average price per bushel?
Ans. to the last question, $815,827'88+.

5. What sum would purchase a horse with 4 shoes, and 8 nails in each shoe, at 1 mill for the first nail, 2 for the second, 4 for the third, &c., doubling to the last? *Ans.* $5,668,109'139.

6. A lady, who was married on new-year's day, received from her father one dollar towards her marriage portion, which he promised to double on the first day of every month during the year. What would her portion amount to? *Ans.* $4095.

Annuities.*

Definitions.

1. An *annuity* is a rent or sum payable or receivable yearly for a certain number of years, or forever.

2. A *deferred annuity* is one which is not to be entered upon immediately, but after a certain number of years.

3. A *reversionary annuity* is one which is not to be entered upon until after the death of some person or persons now living.

4. The *present worth* of an annuity is such a sum of money as will, at compound interest, produce an amount equal to the amount of the annuity.

Case I.

To find the amount of an annuity, which has remained unpaid (or been forborne), for a given time: in other words, to find the sum of a series, when the first term, the number of terms, and the ratio, are given. [See Case III., Progression by Ratios.]

1. What is the amount of an annuity of $100, forborne 6 years, at 5 per cent. compound interest? *Ans.* $680'19.

2. What is the amount of an annuity of $150, payable half yearly, forborne 2 years, allowing half-yearly compound interest at 6 per cent. per annum? *Ans.* $313'772.

3. What would a salary of $700 a year amount to, if left unpaid for 4 years, allowing quarterly compound interest at 12 per cent. per annum? *Ans.* $3527'454.

4. If the annual rent of a house be $200 per annum, payable quarterly, and it remains unpaid 4 years, how much will be due, allowing compound interest at the rate of 12 per cent. per annum? *Ans.* $1007'844.

5. What would be the amount of a pension at $65 per annum, which had been unpaid for 14 years, if compound interest were reckoned at the rate of 6 per cent. per annum? *Ans.* $1365'97.

Case II.

To find the present worth of an annuity which is to terminate in a given number of years.

1. What is the present worth of an annuity of $150, to continue 6 years; reckoning compound interest at 6 per cent. per annum?

Ans. Amount of $150 for 6 years, $1046'297. Present worth of $1046'297, 6 years hence, at 6 per cent., $737'598.

* In most works on Arithmetic, which treat of compound interest and annuities, the student is furnished with tables of the amount of $1, at various rates per cent., for from 1 to 30 or 40 years, and also of the present worth of annuities, &c., in like manner. Such tables are exceedingly convenient in offices where this kind of business is transacted, as the object there is to ascertain values as quickly as may be, and with as little calculation as possible. But they are entirely out of place in a school, where the main object is mental discipline and expertness in calculation. Where the teacher, however, thinks such calculations of ratios occupy too much time, each student can be directed to form the requisite tables for himself, as separate exercises.

2. How much ready money will purchase an annuity of $250, to continue 20 years, at 6 per cent. compound interest? *Ans.* $2867'47.

3. What is the present worth of an annuity of $150 for 30 years, at 5 per cent. compound interest? *Ans.* $2315'96.

4. What is the present worth of an annuity of $500, to continue 10 years, at 6 per cent. per annum, compound interest? *Ans.* $3680'045.

5. A gentleman subscribed $1000 for a college, payable in 10 annual instalments, of $100 each. What is the present worth of such a subscription; that is, what sum of money placed at interest would exactly pay each instalment as it fell due, reckoning the interest at 6 per cent.? *Ans.* $736.

Case III.

To find the present worth of an annuity in reversion.

a. Where the term of the annuity is limited.

1. What is the present worth of an annuity of $100, to be continued 6 years, but not to commence till 3 years hence, allowing 6 per cent. compound interest?

Present worth of an annuity of $100 for 9 years, at 6 per cent., $680'169. Worth of do. for 3 years, $267,301. Difference, $412'868. *Ans.*

2. What is the present worth of an annuity of $320, to continue 7 years, but not to commence till after the expiration of 5 years, reckoning interest at 5 per cent. per annum? *Ans.* $1450'728.

3. What is the present worth of an annuity of $200, to be continued 5 years, but not to commence till after the expiration of 2 years, reckoning interest at 6 per cent.? *Ans.* $749'798.

4. A father, dying, left a will, by which, among other arrangements for his family, he directed that his only son should have $500 a year as soon as he attained his majority, the annuity to be continued for 4 years, to assist him at the commencement of his profession. The son was exactly 18 on the day of his father's death. What sum would be necessary for the payment of this bequest, allowing compound interest at the rate of 6 per cent. per annum? *Ans.* $1454'684.

5. What is the present worth of a reversion of $400 a year, to continue 7 years, but not to commence until the end of 8 years, allowing compound interest at 4 per cent. per annum? *Ans.* $1754'356.

b. Where the term of the reversionary annuity is unlimited.

[Where an annuity is to continue forever from the present time, it is manifest that its present worth will be that sum whose interest for 1 year is equal to the annuity. All that is necessary, therefore, where the annuity is reversionary, is to find what sum of money will produce such a principal at the time when the annuity is to commence.]

1. What is the present worth of a reversionary annuity of $200 a year, to commence in 6 years, and to continue forever, interest being 6 per cent.?

Ans. { Present worth, if entered on immediately, $3333⅓.
Present worth of $3333⅓ due 6 years hence, $2349'87.

2. Find the present worth of a reversionary annuity of $500 a year, to commence in 4 years, and to continue forever, interest at 6 per cent. *Ans.* $6600'783.

3. A man left his son an annuity of $400, to commence on his majority 3 years thereafter, and to continue forever to him, and his nearest heir. What sum would secure such an annuity at the father's death, allowing interest at 5 per cent.? *Ans.* $6910'70.

4. What is the present worth of an annuity of $1000, to commence in 6 years, and to continue forever, interest at 6 per cent.? *Ans.* $11,749'35.

5. What is the present worth of a reversion in perpetuity of $100, to commence in 4 years, interest at 5 per cent.? *Ans.* $1645'40.

PERMUTATION,

OR, CHANGE OF ARRANGEMENT.

Definitions.

PERMUTATION is the method of finding in how many different ways the order of things may be varied.

Exercises for the Slate or Black-board.

1. In how many different positions can 5 blocks be placed in succession?

1st. If the blocks be letters in the order of the alphabet, it is obvious that *a* (the first) alone can only be placed in *one* position.

2d. If now *b* (the second block) be added, the two can be placed in two positions, namely, *ab* and *ba* · or $1\times2=2$.

3d. Adding *c* (the third) we have (commencing with *c*) *cab*, *cba*; (with *c* in the middle) *acb*, *bca*; (with *c* last) *abc*, *bac*: or $1\times2\times3=6$.

4th. Adding *d* (the fourth), it is plain that the 6 changes are increased four fold, since *d* may be placed as the first letter of the series, and also as the second, third, and fourth, making the number $1\times2\times3\times4=24$.

5th. In the same manner the 5th will increase the last product five fold; and, in general, every number added in the natural arrangement of numbers (1,2,3,4, &c.) will increase the number so many times.

All such questions, then, are evidently solved by taking the products of the natural numbers from 1 to the given number inclusive, the last product furnishing the answer.

Ans. to Exercise 1, $1\times2\times3\times4\times5=120$.

2. How many ways may the 6 vowels, *a*, *e*, *i*, *o*, *u*, *y*, be placed, one after another? *Ans.* 720.

3. How many variations may be made in the position of the 9 digits? *Ans.* 362880.

4. Find how many changes may be rung on a chime of 8 bells; by dividing the answer to the last question by a certain number.

Ans. 40320.

5. A gentleman once offered to sell a township of land in Illinois, 6 miles square, for 1 cent for every different order in which the letters of the English alphabet could be arranged. What would be the price per acre at that rate? *Ans.* 1,750,396,969,533,696,000.

PROPERTIES OF NUMBERS.

The following exercises on the properties of numbers will not only assist the student in the art of computation, but, what is of vastly more importance, aid in the development of his faculties, by affording admirable materials for thought.

1. The sum, or the difference, of two even, or of two *odd* numbers will be an *even* number. Why? But the sum, or the difference of an even and an odd number will be an odd number. Why?

2. The sum of any number of even numbers, or of an even number of odd numbers, will be even. Why? But the sum of an odd number of odd numbers will be odd. Why?

3. If one, or both, of two factors be even numbers, the product will be even; but if both be odd, the product will be odd. Why?

4. If any two numbers be, severally, divisible by a third (without a remainder), their sum, and their difference, will also be divisible by the same. Why?

5. If several different numbers be each divisible by any other number, then their sum will be divisible by the same number, and their product by the same number, or by any power thereof whose index does not exceed the number of factors. Why?

6. If several different numbers be each divisible by 3 or by 9, then their sum, and also the sum of their digits (see p. 112), will also be divisible by the same numbers respectively. Why?

7. If any number be multiplied by a number divisible by 9 or by 3; then the product, and also the sum of its digits, will also be divisible by the same numbers respectively. Why?

8. If any number be divisible by 3, by 9, or by 11, then its reverse (the same figures written backwards) will also be divisible by the same numbers respectively. Why? See No. 6.

9. Any number diminished by the sum of its digits will become divisible by 9 and also by 3. Why? Any number divided by 9 will leave the same remainder as the sum of its digits when divided by 9. Why? The same remark also applies to 3. Why?

10. The sum of any number, consisting of an *even* number of places, and its reverse, will be divisible by 11, and their difference by 9. But, if the number of places be *odd*, then the difference of the number and its reverse will be divisible both by 11 and by 9. Why?

11. If the sum of the digits in the odd places of any number, beginning at the place of units, be equal to the sum of the digits in the even places, then both the whole number and its reverse will be divisible by

11; their sum will also be divisible by 11, and their difference both by 11 and by 9. Why? When, in the above case, the number and its reverse (divisible by 11) consist of an *odd* number of places, then the sum of the quotients will be divisible by 11, and their difference by 9. But when the number, as above, consists of an *even* number of places, then the difference of the quotients will be divisible both by 11 and by 9. Why?

12. Any prime number (greater than 3) divided by 6, must have a remainder of 1 or of 5. Why? See p. 88.

13. Every prime number greater than 3 is divisible by 6, if 1 be either added to it or subtracted from it, according to circumstances. Thus 37—1, and 41+1, form numbers divisible by 6, and so of all other primes. Why?

14. Every number is either a prime number, or composed of prime numbers. Why?

15. An odd number cannot be divided by an even number without a remainder. Why?

16. A square number, or a cube number, arising from an even root, is even. Why? See No. 2.

17. The square and the cube of an odd number are odd. Why? See No. 3.

18. If an odd number measure an even number, it will also measure the half of it. Why?

19. If a square number be either multiplied or divided by a square, the product or quotient is a square. Why?

20. If a square number be either multiplied or divided by a number that is not a square, the product or quotient is not a square. Why?

21. The product arising from two different prime numbers cannot be a square. Why?

22. The product of no two different numbers, prime to each other, can make a square, unless each of these numbers be a square. Why?

23. The difference between an integral cube and its root is always divisible by 6. Why? See the synthesis of a cube, p. 165.

24. Every prime number above 2 is either 1 greater or 1 less than some multiple of 4. Why? And every prime number above 3 is either 1 greater or 1 less than some multiple of six. Why? See p. 88.

THE ELEMENTARY OPERATIONS OF ARITHMETIC

PERFORMED WHOLLY BY INSPECTION.

The principal object of the following exercises is to develop the power of attention, to accustom the pupil to restrain his wandering thoughts, to fix them steadily upon one point, to the complete exclusion of all extraneous matters. Before such a power of concentration, all the seeming difficulties and obscurities of science disappear; and that this power may be gained by a *faithful* practice of exercises like these, will hardly be disputed.

Plan of Conducting the Exercise.

Let the teacher, or one of the pupils, write a few of the following examples on the black-board, and number them, and then let each member of the class at once proceed to the computation, in perfect silence, writing the *complete* result on his slate with the appropriate number, as below, as soon as it is ascertained.

Results.

No. 1. . . . .
No. 2. . . . .
No. 3. . . . .
&c. &c.

The teacher, however, should be careful to see that no pupil write down a partial result, nor have any of them ready written on the slate. As the object is to cause the class to go through the *whole* of each exercise mentally *without a pause*, it is easy to see that either of these practices would render the whole proceeding nugatory.

1. *Addition by Inspection.*

All these exercises should be thoroughly practised on the slate before recitation on the black-board, the pupils being cautioned never to use the pencil until they are prepared to write the result in full. All the lessons should be short. They may be practised while the rest of the book is reviewed.

Find the sums or amounts of the following columns of figures, not writing them on the slate till the *whole result* of each exercise is ascertained. Prove by addition in the ordinary manner.

No. 1.	No. 2.	No. 3.	No. 4.
426	124	213	723
373	356	426	857
254	472	718	239

5.	6.	7.	8.	9.
719	912	285	537	738
437	738	437	298	953
256	912	264	451	421
324	326	519	724	698

10.	11.	12.	13.	14.
473	834	279	354	564
912	926	145	747	947
728	276	372	832	328
963	385	864	174	473
465	714	926	913	298

15.	16.	17.	18.	19.
621	137	735	286	882
793	426	912	425	375
426	718	437	938	926
318	942	285	726	378
925	375	942	458	825
514	486	368	921	174

20.	21.	22.	23.	24.
971	128	396	738	929
263	493	248	124	135
485	521	151	689	784
926	436	289	753	933
379	792	827	172	257
634	858	473	938	781
257	976	925	181	324
931	259	126	247	471

25.	26.	27.	28.	29.
1365	3258	8914	6738	1543
4734	1947	3246	9146	2673
9128	2836	9145	2052	7844
6432	6244	2786	7606	9127
5638	7128	3823	3235	8324
1927	9435	1574	9574	7126
3825	8817	9138	3842	8891
1917	9432	5245	5178	9312

30.	31.	32.	33.
73426	28547	44572	98137
84932	32984	96385	24325
34578	17054	10472	87815
91245	36408	67347	91233
37962	91525	91234	54620
18459	71347	32896	87356
30640	36482	88709	32104
84516	93105	21594	66527
93284	71746	38727	93162
13797	81272	21345	45231

The class may now take the first four exercises, and form them into one, with 6 figures in width and 6 in depth, and proceed in like manner with the remainder, thus increasing the exercises both in width and depth to the end, always remembering that the result of each addition is not to be written till the student has it complete in his mind.

Repeat all the above exercises in addition, commencing each at the left instead of the right. For example, in No. 5, say : sixteen hundred ; a hundred and ten, seventeen hundred and ten ; twenty-six, seventeen hundred and thirty-six.

2. *Subtraction by Inspection.*

Find the differences of the following numbers, observing that the subtrahend is placed sometimes above and sometimes below. The whole result should be ascertained before any part of it is written. Prove by subtraction in the ordinary method.

1.	2.	3.	4.	5.	6.
5276	3157	8126	1309	9138	3284
8284	4848	2473	2214	2099	7721

7.	8.	9.	10.	11.
15389	91205	28405	64173	73085
4768	12345	71324	28094	16409

12.	13.	14.	15.
796346	9712645	287638472	8296384
149287	2793487	639724576	5149427

Repeat the above exercises, by throwing two into one, proving as before; and repeat them once more, performing the subtraction by adding the complement of the subtrahend.

Find the differences of the following pairs of numbers, by the addition of the complement, proving by adding the result to the double subtrahend.

1.	2.	3.	4.	5.	6.
326 }	146 }	845 }	378 }	627 }	736 }
721 }	938 }	792 }	624 }	836 }	472 }
245 }	247 }	148 }	327 }	924 }	315 }
378 }	146 }	973 }	415 }	356 }	249 }

7.	8.	9.	10.	11.
1348 }	3247 }	2826 }	1563 }	5257 }
7692 }	1958 }	7331 }	2844 }	4348 }
387 }	637 }	2087 }	647 }	1252 }
6914 }	28 }	639 }	1248 }	4873 }

Repeat the first ten exercises immediately above, first by doubling two horizontally, and again by doubling each pair vertically, placing the four numbers of the minuend together, and also the four of the subtrahend, connecting each set by a brace.

3. *Multiplication by Inspection.*

Find the products of the following factors, not writing them till every figure of the result is attained. Prove by multiplication in the ordinary manner.

1.	2.	3.	4.	5.
3452	6354	8492	2658	4936
4	5	6	7	8

Method of Operation.—[To be read slowly by the teacher, the class keeping their eye on the figures.] No. 1. Beginning on the left, twelve thousand; *sixteen hundred*, thirteen thousand, six hundred; *two hundred*, thirteen thousand, eight hundred; thirteen thousand, eight hundred and eight.

No. 2. Thirty thousand; *fifteen hundred*, thirty-one thousand, five hundred; *two hundred and fifty*, thirty-one thousand, seven hundred and fifty; thirty-one thousand, seven hundred and seventy.

6.	7.	8.	9.	10.
2598	4673	7856	2896	3875
9	3	4	7	6

11.	12.	13.	14.
34625	78452	29867	26548
8	9	5	4

15.	16.	17.	18.
286943	325896	534278	394827
6	5	9	4

1. Repeat the above eighteen exercises, omitting the words in italics: that is, throwing the partial product at once into the general result *without naming it.*

2. Repeat the same eighteen exercises, with an additional figure to each multiplier: that is, 1 ten to each, then 2, 3, 4, and 5 tens to each.

Method of operating with two figures in the multiplier. No. 1, with 14. First by 10, then by 4. 10. Thirty-four thousand, five hundred and twenty. 4. Twelve thousand, forty-six thousand, five hundred and twenty; sixteen hundred, forty-eight thousand, one hundred and twenty; two hundred and eight, forty-eight thousand, three hundred and twenty-eight.

No. 2, with 35. First by 30, then by 5 [$5=\frac{10}{2}$]. 30. A hundred and eighty-nine thousand; sixteen hundred and twenty, a hundred and ninety thousand, six hundred and twenty. 5. Thirty-one thousand, seven hundred, and seventy, two hundred and twenty-two thousand, three hundred, and ninety.

No. 3, with 26. First by 20, then by 6. 20. A hundred and sixty-nine thousand, eight hundred and forty. 6. Fifty thousand, four hundred, two hundred and twenty thousand, two hundred and forty; five hundred and forty; two hundred and twenty thousand, seven hundred and eighty; twelve; two hundred and twenty thousand, seven hundred and ninety-two.

No. 5, with 38. 3. A hundred and forty-seven thousand; a thousand and eighty, a hundred and forty-eight thousand and eighty. 8. Thirty-two thousand; a hundred and eighty thousand and eighty; seventy-two hundred; a hundred and eighty-seven thousand, two hundred and eighty; two hundred and forty; a hundred and eighty-seven thousand, five hundred and twenty; forty-eight; a hundred and eighty-seven thousand, five hundred and sixty-eight.

3. Repeat the 18 exercises with the figures of the multiplicand in reverse order, and with 6, and 7, and 8, and 9, in the ten's place of the multiplier.

19. 1,357,246 234	20. 4,738,324 342	21. 2,854,963 432	22. 7,128,435 243
23. 4,263,065 345	24. 5,849,212 453	25. 3,246,154 254	26. 3,782,465 351

☞ The teacher may extend these exercises as far as he may find it profitable to the class. Some pupils have accomplished the multiplication of 9 figures in the one factor, and 5 in the other, after a very short practice. To others they come hard. But all will be highly benefited by their use.

4. *Division by Inspection.*

Find the quotients in the following exercises by inspection, beginning at the left, not writing them till the whole quotient is attained. Prove by multiplication by inspection.

1. 2)653,492 2. 4)134,684 3. 7)179,426 4. 5)286,942

5. 3)295,464 6. 9)264,834 7. 8)389,264 8. 6)521,424

9. 7)1,356,493 10. 8)7,287,934 11. 6)3,457,836

12. 4)6,237,948 13. 5)2,176,385 14. 9)2,693,754

Repeat each of the above 11 exercises, with an additional figure for tens in the divisor, namely, 1, 2, 3, 4. Then repeat again, with the figures in each dividend reversed, with an additional figure, 6, 7, 8, 9 tens in the divisor. Prove as above.

The teacher can extend this practice by new exercises, increasing the number of figures, both in the divisor and the dividend, till the class has acquired sufficient dexterity.

5. *Evolution by Inspection.*

a. *Extraction of the Square Root by Inspection.*

Find the nearest square root of each of the following numbers: 576; 1296; 1458; 3975; 2482; 9176; 7056; 2209; 20736; 53824. Prove by involution by inspection.

b. *Extraction of the Cube Root by Inspection.*

Find the nearest cube root of the following numbers by inspection; and prove by involution by inspection: 13824; 46656; 53824; 592,-704; 638,576; 2,985,984; 12,487,168.

☞ When a *perfect* cube does not exceed 1,000,000, its root may be found almost at a glance, as follows: Fill up the table of squares and cubes in p. 164, if not already done; and then take notice that the unit's figure in the cubes of 2 and of 3, and also in those of their complements (8 and 7), is the same as the complements of their respective roots. Observe, also, that the unit's figure in the cubes of all the other digits (1, 4, 5, 6, 9) is the same as that of their root. Thus:

Root of 8 is 2, the complement of 8.
" " 27 is 3, complement of the last figure 7.
" " 343 is 7, complement of the last figure 3.
512 is 8, complement of the last figure 2.

While root of 1 is 1	Unit's figure of the power same as the root.
" " 64 is 4	
" " 125 is 5	
" " 216 is 6	
" " 729 is 9	

From the above table it is evident that it only requires a knowledge of the cube of each of the digits, to determine the root of any perfect cube not exceeding 1,000,000, by a mere glance at the second period and at the unit's figure of the power.

Exercises for the Slate or Black-board.

1. What is the cube root of 262,144? *Ans.* 6 is the greatest root in the second period, the unit's figure is 4: 64, therefore, is the root.

2. What is the cube root of 389,017? *Ans.* 7 being the greatest cube in 389, and 3 the complement of 7, the cube root is 73.

3. Find the cube roots of the following perfect powers by a glance: 54,872; 884,736; 185,193; 474,552; 5832; 15,625; 59,319. Prove by involution by inspection.

☞ The extraction of the square root by this method requires more attention; but for that very reason is more useful as a mental exercise. The roots of *perfect* squares which do not exceed 10,000 may be ascertained thus. By an examination of the squares of the digits, it will be perceived that every perfect square ending in 5 has 5 for the unit's figure of its root; and that the squares of 1, 2, 3, 4, and of their complements, 9, 8, 7, 6, have for unit's figure 1, 4, 9, 6, respectively. Thus, having determined the ten's figure of the root by a glance at the second period, we know that if the unit's figure of the square be

5		5
1		1 or its complement 9
4	the unit's figure of the root is	2 " " " 8
9		3 " " " 7
6		4 " " " 6

Exercises for the Slate and Black-board.

1. What is the square root of 5184? *Ans.* The greatest square in the second period being that of 7, and the remainder small compared with 7, the unit must be 2 rather than 8. The whole root, of course, is 72.

2. What is the square root of 1296? *Ans.* The greatest square in 12 is that of 3, and the remainder being great in proportion to 3, must be 6 rather than 4. The whole root, then, is 36.

3. What is the square root of 2025? *Ans.* The greatest square in 20 is 4, and as the unit's figure in the square is 5, that of the root must be 5 also. The whole root, then, is 45.

4. Find the square roots of 1024; 3136; 784; 4225; 2116; 3249; 6561; 2401. Prove by involution by inspection.

SYNOPSIS,

OR, RECAPITULATION OF PRINCIPLES DEVELOPED IN THE PRECEDING PAGES.

I. There are only two operations in arithmetic, *increase* and *decrease*. A number may be increased by one or more additions. It may be diminished by one or more subtractions. Such is the whole sum and substance of arithmetic.

Both of these can be performed by numeration. All other processes are mere abbreviations of this foundation of arithmetic by the omission of steps become superfluous by repeated practice.

II. In adding or subtracting *integers*, the numbers must be of *one denomination*. Four *thousand* added to three *hundred*, can only make *four thousand three hundred*. The 4 and the 3 do not make 7 of any denomination. For a similar reason, 3 hundred cannot be taken from 7 thousand, so as to leave just 4 of any denomination whatever. The latter operation can only be performed by *changing the denomination* of one of the thousands into hundreds, and then subtracting the 3 hundred from the 10 hundred, leaving *six* thousand and *seven* hundred.

Precisely the same remarks apply to fractional quantities. In order that fractions may be united into one number, their denomination must be the same. Three *fourths* and four *fifths*, or three *pecks* and four *bushels*, can be united or subtracted, only after some change of denomination which renders them similar.

No such restriction, however, is necessary in multiplication or division. The product of 4 thousand or 4 hundred by 2 or by 2 ty can be found just as readily as if the denominations were the same ; and the quotient of 4 million by 2 is as easily found as the quotient of 4 by the same number. The reason is obvious. A product is not one factor increased by another. It is the amount of the one taken *as many times* as there are units in the other ; and a quotient merely points out *how many times* one number is contained in another. And this is not less so with fractions than with integers. For $\frac{3}{4}$ can be taken $\frac{7}{8}$ *times*, and we can find *how many times* $\frac{1}{6}$ or $\frac{1}{9}$ can be found in $\frac{1}{3}$, or $\frac{2}{7}$; and the same remarks apply equally to bushels, yards, &c., as to 3ds or 7ths.

But, after all, the difference is only in appearance. Multiplication is, in fact, simply an *addition of identical numbers*, while division is the subtraction of numbers equally *identical*. It is only the *manner* in which the operations are performed which gives them the *appearance of different denominations*. Bringing the numbers to the same denomination is one of the *steps* that become *superfluous*, by the use of multiplication and division in place of addition and subtraction. See p. 91.

III. The *object* of all arithmetical operations is to produce a balance, or, in technical language, to form an equation. Thus, both in integers and fractional quantities,

In addition, we seek to find a number=the sum of the given numbers.
Thus, given numbers 24+37+41=102 sum.
In subtraction, a number that with the subtrahend will=the minuend.
Thus, subtrahend 255+134 remainder=389 minuend.
In multiplication, a number=the product of the two given factors.
Thus, product 864=36×24 factors.
In division, a number whose product with the divisor=the dividend.
Thus, dividend 2432=divisor 19×128 quotient.

In all the *changes* of fractional quantities we seek to place a fraction, without altering its value, in a more convenient or simple and intelligible *form*.

In proportion, the sole object is to change the complete ratio into the same denomination with that of the imperfect ratio, so that the term wanting in the latter may appear. In other words, to find a fraction of the denomination of the imperfect ratio=to the complete ratio.

IV. When figures are written horizontally, or side by side, whether they be integers or decimal fractions, every figure is tenfold greater than the same figure immediately on its right, and tenfold less than the same figure immediately on its left. See p. 114.

V. Every figure becomes tenfold greater by being removed one rank or place towards the left, tenfold less by being removed one rank or place towards the right, p. 115.

VI. Ten units of any one place make one unit of the next place to the left; and one unit of any one place makes ten units of any one place to the right, p. 117.

VII. When there is a separatrix, the unit's place is immediately on its left; when there is none, the right hand figure occupies the unit's place, p. 119. When a separatrix is used, any number of figures or ciphers can be placed to the right of a number, without changing the value of any of the figures in that number, p. 118, l. 27.

VIII. The cipher is superfluous unless it occupies the place of units, or intervenes between a significant figure and the place of units, p. 122.

IX. If equal numbers be added to unequal numbers, their difference remains unchanged, p. 139.

X. 1. If the difference between two numbers be added to the smaller, it becomes equal to the greater; 2. If taken from the greater, it becomes equal to the smaller, p. 140.

XI. The first significant figure on the right of an arithmetical complement is always greater by 1 than if it stood one or more ranks to the left, p. 143, l. 6.

XII. The difference between two numbers may be obtained by *adding* the complement of the smaller to the larger, and diminishing this sum by 1 of the next higher rank of figures than is contained in the smaller, p. 144.

XIII. In multiplication, the number of decimal fractional places in the product is always equal to the number in both factors, and, consequently, as the dividend is the product of the divisor and quotient, there will always be as many fractional places in the dividend as in the divisor and quotient; and, conversely, as many in the divisor and quotient as in the dividend. In the same manner the number of ciphers at the right of a product is always equal to the number at the right of both factors, p. 154.

XIV. An integer may have any number of decimal places annexed to it, by affixing a separatrix to show the place of units, and then adding as many ciphers as may be required. Thus, if 48 is to be divided by ‘4, by changing the dividend to 48‘0, the quotient is the whole number 120, p. 118, l. 27.

XV. 1. The SQUARE of any number of tens and units=the squares of the tens and of the units taken separately, plus twice the product of the tens and units. 2. The CUBE of any number of tens and units=the cubes of the tens and of the units taken separately, plus three times the square of the tens multiplied by the units, and three times the square of the units multiplied by the tens, p. 166.

XVI. In common fractions,

1. The numerator expresses the number, the denominator its denomination, or value, p. 67, 8.

2. The denominator expresses the number of parts into which a unit is divided, the numerator the number of those parts which the fraction contains.—*Ib.*

3. The numerator is the dividend, the denominator the divisor, and both terms taken together the quotient. — *Ib.*

4. By { multiplying / dividing } the numerator, the fraction is { multiplied. / divided, p. 80. }

5. By { multiplying / dividing } the denominator, the fraction is { divided. / multiplied, p. 81. }

6. By { multiplying / dividing } both, the value of the fraction is unchanged, p. 82.

7. By removing the denominator, the fraction is multiplied by a number equal to the denominator.

8. Any number, whether fractional or integral, may be represented in an infinite variety of forms, all differing in character, yet all exactly equal, p. 82.

9. Any number whatever may be expressed by a common fraction, whose numerator (or whose denominator) shall consist of any specified number, whether whole or fractional, p. 211.

10. Any numerator }
11. Any denominator } may be used to express any number whatever.

XVII. An integer or a decimal fraction is changed to a common fraction by *expressing its secondary name* under it in figures ; a common fraction is changed to a decimal fraction or to an integer by *performing the division indicated*, p. 193.

XVIII. Circulating decimals with a simple repetend are changed to common fractions by using the repetend as numerator, and as many 9s as there are figures in the repetend as denominator.

XIX. Circulating decimals, with compound repetends, may be changed to common fractions, by changing separately the repetends and the figures that precede them to common fractions, and then adding them together, p. 198, l. 32.

XX. In changing complex fractions of three terms to simple fractions, when the upper term is an integer, the upper and lower terms are factors of the numerator, and the other term is the denominator ; but, when the lower term is the integer, the upper term is the numerator, and the other two are factors of the denominator, p. 203.

XXI. The form of a fraction can always be conveniently changed by multiplication ; but by division it frequently becomes more intricate, p. 92, l. 10, and p. 200, l. 14.

XXII. The same quotient will result, whether division is performed by a composite number, or by its factors separately, p. 86, l. 24.

XXIII. Addition or subtraction of common fractions is performed by exactly the same process as addition or subtraction of integers, p. 91.

XXIV. *Signs for discovering Factors of Numbers.*

1. Every *even* number is necessarily divisible by 2. For, in dividing the tens, the remainder must either be 0 or 1. If it be nothing, the last

number to be divided will be 0, 2, 4, 6, or 8 ; if it be 1, the last number must be 10, 12, 14, 16, or 18 ; and all of these are divisible by 2, p. 85.

2. If the *two right hand* figures of a number are divisible by 4, or by 25, the whole number is divisible by 4 or by 25. For both those numbers will divide *one* hundred, and consequently any number of hundreds without remainder. If the number of tens be even, we need not look further than the units to decide if the number be divisible by 4, p. 86.

3. If the three right hand figures of a number are divisible by 8, or by 125, the whole number is so divisible. For as *one* thousand is divisible by both those numbers, so, of course, is any number of thousands. If the number of hundreds be even, we need not look further than the *tens* to decide whether the whole number be divisible by 8. Therefore, when the hundreds are odd, we have only to render them even by taking *one* hundred with the tens and units. Thus 953752 is divisible by 8 because 152 is divisible by 8, every *even* number of hundreds being already known to be so divisible, p. 86.

4. Every number terminated on the right by 0, or by 5, is divisible by 5, because ten or any number of tens is divisible by 5, p. 86.

5. If the *sum* of the significant figures of any number, added horizontally, be divisible by 3, or by 9, the whole number is divisible by 3 or by 9. Thus, the number of 6273 is divisible by 9, because, since 1000 consists of 999 and 1, it will leave a remainder of 1 when divided by 9, and consequently 6000 will leave a remainder of 6. For a like reason 200 will leave a remainder of 2, 70 a remainder of 7, and 3 units a remainder of 3. Now, as the remainders in this division are *its significant figures*, this must evidently be so in any number whatever. If, therefore, the sum of the significant figures is divisible by 9, the whole number is so divisible. The same demonstration will answer for 3. Therefore, *every number, the sum of whose significant figures is divisible by* 9 *or by* 3, *is itself divisible by* 9 *or by* 3. It is obvious, also, that every number divisible by 9 is also divisible by 3, though the converse by no means always follows, namely, that every number divisible by 3 is also divisible by 9, p. 85.

6. Every even number divisible by 3 is also divisible by 6. For 2 is a factor in every even number, and $2\times3=6$; and it is evident that the *products* of factors of any number are also factors in that number. For the same reason, every even number divisible by 9 is also divisible by 18, p. 86.

7. Every number divisible by 3 and by 4 is also divisible by 12 ; every number divisible by 3 and by 5 is also divisible by 15 ; every number divisible by 3 and by 8 is divisible by 24 ; and every number divisible by 9 and by 25 is divisible by 225, p. 86.

8. Table of the foregoing signs for the discovery of factors :

Factors.	*Their signs.*
2	Even numbers.
3	When sum of significant figures divisible.
4	When 2 right hand figures divisible ; or one, if tens be even.
5	When terminated by 0 or by 5.
6	Even numbers divisible by 3.

8	When 3 right hand figures divisible ; or 2, if hundreds be even.
9	When sum of significant figures divisible.
10	When last figure a cipher.
12	When divisible by 3 and by 4.
15	When divisible by 3 and by 5.
18	Even numbers divisible by 9.
24	When divisible by 3 and by 8.
25	When two right hand figures divisible.
75	When divisible by 3 and by 25.
125	When 3 right hand figures divisible.
225	When divisible by 9 and by 25.

9. All the factors in an odd composite number must be odd ; since every number is even which contains an even factor.

10. Every prime number, except 2, is odd ; therefore, all the prime factors, in any number, except the twos, are odd.

XXV. In the multiplication of common fractions, the numerators are multiplied for a new numerator, and the denominators for a new denominator, p. 94, 5. When one or more factors can be found in either or both of the numerators, which can also be found in either or both denominators, they may be cast out of both before the multiplication, and thus leave the fractional product in its lowest denomination, p. 94, 9.

XXVI. In the division of common fractions, reverse the divisor, and proceed as in multiplication, p. 95, 11.

XXVII. In a progression by differences, every term in an increasing series consists of the *first term* added to the *common difference*, taken once less than the number of that term, p. 325.

XXVIII. The sum of all the terms is equal to half the product of the number of terms by the sum of the first and last terms, p. 327.

XXIX. In a progression by ratios, the sum of a series is equal to the last term multiplied by the ratio, diminished by the first term, and divided by the ratio less one, p. 332.

www.ingramcontent.com/pod-product-compliance
Lightning Source LLC
LaVergne TN
LVHW010158110826
845151LV00002B/544

* 9 7 8 1 4 2 5 5 3 6 3 4 3 *